"In an age driven by algorithms, understanding their power is essential. 'Digital Shock' sheds light on how these unseen forces influence our decisions and the course of our society."

— Fei-Fei Li, Computer Scientist and Co-Director of Stanford AI Lab

"'Digital Shock' is a thought-provoking journey into the heart of our digital world. It shows how algorithms are shaping our present and defining our future in ways we cannot ignore."

— Reid Hoffman, Co-Founder of LinkedIn and Author of "Blitzscaling"

"As we navigate the complexities of the digital era, 'Digital Shock' serves as a guide to understanding the forces that shape our lives. A must-read for anyone seeking clarity in the algorithmic age."

— Arianna Huffington, Founder of HuffPost and Thrive Global

"Algorithms dictate so much of what we do, from the news we see to the products we buy. 'Digital Shock' is a compelling exploration of the impact these algorithms have on our lives."

— Nate Silver, Statistician and Founder of FiveThirtyEight

"Our reliance on technology is shaping the world in unprecedented ways. 'Digital Shock' dives deep into the consequences of this reliance and is a wake-up call for the digital age."

— Sherry Turkle, MIT Professor and Author of "Reclaiming Conversation"

"Discover the untold stories behind our digital existence with 'Digital Shock'. This thought-provoking book left me questioning my own relationship with technology. Are you ready to confront the truth? #TechnologyExposed

— Kshitij Kumar, serial CDO (Haleon, ex-Farfetch, ex-OneFootball)..

"Algorithms are the building blocks of the digital landscape. 'Digital Shock' offers a captivating exploration of their influence on our society, uncovering both their

Digital Shock

Hari Guleria

We must realize that we live in the best period humanity has ever known. To understand this fact, we must look past Russia's war with Ukraine, China's treatment of Uighurs, and Chat-GPT's and AI-LLM's growing influence in 2023 amid the technocratic audience. Despite all this, the world is steadily improving, and *We the People,* hold the reigns to stand up and fight or succumb to the authoritative power of the few as they have for thousands of years. In the words of Shantanu, the CEO of Adobe, *"When you think about what's happening in the world of collaborative creativity, it's very clear that every single person has a story to tell."* Each story is a personal opportunity; each story can become a digital lighthouse app that can disrupt the greatest of companies in a year. Today, we're living at the cusp of the greatest individual opportunity we will see in our lifetimes. It is up to us to choose the path of becoming the digital disrupter or become a digital victim.

At the current state of the *Humane Use of Our Intelligence,* humanity may, hopefully not, be positioning itself to run out of time. The decisions we individually, as a nation, and collectively make in the next decade and every decade after that will shape the future of humanity on planet Earth.

Suppose this generation loses the definition of right and wrong. In that case, there is a high probability that we could all collectively lose our sanity despite being the only known intelligent species in the universe.

As media, scientists, computer science geeks, and even Elon Musk warn us of the dangers of unleashing AI. In my opinion, our biggest threat right now is Alternative Intelligence being digitally shoved into the minds of unwary citizens.

Fact Check: The US is being systematically destroyed by political trolls and media opinions broadcasting alternative-realities of elected leaders. Our planet is not being destroyed by any rogue Artificial Intelligence robots but by elected leaders. — Author

We are all stuck in the middle of three paradigms:

1.	Einstein once stated, **"Change is the only constant."** So, any person hoping that things must remain the same, is living in an unachievable reality. The more we resist change, the greater the gap between reality, resistance, and existence. One of the three must tear at a certain point, and it won't be an evolutionary change.

2.	Einstein also defined insanity as **"Insanity is doing the same thing over and over and expecting different results."** So, if we keep resisting change and continue doing the same things while hoping that things will adjust to our old ways, then, according to Einstein's prediction, we know that we will have to fight our insanity.

3.	Throughout history, at each significant fork in life, there are always two intelligence paths: adapt or reject. In such instances, people who chose the *adapt* option have realized strategic success. In contrast, those who chose to *reject* the ongoing change have inevitably felt short-term happiness at their resistance but strategically floundered in their ways. This rule applies not only at an individual level but also at a corporate and national level.

Digital Proof: 2022 stats - 1 minute on the Internet, 24x7

1. 694,000,000 songs downloaded
2. 444,000,000 people viewed Facebook live streams
3. 231,000,000 emails sent
4. 174,000 Apps downloaded
5. 167,000,000 videos watched on Tik-Tok
6. 20,800,000 active users on LinkedIn
7. 12,500,000 Shared rides Taken
8. 16,200,000 Text Messages sent
9. 6,000,000 people shopped online
10. 5,900,000 Google Searches
11. 3,670,000 YouTube videos watched
12. 2,100,000 active users on Facebook
13. 2,000,000 Snapchat chats
14. 575,0000 Tweets sent
15. 452,000 hours of content streamed on Netflix
16. 66,000 photos and videos shared on Instagram
17. 46,000 searches on Pinterest

Source: LocaliQ Stephanie Heitman, May 5, 2022

AI and the future

Humans need to write a constitution for AI that no AI program can break. Who is in the best position to lead this constitution? It cannot be China or Russia, and it cannot be the UN. Can it be Silicone Valley, many people may think otherwise. Then Who? Should it be a democratic vote or decided by the already rich and powerful? All AI programming must have implied fences where an AI cannot go out of defined boundaries to meet its stated objectives.

Dedication

This book is dedicated to my late father, a brigadier in the army. He took us across India almost like clockwork every three years while serving across the planet. Its most important contribution was that it unconsciously taught us to adapt to ever-changing environments, i.e., new postings, new cities, new schools, new people, and new thoughts each time. Today, because of him, I stand here as a true citizen of planet Earth.

To my mother, Savitri, who still takes us to India every year. She kept our family together after my father's death, which is the anchor for our family today. She got covid at 99 and went on to cross 100.

To my dear wife, Gauri, my son, Arnav, in New York, working as an investment banker, and my daughter, Iraa, mid-way on her journey to become a doctor, the medical kind at Milwaukee. It is also dedicated to our beagle, Kona, who passed away in 2021

It is dedicated to my siblings — my three sisters, Shakunt, Asha, and Urmil, and my brother, Ex-Col. Virendra, who have collectively molded me, in their respective ways, to become who I am today. It is dedicated to all my friends and customers who helped me reach my social, professional, and digital maturity. At the same time, we together implemented many Oracle, SAP, and S/4HANA projects and a few Digital Transformations, where, one step at a time, we together began to comprehend the overall potential, opportunities, and impact of Digitization.

Most of all, it is dedicated to you, the reader, with the hope that after reading this book, you come out to know a little, or a lot, more about surviving the digital ocean and forest we all find ourselves in. This book is designed to first understand the digital threats, then understand the many more digital opportunities, and finally to help you build your very own path toward Digital Adaptation and social decisions. If it changes your mind or you learn something new, even in a single attribute, then this book has done its job.

My other books: See at the end of the Book

Acknowledgment

Writing a book is never a solitary endeavor; this book is no exception. I would like to express my deep gratitude to every person in my personal and professional life. I want to express my deep gratitude to everyone who has supported me throughout this journey.

Firstly, I would like to thank my family for their unwavering love and encouragement. Their belief in me gave me the strength to persevere and complete this book.

I am also grateful to every one of my colleagues and friends who provided me with valuable insights, feedback, and inspiration throughout the writing process. Sometimes, something as simple as a *Great chapter, but what should I do after reading it,* and how easy or difficult is it to pull one outside this digital shock? Thank you for your intellectual generosity and willingness to engage in fruitful discussions.

I would like to acknowledge the numerous experts in the field of digital technologies who shared their knowledge and expert opinions with me. Their contributions have been invaluable to the accuracy and richness of this book. I would also like to express my heartfelt thanks to my editor, who provided me with guidance and support from the beginning to the end of this project. His meticulous attention to detail and commitment to excellence made this book possible.

Finally, I would like to thank the readers, as without you, this book is just like a wink in the dark. No one except the author will know that it even exists. Most of all I would like to thank Open AI and ChatGPT for creating so many millions of doors even where we knew none existed. The greatest opportunity for a thinking mind

Thank you all for your contributions and support.

Note: This book was started in 2014-5. It was rewritten in 2020 when Covid-19 made all my older writings became redundant with the advent of global digital transformations. Due to this some of the dates may seem old but they are the dates at the time of that paragraph. Thank you.

About the Author

Hari Guleria is a multi-faceted, leading figure in enterprise Business solutions. He has been a division head, business coach and a Customer & User experience designer. He authors articles around his profession of SAP HANA ERP, and in the world of astrophysics, with speaker sessions, and is a prolific author of books and articles.

Professionally, Hari has worked for SAP as Principal Architect, at HCL Axon as Director of SAP HANA for North America, at HP as Director of SAP HANA, at Pridevel/Cisco as Global VP of SAP HANA, and most recently, leading SAP HANA implementations, upgrades, and PMO support for US unicorns, and new SAP customers. He aids enterprise customers who do not see light t the end of their SAP implementations.

In SAP HANA ERP, he was one of the few leaders when HANA was launched in 2010. He has over 30,000 followers on LinkedIn. He has become a name synonymous with delivering success in SAP HANA implementations. In 2011, he published a book, 'BI Valuenomics — The Story of Meeting Business Expectations in BI,' a book more relevant today than it was a decade ago.

In the world of theoretical astrophysics, Hari has published a groundbreaking *Beyond Drake — No Earth 2.0*. Here, Hari fills two gaps: The first is updating Frank Drake and Carl Sagan's 1961 equation on the search for alien Intelligent Life in our Milky Way. Hari added a new, mind-blowing, philosophical set of KPIs for resetting our human intelligence parameters as our only quorum of one while we search for life outside planet Earth.

Hari has been a speaker in over 50 sessions at global conferences on SAP, SAP HANA, and Big Data and Analytics.

Hari is now focusing on authoring his global experience and sharing what he has learned in this book, *Digital Shock*. This book is the first of many to come. His next book, already in the pipeline, is an attempt at science fiction.

Hari lives in Cupertino, California, in the heart of Silicon Valley and the global HQ for Apple.

Hari may be contacted at harig.auth@gmail.com. Please start with 'Digital Shock:' in the subject line if it is regarding any comments about this book.

To get access to the charts, and additional comments on this subject please go to my Blog on 'Digital Shock'[i]

Preface

When we refer to digital addiction, we're normally talking of the smartphone, smart TV, or networked laptop. However, look a little deeper, and we discover the addiction is not to the device but to the content that social media throws at an individual level based on their unique fear tolerances and preferences. If you think that this is science fiction and somewhere in the far future, then just read on.

When we talk of digital opportunities, we think that is for Jeff Bezos (Amazon), Mark Zuckerberg (Facebook), or Richard Branson (Virgin). However, look a little deeper, and we quickly realize that we are at the beginning of the greatest continuous gold rush in human history with many self-made billionaires between the ages of 12 and 30; 4 of them are, i.e., 57%, are Stanford dropouts. Overall, there were 2,668 billionaires on June 1st, 2022 across the planet. Each with more individual wealth than a single individual owned ever before. These are billionaires who have made their wealth on their own steam and lifetime, i.e., not from their parent's wealth handouts.

So, if you take just three things from this book toward your success, it is:

- **Stop pleasure voyeurism** on Social Networks, Media Channels, TV programs, and WhatsApp, as these are designed to distract your true potential by providing delight in seeing how good life is for someone out there or how bad yours could get very rapidly. Replace all this with reading books — learn from the experience of others.

- **Start learning** from others. Find Mentors, read biographies of great people and learn from their experiences. Find a passion and attend a micro course. Delve into the briefs at the start of each chapter in this book on how someone like you and me succeeded in their lives using digital assets. Then, go make a change that you believe in. Replace thinking with thinking big. Then go and think very big indeed.

- **Believe in your good self:** You can go and change the world. Find a gap in what you need. The opposite side of that need is your solution. Then, go build a digital solution and make it easy to use. There are thousands of people providing microservices that people need. Some are earning more than before, while others are taking their ideas into global solutions.

- **Plus - Focus on the 'P' factor:** 'P' stands for 'we the People' the center of the known intelligent universe, it also stands for 'positive' impacts. This book focuses on what can assist 'we the people' and looks more at how ML, AI, Algorithms, and also on how bots can be made to make our individual and collective lives worse.

This is done for awareness. However, the reality is that collectively, we need to focus on the 'P' factor or the positive things "We the People" can individually and collectively develop and contribute toward in order to make this a better planet. While a bot under the control of a Russian troll can be a global political problem, a similar bot under the control of a civilian Ukrainian can blow up an advanced Russian airplane in a Russian airport. What we discuss here is the science of 'under the control' of an intelligent mind that can be used to develop, spread, and deliver efficiencies that we cannot yet imagine. In each chapter, we will discuss the 'P' factor at the end as we will discuss the individual, an enterprise, and a nation-state. Be aware of the bad, guys, but remain focused on the good.

This book is a very non-technical roadmap on how individuals, companies, and nations can capitalize on the current digital shock that has created a digital fog in the short-, mid-, and long-term plans of all three stakeholders mentioned here. Individuals, Companies, and Nations are all at a critical inflection point. The pace of digital change has far exceeded human change adaptation capabilities, and the ramifications of this gap can be catastrophic if not controlled and structured. The solution is with you & I — the very persons who we can change as they adapt.

According to Harvard Review:

1.   92% of enterprise executives confirm that now is the unique opportunity for all companies to capitalize on their digital adaptation capabilities

2.   78% of digital product and transaction users have formed lasting habits during the COVID-19 pandemic lockdowns. They have also confirmed that the individual's digital adoption has accelerated to a point, whereby turning back to a pre-2020 level of non-digital interactions is no longer possible.

3.   58% of enterprise stakeholders use analytics for tracking and measuring the impact of overall digital transformations. Topping the global list of surveys, web analytics, and business intelligence initiatives.

4.   78% of critical stakeholders consider Informatics as the 'Center of the Digital Transformation' universe.

The three main participants are the individual, the enterprises, and the nation states with each of their leaders. It is generally agreed that at every level, these three participants need to adopt a digital-first mindset to redefine their roadmaps and meet their customer requirements. For an **individual,** their first customer is themselves, their family, overall survival by making the right decisions, and their work. For a **company,** its customers are their internal business users, and the end customers of their services or products are its external customers. For a state or **nation**, their customers are all other states and every single citizen that makes up their nation.

It is important to accept that this digital disruption and adoption will continue to accelerate at unprecedented rates. According to IDC's FutureScape report, it is prudent to plan that by 2023-24, more than 500 million cloud-native applications will be created and deployed. One or any of these could be yours.

"Creating successful digital business hinges on the individual, the company, and national abilities to firstly meet, then exceed their customer expectations." — **Alex Paleologos**

Contents

Dedication i

Acknowledgment ii

Preface iv

Introduction 1

Chapter 1 — The Seventh Wave 9

Chapter 2 — Global Digital Disruption 24

Chapter 3 — Information: The Center of the Universe 34

Chapter 4 — Face of Change 58

Chapter 5 — The Digital Opportunity 73

Chapter 6 — The Digital Society 90

Chapter 7 — The New Digital Huma 109

Chapter 8 — The New Digital Enterprise 140

Chapter 9 — The New Digital Nation 162

Chapter 10 — The Digital Segmentations 177

Chapter 11 — Fracturing Humanity's Mind 202

Chapter 12 — The Realms of Multiplicity 210

Chapter 13 — Mining the Restless Individual 225

Chapter 14 — Physical and Neurological Magnitudes 238

Chapter 15 — Digital Shock, Mental Adaptation 269

Chapter 16 — Navigating the Digital World 283

Chapter 17 — Some Taboo Topics 301

Chapter 18 — Clearing the Stormy Haze: Risking Small Predictions 425

Post log 439

Bibliography 449

References 450

Introduction

At the very beginning, it is critical to remember that the total digital revolution is based on a simple Equation. This equation evolved to become Math, Algebra, Calculus, Linear Algebra, Probability Theories and the foundation of programming, ERP, predictive analytics, Coding, AI, ML, LLM and now Chat-GPT. This proceeds to evolve from manual calculations to automation in computers. The simple equation is still the base logic inside computers. These then evolved into software robots that we call bots.

This book is designed for a curious, non-technical reader; thus, I keep as far from being technical as I can. My first book, published in 2011, was the result of a question by a CFO from a major US Biotech company. He asked if there was a book on Data and Analytics for a business stakeholder; everything he saw in the market was highly technical. I at once saw an opportunity. After 6 months of furious writing, when I enthusiastically presented him with my draft, he read the first 10 pages and told me that I had not understood his needs and he could not read it. He added, "When your 14-year-old son can read and understand every sentence. Only then the book is ready for me." This fundamentally changed my writing style as I had to restart that book from scratch. His advice resulted in *BI Valuenomics — The Story of Meeting Business Expectations in BI*.

Fast forward to today, and our subject of Digital shock, digital competency, and disruption is so huge that one can write a book a hundred times this size on each technical topic. So, considering my target audience, I have taken a deep dive into the world of equations as the atomic brick in our digital world and tried to explain this to 3 prime audiences, with 1 single principle.

1. A curious teenager.

2. A 20 to 60-year-old enterprise manager.

3. A 40 to 90-year-old nationally elected leader and the 4th one will be exposed toward the end of the book.

4. *Basic rule: Keep it non-technical and write it for a 14-year-old teenager*

Until Jan 2023, my world was surrounded by technical coding competence, and then ChatGPT happened. Now, even technologists are in a state of digital shock and living inside a global fog. Most people, technical and non-technical, I talk to, seem to be surrounded by a fog of a dystopian yet bleak future, in one form or another, and perceive darker times around the corner, while only a few are ecstatic about our digital revolution and the opportunities it presents. It is my intention to make every reader into an enthusiastic digital disrupter and expose to them the infinite opportunities this digital disturbance has presented at our doorsteps. Reminder — "In life, opportunity knocks but once, while temptations lean on the doorbell". It seems that we are getting closer to the total eclipse of the world order under current strains in almost every fabric of our society. We all need to first find out whether this is because of the post-COVID-19 trauma, or when political parties have gone amuck in their hyper-polarized messaging of *Us vs. Them*, which has collectively resulted in an overflow of dystopian futures that are dominating our minds.

In my mind, there are two futures we have to choose from. The first is the greatest opportunity we humans have ever been offered. This is the time of the greatest peace, believe it or not. I hope by the end of the book, you see the light of this statement.

And the second is the constant negative barrage of our dismal future that our media, religions, children's books, and politicians talk about every day. We touch briefly on why they actually need to do this. This is the only time in history when a dedicated person could become a trillionaire in their own lifetime. This is the first time when most people in developed democracies today when the only barrier between your dream and what you can achieve is — You. My next micro book will be on Quantum Thinking, as I believe that is a timeless reality that we have not been able to quantify so far, but that is altogether another story altogether. Your first and only takeaway has to be to think positively, believe in your success, and then think big. Once you have thought how big, now multiply this a hundred-fold, and that is your starting point. On the dismal side, there are more than enough people who want to collectively take us down this rabbit hole called the beautiful past. The problem with that is the past was never either like the stories and legends or like the Hollywood movies we think represent it. It was never like proponents of the past wanted you to believe. Women, barely a hundred years ago, could not vote, nor did they have a voice in their man's world. Some women do not have that even today. Some of our great-grandmothers were slaves that could be locked, raped, or worse, at the will of their owners. Humans were property till very recently. How could that be better in any way for any woman? Some men a hundred years ago were mostly owned as labor or slaves to rich, powerful kings, dukes, and landlords with little to say who they could fight or die for; in some nations, that is still prevalent today.

On the bright side, the world has never been better than it is today. Despite what the media and news channels tell us. As our examples will demonstrate, as of today, there are no age, sex, location, or color limitations to curtail your ideas. For the first time, an individual can start a global corporation and become the richest company on the planet in their lifetime. The only barriers and fears are between your two ears and, of course, the nation you live in. For example, if you live in North Korea, then you cannot have your idea. If in China or Russia, then you have to follow Xi's commands, and he/they could take over your company at will. For example, Putin is probably the richest person on the planet as he keeps locking their successful business leaders and probably owns their businesses. On the other side, Apple, Tesla, Facebook, Amazon, Airbnb, and Uber are a few of the thousands of examples that would have been impossible just twenty years ago.

The Republicans in the US want to take the US back to 1864 and some even back to the times of Christ. So they could rule over the Black people and currently deny them the right to vote, and the second because they want to implement rules they thought existed at that period.

So why would anyone want to go 100, 500, or 1,000 years back? Things only get worse as you go back in time, and as you get forward, they get disproportionally better. Our greatest risk in the future is a nuclear Armageddon started by a single autocratic lunatic. We have plenty of them across the planet.

The only person stopping you from being rich and comfortable is probably just you.

It's time for a change. A very big, digitally-connected, intelligent change. Only this time, the connections are to become global.

Before 2020 and the pandemic, everything was running super-fast, and no one had time to prune their thoughts or reflect on anything. The pandemic not only locked humanity into their homes but also made almost everyone visit their minds. It made people question their choices. After a lot of introspection, some decided it was time for a change in how and what they did, while others started to follow some pied piper that entered their homes and influenced the circle and made them become a part of a tribe of one sort or another. For one lot, it became the greatest opportunity and changed for a long time, while for others, they found, participated, and became part of a tribe of their self-induced or polarized digital addictions.

Almost every person I have talked to, and many I did not, in the last two years, man or woman, Black or White, Democrat or Republican, Indian or Chinese, Russian or Ukrainian — all have one thing in common — a firm belief that only darker days are ahead. Many of us believe and resonate with the embedded beliefs that the world is going rapidly south. The big message out there to grab your attention is, 'How long before this civilized world, we know, collapses? The strains are many as the world has taken a 180° turn since the beginning of this century when it seemed that the planet was heading toward the top of the mountain.

Then, around 2005, just as we thought the planet was getting into a mode of perfection, things started to crash all around. Out came the powerful demons supported by media and politics, both of which thrive on fear. Suddenly, there is this explosion of political distortions, fake news, climate change or not, overpopulation or not, China a friend or foe, North Korea a who is an improbable Trump friend or a missile shooting enemy for the US, and a host of other fractures that suddenly seemed to come out from under the dark rocks in the swamp. Then came the crash of 2008 and the subsequent fall from grace. By the time the world started recovering, we all got hit by the non-stop barrage of Covid-19 that grounded the whole planet to a very rapid stop. In the US, this was followed by a total political fracture of the US trust, belief system, and the American psyche, no matter how we look at it.

To most people now, it seems just a matter of time before the world falls apart. However, the truth is that reality is far from heading toward the end. The reality that media and politics do not want to share and something that I firmly believe in is that despite our highs of the early 2000s, we could not be in a better state of global opportunity. Take out the media, the trolls, and the political feeds, and we may realize that the world is heading in the right direction and the world is full of unheard-of opportunities.

We are currently entering a state of a total surplus of all times, unheard-of opportunities, and a world where the big boys are desperately getting bigger, but the world of opportunities for the common man has never been better. We are collectively sitting right in the middle of the greatest opportunities for a bright and innovative future, but only if we try and get our heads out of the rabbit holes of social dystopia.

Each time in the past, when the change happened, wealth distribution exploded. Current changes with the digital core and cloud deployments are opening unimagined opportunities that are now available for thinking. Change at the speed of thought is the new roadmap to salvation. These digitally enhanced opportunities are transforming societies, opening global communications, allowing face-to-face meetings, disruptive thoughts, and almost everything one can imagine.

This is the first time in our history that individuals have become trillionaires in their lifetimes.

The computing power of the laptop in your hand is very close to overtaking that of the human brain? The big question is, what happens when your laptop is smarter than you? I think mine already is. Cars are going electric. My favorite, Elon Musk, has challenged established industries and government institutions by sheer imagination and confidence in his convictions. He is our model child for our digital future. He built the first electric car and not only became the top car manufacturer but also changed the total gas car industry. He then built the first private rocket, which became the preferred rocket for the US government and NASA, which had become dependent on Russia to fly their astronauts to the space station. He has now proceeded to build the first human-to-computer interface with Neuralink, a new company, which, by the way, just got FDA approval for human trials.

He is just one human; you could be the next.

The new access via the internet is allowing students in India to attend MIT courses in the US. A consultant in Poland can configure a hi-tech system in the US. Genetic engineering scientists are collaborating across the planet to build genetic scaffolds that will build the future artificial hearts and new body parts built from your own stem cells. The world is full of opportunities, and the only limitation is your imagination and your will to do something. Barriers of time, age, and geo-locations have disappeared.

These societal changes and digital innovations have finally been placed in our hands — tools that can democratically change not only our individual lives but also the planet and make our global society a better place for all. Hence, we can join hands and collectively do something about this.

On one side are our unlimited opportunities that are so abundant that most of us cannot comprehend them due to the fog created on the outside.

On the other is our human brain's survival architecture, which can work for us or against us. Our psychological architecture is built for survival, but over the centuries, our learned colleagues have found newer and newer ways to hack into our weaknesses. Survival, as we all know, is a primary instinct of all living things, the most one can say. Anything that interrupts our survivability gets our attention very fast. The way to attract any human is to shout, *"Watch out."* Everything in our mind shuts down, our survival instincts take over, and we get into our well-known fight-or-flight mode. Media, politics, and trolls work incessantly on this one attraction and feed us daily and hourly with the bad news that addictively attracts us. What it does, in the long run, is that despite our lives being better, more comfortable, and more fulfilling than ever before in history, we have a collective pessimistic view of the world around us.

Today, toward the end of September 2023, it is hard to look at the future and not consider the Putin attack on Ukraine, China backing Russia, India abstaining, the Chinese treating the Muslim Uighurs worse than slaves, and the US supreme court proposing to the ban-abortion in the US. By the end of the first quarter of 2022, our world seemed to be headed toward ultra-polarized views on critical areas like abortion, climate change, global warming, wars, nuclear threats, internal and external terrorism, economic crisis, and the resulting food and other essential shortages. All are projected to only get worse.

There are three culprits for this incessant messaging; the first two are media and politics, and the third we will discuss at the end of this book.

Media collectively tend to push their pessimistic interpretations of all events as Main Street facts. The reason is that humans are more attuned to events that simulate crisis events, which then triggers our survival instincts. So, with a clear intention to get more readership, they routinely create a mountain crisis event in every molehill they can find. These end-of-existence positions not only attract the viewers but keep us glued to the sequence of these events. As one event starts to die, it will push a set of other events where you can choose which one is a greater threat from your point of view. This generates viewership, which generates revenues, which generates higher ad revenues and, thus, profits. Media has access to micro-segmented tribal members, and it already targets messages based on individual households.

Politics rides on a very similar wave but with a slightly different method of operation. They work on the principle of *Us vs. Them* with a communication philosophy of "It's not your fault you're in this mess; it's all their fault they did this." The reason for this is to form bigger and more loyal tribes on one side and alienate tribal members from competition. Here, the aim is to get more loyal voters, again via using media to create the interpretation of self-defined crises, often with a singular political divide 'point of view' interpretation based on their followers and tribal addictions at an individual level. Politicians leverage the power and segmentations of media, enhancing that with content providers like Comcast and AT&T to further microsegment their targets with finer digital products while using providers like the banned Cambridge Analytica to provide hyper-focused communications at an individual level.

In addition to the above two influencers, our politicians are their buyers and funding partners. The media is the biggest troll that delivers their divisionary polarized messages directly into your homes, in your living room, while your mind is relaxed and wide open.

But why does this work so efficiently?

The first gateway filter to all sensory inputs is our amygdala, which triggers all kinds of survival chemicals in our body and brain. It's always on alert and, for thousands of years, focused on triggering the fight-or-flight response. In the ancient days if you saw a tiger you stoppend thinking of everything else as you ran for survival, or fought for it if you had a weapon in your hands. Today, when you read of a very targeted threat on your TV, something that seems closer than some Climate change that will hit humanity after hundreds of years (fake), or even a war in faraway Ukraine, or China mistreating people in Tibet. All of this gets sidestepped when your brain is made to believe that your imminent danger is people of the opposite political tribe threatening your life. This fake distraction seems immediate and is made to be a bigger threat than all those far-away events. Just like, if you have a person with a gun facing you and asking you for money, it does not matter if the economy is nose-diving, Russia released a sub-nuclear bomb in Ukraine, or they just proved that climate change will hit us in the next 5 years, or even that your bank is being robbed. Your amygdala will ensure your focus only on the gun and the hand holding it — all else will blur into your foggy mental background as if it does not exist. Your brain does this as a survival imperative. You and I are here because our ancestors listened to this instinctive warning; now, it is being misused by trolls and influencers.

The second is the kind of information we choose to receive and slowly get addicted to. Research establishes that positive news is not as attractive to human eyeballs, nor as addictive to our minds, as threatening news is. Same reason as above. We are surrounded by "If it bleeds, it leads" or "If you think it will hurt you, you will get addicted to it," so each media provider tries to outdo the other in their battle for your attention. For people glued to their preferential media, the information fed into their blotting papered brain is simply snowballing off each additional rendition to what is grabbing the most eyeballs right now. More and more people are listening to fewer channels as they get more polarized and start to believe in common tribal themes.

However, if we look at empirical data over time, we see that the world we live in today has never been safer, wealthier, healthier, and less hungry than ever before. On one side are facts, science, and the truth establishing the world is indeed getting better; on the other are media and politicians running us down rabbit holes of panic, despair, and depression — when they seem to be our only salvation.

The fact is that the future is far brighter than our brains, media, and politicians would have us believe.

The new world, along with its opportunities and threats, is our digital world. Everything is getting digitized, even voice and pictures, and all this data is being stored in what we term big-data storage devices. To get highly focused information, we are segmenting all this data into ever smaller groups. The goal is 1:1 grouping, i.e., all the analytics we need at a single individual level.

Because there are so many areas we can talk about and so many layers. This book is designed for a reader from 9 years to 110 years of age.

Digitization impacts almost every person, every company, every process, and every nation. It has become all-pervasive.

If my book focuses on my experience areas, i.e., the technical details of digitization, then the book could get extremely boring, become applicable for a narrow audience, and become highly technical, so I'm keeping away from that.

The book right now is written for the general audience.

Once again, if I try to cover too many areas, the topics will again get confusing as we will not be able to cover even a few aspects of digitization in a single book. So, I have aimed this book, which a 12-16-year-old reader should be able to read and comprehend, with a deep dive into history and a focus on a few aspects of the digital world we live in today.

So, the resultant book is a brief history of humans culminating in our current digital era. While all along with a deep focus on one of the most important opportunities and threats to our new and evolving digital future. This singularity is the algorithm. Using this single programmatic bot, I have tried to simplify its use and misuse on individuals, groups of individuals that I refer to as digital tribes, companies and enterprises, states, nations, and politicians, and some of the various nuances that exist in between.

By 2021, our planet has entered a world of distrust. This distrust has been brought about by the fracturing of the human mind by powerful AI, artificial intelligence, and algorithm-driven bots that are creating the artificially induced *Trust Gap*. The goal of the controllers is to create a *scarcity mindset* in their followers, while the reality is that there is an opportunity to realize the *abundance mindset* that can take humanity forward. Saving Humanity must become our prime goal as we go through this book, realize the way our minds are being manipulated, and come out by the end with a pure and empirical mind that works using your biological intelligence and your desire to save the planet and thereby save humanity itself. Do not believe that your nation, politics, and religion sit outside the normal rules, scientific facts, cultural or social rules. That is the foundation of *Us vs. Them*.

Chapter 1 — The Seventh Wave

Garett Gee turned a lucky guess into a business opportunity when he was still at his university. When iPad2 was about to be released, he realized there were a lot of people who wanted to be ahead of the line to buy one. As a student, he also realized that as the iPad2 was released, there would be a blog post somewhere listing the top 10 apps for the new iPad2. He began looking for one and could not find it — this presented him with a unique opportunity to create one.

While most kids focused on games, Garett noticed that QR codes were popping up everywhere in shops, items, cereals, and even business cards. In his research, he found that current QR Code readers were clunky and slow. After realizing there could be an easier way to read a QR code, he made it his mission to design a new QR reader ad app. His goal was to become the first QR app for iPad2.

Firstly, he stood in a long line to pick up the first release of the iPad2 and put it into the hands of his IOS developer expertise. After two sleepless nights, very much like a hackathon, they had accomplished their goal.

His guess about the blog also proved to be correct as it was released soon after the launch of iPad2. As his app was targeted for launch on the new iPad2, he made it into that list. That blog was read by millions, and he rode on the crest of that blog's wave.

He then recruited two of his classmates. Together, they launched Scan in February 2011. Their company raised $1.5 million from venture capitalists; one of the investors was Google Ventures. In the first year, Scan accomplished 10 million downloads.

The number of scans kept growing, and by October 2011, they had reached 21 million. Initially, his app was for free, and he made his money in click revenue from embedded ads. His next step was to offer a paid version with a low annual fee and no ads. 30% of his customers took that offer, and the company continued to grow. The company was acquired by Snapchat in 2014 for $54 million.

"Every act of creation starts first from an act of destruction." — **Pablo Picasso, artist**

A billion years ago, when the earth had cooled down, the oceans had formed, and there was life in the oceans; our planet was a violent place of volcanoes and earthquakes. The phenomenon of continental drifts would frequently cause huge earthquakes, which in turn created tsunamis, both large and small. These tsunamis carried plants and fish from the ocean and left them on high ground as the waters receded. As the frequency and power of these earthquakes decrease, the planet slowly stabilizes to some extent. However, there were small upheavals that created high waves and tides, which threw some of the sea creatures and plants at a higher frequency.

These unplanned visitors were mostly left on dry land as the waters receded, where their struggling lungs gasped while they rapidly and later slowly died.

Over millions of years, some of the plants, due to repeatedly being thrown onto the ground, evolved. Over shorter times, some of these ocean creatures and plants began to adapt and evolve. Then, after a billion such cycles, some plants put roots into the grounds, while some of the creatures became amphibians who could stay longer and longer out of the oceans. Until one day, a single species found they could survive on land without going back into the ocean. Thus, they began the greatest adventure, the adventure of life, as they walked this alien land where they flourished, adapted, and evolved. From that point, I began this unique journey at the end of which you and I are now. Not just us, but all the life on this, a mere *pale blue dot*.

Around one to five million years ago, we believe that somewhere in current South Africa, an unnamed humanoid was most excited when they found that two special types of stones when rubbed together, create sparks at night. On a very dry night, they huddled in the dark to play with their magical stones; one of them rubbed the stones rather vigorously while sitting next to some dry grass.

Sparks fell on some dry grass, and the enthusiastic person kept rubbing the stones with his hands. Suddenly, there was a whiff of smoke, and they all pulled back in astonishment. Then, as our ancestral enthusiastically continued with more vigor, somewhere along the way, the grass caught fire. Initially, it was a moment of great fear. But later, it converted to excitement.

Some tried to catch the flame and burned their hands, and the first lesson was learned: fire is hot. Over time, they learned how to start a fire on demand, and that was the greatest evolutionary step for mankind. Humans could finally build a fire at will. This, some believe, was the birth of intelligence.

Our humanoid ancestors had discovered the art of fire.

This is the greatest launchpad of humanoids that took us on a long journey from simply lighting a fire to fire for security and somewhere along the fire for food, warmth in the cold, and light in the night that kept animals away from the fire. The evolution of these fire-bearing people was very rapid, and they were viewed as magicians or superior by most other humanoids. It was a secret some tribes carried for many years. We, today, are the result of this greatest discovery on a long journey in a time tunnel as we continue adapting and learning across time and space. Each evolutionary step has been a discovery, and with every discovery, we came to new physical, intellectual, and now technological forks. Each fork takes us in new directions. At each fork, some people feel secure with the warmth of the status quo, while others flow with the change and take their group to new heights or even extinction. New heights if the path was a successful road, and extinction if it led to disaster or even death.

Somewhere in Africa, around 1.5 million years ago, humans discovered the art of lighting a fire. We know this today as we can find microscopic ash that we can carbon date and thereby confirm that this was approximately the first wave when humans discovered the art of making fire. This singular discovery altered humanity for close to ¼ million years and helped take humans from nomadic apes to domesticated humanoids who would slowly evolve and adapt to become you, the reader, and me, the writer of this today.

Never forget we are the result of millennia of humans that have evolved, survived, and lived to pass on their genes to the next generation.

The digital disruption — "Disrupt or be Disrupted" was the advice we were giving in the 2014-16 period; by 2017, this had morphed into "Become the Disrupter."

Disruptive technology is a revolution in the world of technology that substantially changes the way that consumers, industries, or businesses work. Disruptive technologies move the methods, techniques, or practices away and replace them. It does so because it has qualities that are far superior and prominent as compared to the old technology that was in use.

This is a revolutionary invention that creates a new market and value network. This, in turn, eventually "Disrupts" the already existing value network and market. This helps in displacing the pre-established marketing leading companies and businesses, goods and products, and alliances. It is important to note here that while all innovations are revolutionary, not all innovations are disruptive.

An example of this can be seen in the form of the introduction of the first automobiles. While it was an innovation in its own right, it was not a disruptive innovation. The first early automobiles in the 19th century were expensive luxury products. They didn't disrupt the market and compete with horse-drawn vehicles. It wasn't until a lower-priced automobile was introduced into the transportation market that the market saw a disruptive innovation.

If we look at history closely, we can understand that certain waves have disrupted the course of history and changed the world and the way things were done in the past. There have been many disruptive innovations, but we can divide them into seven distinct parts or waves. Let's understand disruptive technology with the help of examples from the past.

It is understandable that with each passing wave, more and more power is given to those who adapted to the new wave. Moreover, each wave represents the desires of humans and speaks volumes of what it was that they wanted to achieve.

The **first wave** of disruptive innovation came with the discovery of fire. Possibly the greatest disruption in the history of mankind. With humans learning to light a fire of their own free will, humans separated themselves in their ability to adapt from every other animal on the planet. and were able to proceed further into this world. They were able to progress and move further in life. This allowed humans to light a fire that allowed them to see in the night, go out hunting, and keep wild animals away. Moreover, it also lets them cook food. This was the first leap that helped the homo sapiens to move toward a more civilized living society. This singular discovery allowed people to stay in one location for longer periods. It allowed humans to eat cooked meat for the first time.

The **second wave** of disruptive innovation came with the discovery of planting seeds for food or agriculture. This started early with food but evolved into the ability of humans to build permanent settlements. The result of fire and agriculture was the birth of villages, towns, and cities. Followed by landlords, Kings, and Emperors. The land became a symbol of wealth, and ownership brought about power. During these times, landowners had power as their land became the breeding ground for agricultural innovation.

During the 1st and 2nd waves, there wasn't a lot of importance or stress given to education. Gaining knowledge and acquiring an education were things that often took the backseat. And so, knowledge and wisdom during that time were under the control and influence of the Church. The Church and other religious figures came together and wrote books at that time, and these books were passed out to people of power. For this reason, during those days, you wouldn't find books lying around in the houses of common citizens. So, not only was knowledge limited to the Bible and other related scriptures, but it was also restricted in its usage, with only the upper-class being allowed to read it.

The **third wave** came about with the discovery of printing. Printing allowed the spread of unfettered knowledge. It broke the barrier between priests and God by bringing God to the common human with religious books finally written in the language of the common person. This helped in changing the dynamics of the world of knowledge. It made acquiring books of information and wisdom easier and more accessible. It changed the paradigm of knowledge itself and allowed the common man to become acquainted with books as well.

During the mid-1800s, we collectively launched into the **fourth wave** when the steam-powered industrial revolution took place. This

4th wave of disruptive innovation changed and revolutionized the world once again. It allowed for a swift shift between power and money. Landowners were no longer considered to be important enough. Those who owned and operated industries successfully (such as the Moghul's industry) soon became the real holders of power. Power went from the hands of landowners and landlords to factory owners during this industrial wave.

This wave lasted until the early 1900s when we went into Ford's production line assembly and the science of manufacturing. This represents an extension of the 4th wave. Eventually, the **fifth wave** came into play, and it brought along the first usage of disruptive technology and computers up until sometime around the 1960s. During this era, those who owned the latest gadgets and knew how to operate them were the people who were considered successful and powerful. Before this wave, it was the industrial engineers who were the arrowhead of success. After this point in time, i.e., the 5th wave, it was the technocrats who became the new Knowledge Workers of our world. These were the people with the highest wages and the most power. These people had the power to influence the decisions of other powerful people as well. Wealth during this wave shifted from the industrialists to the technology manufacturers like IBM and Univac that dominated wealth.

The **sixth wave** came with the rise in digital availability and usage of information. Information was made easily accessible during this age, and so it allowed everyone to use information freely. This can be seen with the increase in the usage of search engines, like Yahoo, and then specifically perfected by Google. Almost overnight, encyclopedias, Britannica, and other informational banks collapsed and were thrown out of business. Almost overnight, people could verify false statements from reliable resources. This is till companies like Cambridge Analytica made it their business to generate, distribute, and flood the digital dust with fake neuro-attractive emotional data bits. Information became synonymous with power and wealth, digital information, that is. In my opinion, this wave democratized information — both for the global good and bad.

With this wave, information became portable, easily accessible, and reachable, literally in your pocket. We carry more information in our hands than a bunch of scientists barely two decades ago. This wave has enabled all of humanity to use information easily. It has allowed all of us to become researchers and fact-checkers. Unfortunately, it also began a phenomenon of neuro-hacking or the technology of 'in through the outdoor' subliminal persuasion that two decades later would lead to critical hyperpolarization across the planet.

The **seventh wave** is the one where we are currently in, right now.

This is the wave of digitalization, which has leveraged knowledge from prior waves and used digital technology as a catalyst and a foundation to succeed in this wave. This wave has considerably revolutionized the world and the way things were done previously. Today, we know a heavily digitalized world. We have digitalized everything to the extent that even the things that couldn't have been digitalized previously have now been digitalized to some extent.

One prominent example of this can be seen in the way we handle music and photography now. Previously, these were two things that were always considered analog. However, contemporary society has changed things drastically. We have incorporated digitalization in photography, from the way pictures are taken and shared to the way they are edited and uploaded. It is the same way with music, as we can now place notes in a song and share it digitally, as well.

It is also astounding that technology can be used to identify or find details about something. One good example is the famous app called *Shazam*, whose main function is to detect songs just by listening to their tones. All users have to do is to play a song on *Shazam's* detector and they will get the song's artist and complete information in just a few seconds. Therefore, the app offers a real-time solution to a common problem of hearing a song but not knowing its name.

Our cell phones have been a tremendous help when it comes to technology and using digitalization as a way to move forward in this society. Our phones can help us connect with people and share our ideas, thoughts, pictures, music, and so on. Moreover, our cell phones have also helped us with the usage of the digital core.

Along all these waves, there was a definition of time during these times. We started with **disorganized times**. During this time, humans had no concept of time and did things when they pleased. Soon, time became binary with just the classification of light and darkness. This basic cycle brought about **cyclical time** that started with day and night, then further split into seasons and still further into years. The cycles were defined by crops and seasons that impacted our agricultural evolution. Soon, humans started recording events, and this brought about **linear time**. This was a longer duration of time, firstly based on rulers, kings, and emperors and then into dynasties. Over time, linear time, along with writing, became dominated by the written word. The written word was written by the winners and the powerful. This led to the concept of history and the art of learning from the past, but buried in this historical art was the reminder of the rich and powerful and their interpretation of the truth. Just as an example, if Germany or Japan had won the Second World War, our history, our heroes, and our modern kings and queens would have been quite different. History is the interpretation of the past as written by the victors. We are now entering the time of Fufactology, the data-driven science of learning from the future. A subject and topic that we shall discuss toward the end of this book.

Computers

Long before computers came into existence, the English language used "*compute*" for quite different reasons. It was used by the Syrians to calculate the grains their collectors received, on to in 1660 Samuel when Pepys wrote of a morning "computing the 30 ships' pay... and it comes to £6,538. I wish we had the money."

Compute was the science of calculating, while a "computer" used to be a person who did calculations. In 1731, the Edinburgh Weekly Journal advised young married women to know their husbands' income "and be so good a computer as to live within it."

It was very common for companies and government departments to advertise jobs as "computers" — right up to the time when the word was used for early electronic devices, and in some cases, not until the 1970s.

My personal favorite factoid is when Harvard University started mathematically proficient women as early as 1875 for very complex astronomical calculations under their director, Edward Pickling. This started as a glass plate variable in stars and then adapted to complex orbital and astronomical calculations. Several women from this group became famous astronomers. They inherited the name Harvard Computers.

The first digital computer was invented in the mid-1800s. However, at that time, technology and resources weren't as developed as one would have hoped for them to be. Therefore, the invention was overlooked by the public at large.

It took a long time for the resources and technology required to become available finally. This allowed people to work on the ideas from the past and develop them into something better. Finally, in the 1960s, the first supercomputer was invented and introduced to the world.

Supercomputers were large enough to take all this digital information, gather it all into one area, and very rapidly align it into the form of information required. This was the sixth wave requirement, and Supercomputers met them easily. Moreover, this helped make the digital core the cloud, the hardware, and the infrastructure that distributed databases all across the planet. This new piece of technology took the world by storm and radically changed things and the way they were done.

Had this occurred at any time before it actually happened, this could and probably would have been a disputed topic. It would have invited critics to tear the ground-breaking invention apart. However, Supercomputers were introduced at the right time, during an era that was able to appreciate what had happened. This helped in creating a great digital disruption.

Digital disruption is a disruption effect that radically changes basic expectations and behaviors in the culture, market, industry, or process. This can be caused by or expressed through digital capabilities, channels, or assets. The word in the market and the world right now is, "Either you get digitized, or you get disrupted." This highlights just how crucial it is for people to adapt to the current trends and needs and get digitalized.

This can be seen through various real-life examples as well. Blockbuster was a complete hit, and it had been a people's favorite. However, it was adamant about sticking to the traditional ways, which is why it lost the race. Today, Netflix has triumphed over Blockbuster and has become a household name. Another example of this can be Whole Foods Market, a top company that has been doing well in business terms. However, its refusal to digitalize the business structure resulted in it losing its customers and market base. Amazon adopted this approach, which is why it was able to take over. Other leading examples of this include Spotify, Airbnb, and Uber.

The more we look at the history of disruption, the more easily we can see how it is causing a shift of moving money, power, and influence from the hands of a few to whoever has a new idea and a desire for change. It has been made clear that those who bet and put everything at stake often win if they are confident that their product or service will create a disruption.

Currently, it is shifting a lot of wealth into the hands of a few, but in the future, clearly, it will shift a lot of wealth into the hands of many as some of the few super-wealthy start giving their wealth toward 'Saving Humanity' endeavors.

New ideas need to be filtered in a structured process; we need to look at the 1 + 4 pillars of digital success.

- User experience

- Data

- Digital

- Higher Quality

- Lower Cost

Out of all these, the most important attribute is the User experience. We can no longer aim for satisfied customers or happy customers. We now need to 'delight' and 'wow' the customers with our products and services.

We need to create and develop sticky products that are so easy to use that the user becomes the most prominent advertiser by sharing them with their friends.

Now, these are all very important for an innovative new product and service to thrive. As someone who had to travel frequently, it was often a hassle for me to get to the airport. Normally, I used to call agencies or companies that would provide me with cab services. I would ask for them to send their drivers at a specific time to catch my flight on time.

However, it wasn't always easy or simple. Usually, the drivers were on time, but that wasn't always the case. I remember that there were instances where the drivers would be late in picking me up. Other times, they wouldn't even show up. They had their own reasons for that, of course. Sometimes, there would be an accident that would block up the road. Other times, the traffic would be so bad that it would take them hours to actually get to me. Either way, it was bad for me as it resulted in me missing out on my flight several times. I am sure many people can relate to this and share my sentiments about the topic.

All of this changed greatly with Uber. Uber came into this world and set about a new trend. Previously, companies would buy cars, get their insurance, and hire drivers to drive those cars around. Uber challenged this because it neither owns the cars nor the insurance of the cars.

The company thought out of the box, and it relied on the three pillars of digital success. The evolution of information was in the form of reports that were analytic. However, Uber introduced a new concept called *informatics*. This helped in visualizing nothing more and, yet, nothing less than what the customer required or needed.

So, with informatics, Uber was able to provide us with a lot of useful information. They took the information about our car requirements, our location and our drop-off location, and time and fair specifications. Then, it turned that information around and informed us about the car (company, model, color, and number plate) and the driver (name, picture, and contact number) that would be coming for us. Moreover, it also informed us about the time it would be coming to pick us up, the route it would be taking to pick us up, and the estimated fare. In my experience, Uber has been very accurate with the information it provides to its clients and customers. From the car and its color to the car's driver and their name, everything has almost always been spot-on to the estimated arrival time. This is something that hadn't been possible before.

So, how does it rely on the four pillars, you may ask? Well, there is data there in the background that helps them. As I create my profile, I enter my data, including my name, number, location, and credit card details. This makes Uber recognize me as a customer. When I enter information about where I want to go or when I wish to leave, Uber takes in more data.

Moreover, they deliver high quality at reduced costs. For any digital company to become successful, balancing these two pillars of digital success is of utmost importance. These are the primary requirements of customers, and therefore, digital success relies heavily on a company or a service meeting these two areas. Many new companies fail because they have been unable to rely on the right pillars of digital success.

There are many opportunities for digitalization. A few samples for this:

- **Micro-segmentation:** It is useful to break things down into ever smaller segments and extract more and more atomic information. For example, back in 2004-5, we helped a US Telecom company, let's call it company A, collect data for their market as they used to split their customers into 5 segments, i.e., under 15, 16 to 30, 30 to 45, 45 to 60, and above 60. Most large global companies created almost similar market segmentations. Around 4 years later, we were conducting detailed TP+CO-PA (Trade Promotions plus Cost & Profitability Analytics) analytics of over 32 million customers. Business Pain: The TP-CO-PA run for this company used to take around 23 hours to run. Due to this reason, they ran this audit once a month on the weekend after the month-end close. One in 4 times, the program was unable to finish in a timely manner, so for these months, the company ran blind till the next month-end weekend. In 2015, the company installed a new, blisteringly fast columnar database from SAP called SAP HANA, and we started with the legacy five segments and, with three months of tweaking, the report ran in 23 minutes. Having achieved this huge optimization, we proposed a more granular analysis by dividing the customer segment into 182 segments, i.e., one for each year from 10 years old to 100 years, and a further segmentation for males and females. Once again, with a little tweaking, this report ran in under 11 minutes. In this example, we went from a marketing and promotions reset time of three to four months to once every two weeks. In 2016, we went live with this report running every weekend. By 2019, I heard the company had dropped the run time from 11 minutes to under 3.5 minutes and started running this report every weekend. Now imagine two telecom companies in the US. Company 'B' runs a 5-segment TP-COPA analysis once a month and takes 3 months to react to any market change. Company 'A' runs a 182-segment TP-COPA analysis every weekend and fine-tunes its customer taxonomy and promotions with surgical precision every two weeks or so. You decide which company will win the digital promotions battle, and you will be right.

With this new digital core, computer companies can conduct analytic runs of 100 billion records in under ten seconds when the paradigm of analytics itself changes. This led us to analyze how an eighteen-year-old boy's choices were different from an eighteen-year-old girl for example. This is a very powerful marketing tool for hyper-targeted digital promotions to a very focused audience. This granularity allowed us to look into the matter of how we can satisfy an extremely focused age bracket in ways that were not possible before. We finally understood that the behaviors and preferences of all people at different ages (micro-segments) are different, and so we should treat them with separate digital messages and enhanced user experience components.

This resulted in a paradigm shift in analytics, marketing, and influencing the minds of customers. Within a few quarters, we found that this was just the beginning. For example, an 18-year-old White girl had rather different preferences from a Black, Mexican, or South Indian girl of the same age. And this went on by nationality, color, religion, and so and so forth.

Such micro-segmentations resulted in a shift in the paradigm of customer analytics completely.

Previously, companies would create trade promotions or marketing materials following these micro-age segments with gender identifiers. They generalized and treated people with the same broad brush. This was either a hit or a miss, and so businesses were unable to capture the full attention of people properly.

Now, businesses entice customers one on one. This is more direct and swifter. It is swift and helps in attracting the right audience to the right product.

- **Micro-adjustments of Addictions:** Humans, by evolution and survival instincts, are more attracted to fear and alarm than to good news. In movies, we're all more attracted to the Joker than Batman, to Loki than to Thor, and to the bad guys than do-good people. Media, politics, and religion all work on this one singularity. We can see that since 2019, media and politics have played a big role in defining the polarized citizens in almost every nation with smartphones, WhatsApp, and smart TVs. The power players proceeded with the divisionary creation of 'Us vs. Them' more effectively than ever before in history, with the addition of it's not your fault, it's theirs. Every divergent political party over the past few years has become a specialist in making daily end-of-day depressing predictions, sometimes even if they have to cook the facts from fake sources or because some powerful person gains from that bit of news somehow.

This fear within people can be activated like a neurological chain-and-ball around their ankles. Media, along with politics, has been capitalizing on this for decades, and now new participants have started entering this domain via the digital network we call social networks. Within these social networks is a mammoth attempt to discredit normal media sources with the 'same '*Us vs. Them*' statements but focused on satisfying an ever-smaller group of like-minded trolls and followers — the micro-segmentation. Like most things — social media can be a boon to good folks but also a beast when there are institutional or paid trolls on the other side. This can be a difficult combination to work with, but it has become a blessing thanks to digitization, both for the good and the bad guys. What we have to do as individuals is separate the false from the truth and separate the influencers from the fact tellers. This may sound logical and easy, but it is becoming more and more difficult to separate facts from spin. Almost 50% of polarized US citizens today believe along political lines with little loyalty to the nation or the truth. Both sides blame the other, and, in the end, who suffers is the nation and its people.

More and more successful companies now know, rather accurately, how to target/attract customers. We have successfully conducted neuro-mining of people's fears and desires, which has proven to be rather profitable for specialist providers. This is because social trolls feed people the information that the people themselves desire to see and hear. This makes recipients attracted to the source of that information or that specific product or service the business is providing them with. For example, if a Republican is very dissatisfied with his or her life, they can be weaponized like a human weapon in a matter of a few months. Remember, we used to see movies of Russian sleepers who could be activated by specific commands. We can today create thousands of open-eyed sleepers by neuro addiction to specific likes and desires that can be targeted today at an individual level. Different messages to each individual but for an overall similar purpose. We then end up with a January 6th coup where one side wants no inquiries on their patriots while the other side thinks that such an event must never happen again in the US. Yet, the polarized battle of the addicts continues, with no one knowing who the puppet masters truly are.

- **Individuality Explosion:** We are living in a time where we are becoming more and more individualized than our past persona, even as early as 2015. In addition, we are also becoming more and more polarized by each other. Every individual is being exposed to social network messages that amplify their individual uniqueness. This is great news as unique attributes like LGBT, BLM, White oppression, race, color, and other small groups that used to feel alone and now have become part of the digital, and thus social, mainstream. Just like companies are micro-segmenting, individuals are exploding their micro-interests into the global digital planet. Some parties, unfortunately, have decided to take it a step further by forcing their individuality down other people's throats.

Now more than ever, there has been a huge individuality explosion. Our individualistic echo chambers explode our individuality and sometimes force us to not accept other points of view. We have lost patience, and we refuse to listen to other people who may have different opinions or suggestions. This issue has grown to such an extent that it plagues families, cities, and nations alike.

This has resulted in individuals being targeted based on their opinions and the things they have been sharing with other people. Different people get noticed and highlighted based on their conversations and communication. While some countries offer privacy, others don't. This either suppresses or encourages individuality explosions.

- **Disruption by Simplicity:** With the advent of micro-segmentation and intelligent bots, we don't have to give one message to a nation, a state, or even a zip code. Digital trolls can simply broadcast an individual message tailored for an individual citizen based on their likes and dislikes and with an aim to make them future potential targets. As we evolve through the ages, in our interactions, the goal has consistently been about making things simpler for the end-users to accomplish their tasks. For example, during the agricultural era, our collective focus was on making more and more efficient user tools. In the middle of that evolution, we today have our smart seeders, tractors, and watering systems that are run by a complex matrix of sensors and decision processors. The goal is to make it task-friendly and more efficient in all key inputs and outputs that are rule-based. Extrapolate this to the future, and we will have entire mega-farms run by robots that will plow, seed, and harvest perfect crops in the most efficient manner. The farms of the future might be in the basement of the very building you live in (Already happening in a Chicago O'Hare Airport, where massive farms are being planned under the runways and buildings so all the fresh food served at the airport comes from within the airport) The same has happened to the horse-drawn cart that has over time evolved into the Tesla today. Extrapolate this to the future, and we head in the direction of autonomous vehicles that will need no drivers, no garages, or single ownership. The goal at the end of these tunnels is to configure all the complexities in the background using digital sensors and configurations to make the experience of the users exceptionally simple.

As you can observe, the evolutionary and transformational waves have been coming in at a faster rate and changing the course of history in more significant ways each time. Previously, what took centuries now only takes a few decades. With the seventh wave in full swing and already influencing the world differently every single day, one can only wonder when the next wave shall come and what it could unfold.

Take Away

1. **Individual:** *Individuals, being the smallest component, can and have to adapt the fastest. The opportunities for individuals are the greatest. Through the ages, success has been defined by life forms that adapted. Keep life simple — flow with the change. While Covid-19 has accelerated digitization, it is important to realize that we're right now in the golden period of digital opportunities. With the right mindset, if children as young as 9 years can succeed in this new digital economy, then no human on the planet can hide behind any excuse of 'Us vs. Them' whatsoever.*

2. **Companies:** *Companies are a collection of individuals. The larger the company, the slower it will be able to move or adapt. Companies have been and will continue to fall like dominoes unless they adapt to digital changes. In the last decade, the message has shifted from Disrupt or Disrupted to Become the Disrupter. No company is too simple, too small, or too large not to disappear in the blink of an eye. We're surrounded by companies going bankrupt, so learn lessons from other companies and plan your digital transformation.*

3. **Nations:** *Nations are the dinosaurs of the pack. Leaders of most nations are old dinosaurs, too. More and more older people with power and money are taking over the reins of control. More and more leaders are heading down the path of autocracy. It is in our hands to control all this. Nations are harnessing both the good and the bad of digital disruption. On the good side, nations are building digital highways; on the bad, they are using this to curb individual freedom. Digital connectivity gives autocrats the power to simply shut down entire cities or specific attributes with the flick of their wrists. Countries have to be very careful as leaders are rapidly traversing from democratically elected to autocratic presidents for life. At this rate, we will rapidly devolve back to hereditary autocratic kings as a form of divine right in the garb of elected presidents like it is in North Korea today. One can see this happening in Russia and China. It almost happened in the US and still could unless we correct our course rapidly.*

4. **The 'P' factor:** *This digital disruption has enabled individuals to build simple applications for iPhones, Androids, or bots that have provided the most needed information at the tip of our fingers. At the enterprise level, it has empowered companies by upgrading to a digital core and conducting global businesses under a single database. I worked for a pharma company that had 14 ERP systems worldwide. We helped them drop it down to two. One for the headquarters and the other for all the rest of the subsidiaries on the planet — a saving of 68% in global IT costs and an increase of 33% in globally harmonized data. At the national level, there is an opportunity to standardize every state's federal and state*

financial KPIs at source and then consolidate them all back at the federal level. This was a project we started at the President's office, but then it kind of faded away due to a lack of commitment at the top and a lack of ownership at the individual state level. Now imagine if the UN made it mandatory for their members to specific global KPIs, some of which they already do, but a lot of it is either missing or not provided by member states. Then we can have a global scorecard that is more actionable. Focus on the great things this digital disruption can do for each layer of humanity, as it will then roll up to goodness at a planetary level.

Chapter 2 — Global Digital Disruption

Right now, there is no age limit to becoming a digital success. You can be 9 years old or 90 years old. This is because none of the agrarian, industrial, or other barriers apply to you. The biases are only in your mind, as there is no age limit on becoming an entrepreneur. Just as a bumblebee can fly because it has not done aeronautical engineering, so can a 12-year-old become a global success all on her own, with a little help from her parents. However, there is an age requirement for forming a legal business entity in the United States. Due to contract law, a person must be at least 18 years of age to form a legal business. Fortunately, parents and guardians can file on behalf of minor children. At age 15, Hillary Yip is the youngest CEO in the world. She founded and runs Minor Mynas, an online education platform for children. She began her journey into entrepreneurship at age 10, dabbling in the tech sector, and now sits at the table with some of the world's most renowned tech geniuses. If a 12-year-old can take an idea to success, what is stopping you?

The world has started moving toward a disruptive global digital disruption. We have all witnessed a very rapid digital transformation all around us, and it has completely changed the dynamics of our thinking, interactions, and society.

When we look at history, we notice the significant shift in attitudes and thinking over time. Previously, people preferred to read paperback books. Now, everything has shifted toward a digital approach. At this age, people prefer to do the reading online. I do most of my reading via audible.

It has been the same way for pretty much everything else. Previously, everything was brought from a local physical brick-and-mortar store. Now, we buy from global digital providers. We can buy things online without ever stepping outside our homes. We can even get things shipped from different parts of the world without seeing things beforehand.

However, as we can see, the speed of change is happening at an ever-faster pace. Thus, to remain in balance, we need to adapt ever faster too. In the early days, things wouldn't move so fast. Changes as significant as this would take much longer to come by. Now, rapid changes have become the norm.

Technology has become the catalyst for all this change, and technology can shift much faster than humans ever could. Everybody had their own predictions and assumptions about how far the world would be digitalized by 2030 or 2050. However, with the COVID-19 pandemic, things took a swift turn. They moved forward at a much faster pace. Everything that was supposed to happen decades later started happening at the moment. In fact, I had been writing this book for 3 years, and within 6 months, my book became obsolete, as everything I was predicting would happen around 2030 is already happening. So, I had to rewrite this book from scratch again.

Previously, for most people, the idea of working from home had only been a dream. However, this fantasy was forced upon the whole planet as soon as the COVID-19 pandemic hit us. With everyone forced to stay indoors and strict limits on any physical contact, there weren't too many options available. People moved online, businesses moved online, governments moved online, and the work that needed to be done shifted from physical to digital infrastructure and online platforms. Only this allowed a minority of the people to work from home. The rest of the people, businesses, and governments that did not have a digital infrastructure went into digital night blindness. Some never recovered, others barely survived.

Many businesses and companies have stubbornly refused to go along with this digitization idea. They refused to agree to an online platform, and Covid-19 made things worse for them and their own companies. At this same time, other companies were unable to establish proper systems in place that would allow employees to work efficiently and smoothly from their homes. For these reasons, many businesses failed and became another one of the COVID-19 fatalities. While others bloomed. The single point of truth — digital adaptation.

Businesses have developed and perfected strategies that allowed them to move forward, even in this pandemic. These are the businesses that did not need workers to be physically on-site to conduct their work. Such businesses allowed their workers and employees to work remotely and online in the perfect manner. Their entire workforce, limited in other cases, conducted regular meetings online with employees while continuing to hold online seminars for clients. Many businesses have thrived in these digitally remote times. These are the COVID Winners. Many individuals have thrived too. These are the digital workforce.

The aftermath of this has been that many businesses have realized that working online will save them costs in the long run. For this reason, many companies have started closing down physical offices or moved to smaller spaces. This remote workforce cuts many overhead costs, which is again a plus point both for businesses and workers.

Moreover, some companies have given their employees the option to work from home forever. Many employees have accepted this offer as well, as it saves them a lot of time as well. In addition to that, working from home often provides staff with flexibility in timings and hours, which has been a bonus in these uncertain times. This leaning toward the online platform has jointly helped the employer and the employee.

Working from home also changed a lot of things. People lost track of time and began to put in many more hours than they used to previously. It affected and altered their working hours and also put a strain on employees. People shifted from the eight-hour shift to a shift that would allow them to get their work done. Work shifted from being hour-based to becoming task-based, where people were expected to complete the tasks during the day, in whichever period they preferred.

Everything else has moved online as well. Schools shifted online, and virtual classes became a revolutionary technology that came to save the day in our time of need. Students not only took classes online; they also gave their exams online.

It has been the same way for many other businesses as well. As gyms shut down, the trainers worked on their clients remotely and online. Offices shut down, and so salesmen began selling their goods and products online. All physical activity was restricted, and so, everything turned to the wondrous world of digitization. The people and companies with an adaptive mindset thrived. While those that fought the digital adaptation found solace in digital messages on politics, religion, on discords of any sort.

Previously, we used to interact with people physically. Now it is being done virtually. We meet people online and prefer to communicate with them through our gadgets instead of meeting them. Whether it's someone in our professional lives or our personal lives, we tend to lean on the shoulders of digitized connections and meet and interact with them virtually. This has led to the collapse of physical interactions.

Thanks to the COVID-19 pandemic, this swift change is what I like to call the "Leap-Frog Effect." It has brought advancement far too quickly than any of us expected or imagined for it to come. The world wasn't ready for it, but we managed to deal with the pandemic and make the most of our circumstances simply because the world had already normalized digitization. What COVID-19 did was bring forward global digital disruption sooner than we anticipated.

Altering the Concept of Time & Place

COVID-19 has affected many different areas and industries along with it. One of the most prominent examples of this is the food industry. Previously, the concept of outdoor dining was frowned upon and thought of as a European concept. However, the pandemic quickly changed that. Some restaurants are now deliberating over keeping outdoor dining, so far, a very European option, now as a permanent option in the US. Many streets have closed for good to help support this outdoor dining experience.

Another effect of COVID-19 was something that I witnessed and experienced firsthand. Back in January 2020, I worked in a well-reputed company, and right before the lockdown took place, we surveyed our infrastructure. We needed to determine if it needed to be updated, and the consensus was that we were good the way we were.

When the pandemic hit us with full force, literally overnight, there was a total lockdown announced. When that happened, all employees were sent home, and there was confusion about what was supposed to happen next. As the whole work-from-home concept came into play, there was a constant need for updating work infrastructure. The number of internet devices that were sold, the switch to better quality devices and packages, was extremely high. We had hoped to stay the way we were, but the lockdown forced us to change, adapt, and evolve. It was a change no one expected to make a few years earlier, but the various disruptions, including covid-19, brought it forward at a much faster pace.

With digitization, the duration of time-to-change has altered significantly. Suppose we forget about COVID-19 and go back to the seven waves of disruption. It is safe for us to assume that there wasn't a lot of priority given to time in the pre-industrial years. This is because there was no proper concept of time nor the need to manage workforces in a time-bound manner.

Thus, we can safely state that in the pre-industrial days, there was no real concept of time other than that it was either day or night. However, as time went on, when the Industrialization wave hit us, the concept of hours was introduced. The system set up hours for everything. Hours when workers would need to go to work and hours when they would go back from work. This need for industrial workers changed education as a place for preparing the industrial workforce on daily hour-based schedules. Schools thus became the tool for the industrial giants to prepare their future workforce for working in their industries. The people in power modeled the schools for industrial production, and the US school systems were allowed to model their institutions accordingly. There were hours for when children would go to school and when they would come back.

Later, with the next industrial wave, time was further sliced, and the concept of minutes was introduced. This helped people keep better track of time and stay in touch with all that was happening to them. Today, it isn't uncommon for people to keep track of seconds and even nanoseconds because, as the adage goes, *time is money.* In fact, even seconds are further divided into a million and billion digitized parts.

As the concept and importance of time changed, the intervals of time changed. And as this changed, the duration of change also changed swiftly. For example, back in the agrarian days, it took weeks and months for news to travel around. Even then, the distance the news would travel would be short and limited. Now, it only takes a few seconds to reach the entire world. The compression of time, and space, has accelerated, and it is completely linked to the speed of change.

Just as an example — when a White policeman, Derek Chauvin, gleefully placed his knee, and his full body weight, on the neck of a Black man for over 9 minutes, casually with his hands in his pocket — this image that reverberated across the US within a few hours and across the planet in a few days. It resulted in this White policeman being convicted of the murder of George Floyd on all five counts. What sets this case apart was a flawless, undisputable optical witness, such as this nonstop video that ran for over 10 minutes. There was a very brave bystander who recorded the entire execution for all of the 10 minutes as evidence that the police could not deny or alter, and the world could see for itself. This is the power of the new digital world that today sits in each of our pockets in the form of a camera that is better than what professional photographers carried just two decades ago. People joined the BLM (Black Lives Matter) peaceful protests worldwide. When some of these protestors were attacked, that too became worldwide news in a few minutes, and the world watched the racial inequality of the untouchables in the US. The videographer won the Pulitzer prize for her video.

Another example is the attack on the US Capitol and the 2020 election, an event that took place on 6th January 2021 — the attack on the American capital became news within seconds nationally and minutes internationally. The world is still watching the USA — the so-called citadel of democracy and the bipolar politician's viewpoints on truth vs. politics in a revolt that was prompted and supported by a defeated president trying to discredit the elections. The bigger impact is on how the two political parties once again have a hyper-polarized opinion of what happened that day — despite hours of videos and proof of the event that left policemen dead and senators running for their dear lives at the time of the attack. Although one can write an entire book on Jan 6th, one thing we can all take home is that the social network, religious institutions, Russian trolls, US politicians, and the ex-president have collectively divided the US mindset to a point where the democracy itself has fractured into unsurmountable differences of right and wrong. The biggest concern for democracy should be the attempt of the senate to *White-Wash*, no pun intended, the entire coup as if it never happened, and their direct refusal to hold an inquiry on the attack on the very people who are in this decision.

Digitization has resulted in what I like to call the global Amazon effect. On one side, we can lock up a White policeman for a murder of an individual. While at the same time, we can have a US political party deny an inquiry into an attack on the Capitol and attempted murder of the US democracy itself with 70% of 47% of the population wanting to both deny the facts and at the same time hide it under the carpet. This is an effect we are starting to see at all levels, from an individual, a company, or an enterprise, and now as a nation. The effect is total digitally induced bipolarity right down to the individual's subconscious mind. Amazon came into existence with the sole purpose of selling books online. Very soon, they realized that there were always many readers interested in reading; most knew exactly what they wanted to buy and didn't always prefer going to the bookstore to buy their books. They decided to capitalize on that and started selling books to such people.

At the center of Amazon's now well-known philosophy is customer experience. Not only as a statement on some executive strategy but something they have always stayed focused on in every endeavor they launch. Amazon is customer–centric and has been responsible for the era of customer experience. It is not surprising that with this customer–centric approach, they found more and more areas where they could use this philosophy for many other things out, areas where this focus was missing. Very soon, Amazon went from being a dragon to becoming a mama dragon; they invested millions of dollars in new ideas in identifying services and products that people were willing to transact online. One acquisition after another, they continued to acquire one company after another.

Thus, Amazon began to expand. It reached the billion–dollar milestone in a short period, earning it the well-deserved title of *Unicorn*. This title is awarded to anyone and everyone who expands and reaches the 1-*billion* milestone quickly. However, very quickly, it went from unicorn to *Dragon* ($100 billion).

Following this successful step, they started this unique trend of becoming a "Mama Dragon," which is a term I like to use to describe what Amazon accomplished. As they are so focused on their customers, their needs, and their experiences, they were able to expand and become bigger and better. They have consistently worked on elevating the sticky experience that they provide to their customers, and so they became the first ones in the industry to offer free shipping, along with a 100% no–questions–asked return policy, etc.

Initially, I was hesitant to transact with Amazon and this concept, but over time, I have accepted this. I have seen my daughter order various things and get free delivery on them. I have also seen the returned items that she wasn't too happy with, and I have seen her get free shipping and free ship-out for it all. This changed my mind and convinced me personally about Amazon's credibility. It is my own experience that has thrilled me the most.

I think the real game-changer for Amazon was Amazon Prime. This created a loyal tribe that is well regarded. Tribal loyalty toward a company is important, and Prime helped establish just that. This increased their tribal following to such an extent that no other company could keep up with it. Even Costco was unable to raise such a tribal following, which is why Amazon flourished. This has highly influenced the retail world all around. At the same time, it is important to note that Costco, too, flourished as more and more retail stores started closing.

The fact that Amazon offered so much on its digital platform that everything was just a click away became a game-changer. Giving all your services away on a smartphone or laptop is one of the biggest forms of digitization. It is digitized technology influenced by different parts of the world, including Russia, China, and Iran, and it has changed the course of history in more ways than we realize. Some countries, like China, created their versions of Amazon, Ali Baba, for instance. Even beating Amazon locally by prohibiting them within China or creating barriers to their operations. Digital excellence can leverage the impact of robotic technology, machine learning, and artificial intelligence put together. An example is Tesla Inc., which manufactures cars that use electricity and clean energy to function. I remember visiting the Tesla factory with my family, and it was beautiful. It made use of digitization in such a perfect way and displayed the finer details of their creativity in such an intricate way that my family and I were all spellbound. It exhibited the work of a genius that had the absolute zeal and passion to create something so fine and convenient. The creations presented by Tesla do not introduce a Disruptive Innovation phase in digitalization, but they also set the scene for every industry, enterprise, and service on the market, as I had detailed here[1]. Unlike other automotive companies like Ford, Tesla is focusing on the right side of the market, considering how it is the age of digitalization.

Many other cars have followed in the footsteps of Tesla. However, Tesla Inc. remains to be the real trendsetter. A long time ago, I predicted that there would be cars that would simply run on technology. I believed that they wouldn't need their steering wheels to be touched. Tesla made that prediction come true, although, as of yet, only partially. I still have faith in digitization and its process, and I know one day, we will have such advanced cars, even cars with no steering wheels.

Moreover, I firmly believe that technology will slowly become so advanced that it could accomplish many other seemingly impossible feats in the automotive world one day. Just as we moved from manual cars to automatic cars and then started opting for electric cars, it is only a matter of time before we will start relying on self-driven cars

---

[1] Guleria, H. (2017). *IOT-4-automotive 6 of 5 bonus: The digital ripples of Tesla model 3*. LinkedIn. https://www.linkedin.com/pulse/iot-4-automotive-6-5-bonus-digital-ripples-tesla-model-hari-guleria/

that use technology and digitization to function.

A peek at the future of cars: Imagine the changes using such self-driving cars would introduce to our world. We would no longer need traffic lights or speed limits, as these are designed for humans who tend to be unpredictable and reckless, but artificial intelligence and technology aren't. Moreover, it would demolish the concept of *Driving Under the Influence* and the charges related to that. It would significantly bring down the rate of accidents as well and ensure the safety of the car's passengers. Moreover, it would also reduce traffic, as cars would have enough intelligence to know which routes to avoid and which ones to take at rather high speeds.

It is already being influenced by AI. New artificial intelligence chips are being created for neural networks, deep learning, and computer vision. The AI hardware will come with GPUs instead of CPUs to accelerate decision workloads, special purpose built-in silicon for neural networks, neuromorphic chips, etc. It is an example of how fast digitization is beginning to change the world once again. AI today controls much of the digital entertainment industry as well. Today, apps can figure out the music being played on every device. They can figure out the exact movie you have been watching as well. You have apps that can connect to other devices and play music, videos, movies, etc. Your music providers and your entertainment providers already recommend music and movies that you might like based on your listening and viewing trends.

Digitally induced AI can also influence politics in countries, which is what reportedly happened in the 2016 US elections. Previously, there was a lot of speculation about interference in the elections. As it turned out, there had indeed been outside interference in the elections, and it was only possible because of digitization. The culprits range from Cambridge Analytics and Russian Trolls to other enemy state players. But, the most influential were the trolls and the buyers sitting inside the US.

Like any other thing in the world, digitization comes with its pros and cons. It has a majority-itemized list of advantages that cannot be denied. However, it also comes with certain disadvantages that cannot be overlooked. The thing to understand is that digitization is quickly taking the world by storm and affecting anything and everything. There isn't a single area or sector left untouched or uncovered by digitization.

Take Aways:

Individual: You and I are living in the Wild West of a hundred years ago, but this time a digital wild west. Then there was the metal gold. Today, there is digital dust. This digital dust is of immeasurable value. It is more valuable than gold or oil. Gold and Oil can only be sold once, but digital dust can be sold again and again without losing its intrinsic value. You can be the next Facebook, Amazon, or Spotify with your unique idea. You can be the next millionaire around the block by simply copying successful US digital ventures locally in your country. The opportunities are endless, and so are the excuses for failure.

Companies: *Digital disruption is not an iterative process but a total transformation of ideas and ways of working. The first step is to have a unique idea and then put all your passion and life behind it. Every problem your company faces in your day-to-day life has an opportunity written on the other side. Don't just plan to be digital, but plan to be the disrupter in your industry.*

Nations: *As a nation, some folks see no light at the end of the current global tunnels, while I see a bright light. Some say the US has taken a nosedive because of the disruptive digital opinions that are being unleashed by the two prime political parties, the ex-president and his political appointees. Our elected appointees are today spending more time criticizing the opposition presidents and their work than doing what they were elected to do — serve the people. We are starting to see an erosion of a constitution and the American 'we the people' foundations as individuals and party priorities are willing to sacrifice the greatest democracy for their petty gains. It is 'we the people' that need to change this path or succumb to the will of the future we vote into office, along with whatever gets the American people.*

The 'P' factor: *However, there is a saying that out of the chaos, the structure will evolve. It goes as far as that without chaos, we would never build structure. So, we were at that cusp. We have been handed a very powerful tool, and Pandora's box is now open for anyone to use. Initially, the good and the bad folks will use this technical accelerator; then, as a planet, we'll use our intelligence to curb the negative and reward the good. The potential of good in humans far outweighs the bad that a few autocrats and bad players can accomplish. We have seen good triumph over bad repeatedly century after century, so read of all the bad to watch out against, but keep your mind focused on the goodness ball that we each need to carry in this game of technical evolution.*

Chapter 3 — Information: The Center of the Universe

Everybody knows of Facebook, and who does not know of its founder, Mark Zuckerberg. He is possibly one of the most famous entrepreneurs who, at 19, co-founded Facebook. If Facebook was a nation, it would be the world's largest nation. In the same league with Mark is Dustin Moskovitz, co-founder of Facebook, and Evan Spiegel and Bobby Murphy, co-founders of Snapchat. It is most important to remember not all young entrepreneurs will reach the status of Facebook, Uber, Airbnb, or Snapchat. In the same breath, it is also important to remember not to give up hope. It is equally important to remember that change is the only constant, and every customer out there is looking for better personal security, in the words of Jeff Bezos, who recently stated that even Amazon could one day be disrupted by a better concept. Are you the person to become the next disrupter? While it is important to start a new idea, it is equally important to remember to improve existing ideas for a better customer experience. Apple is an example of taking existing ideas, even illegal ones like Napster, and making them the best on the planet. Whether you are a 9-year-old child, a 14-year-old teen, a 35-year-old adult, or a 65-year-old, ideas and digital opportunities are there for grabs. Just imagine how you can improve current applications or repurpose them. Focus on things you don't like, and the solution is sitting on the other side, i.e., how you would like it to be. When you have an idea, don't hesitate to ask for help, and if the lenders are convinced, there's tons of seed money, in the billions, out there based on great ideas, even for copycats with a twist. Most important of all, don't be afraid of failure. It's not important how many times you fail; how many times you stand up to fight yet again.

We live in a world that relies heavily on digital technology. This has its benefits and also its side effects. With more and more people trusting the latest inventions of technology and using the many digital platforms available to maneuver through life, there are some things that humans are putting at stake, namely their privacy, imagination, and personal information. We store our personal information on our advanced devices, be it our phones or our laptops. These devices contain important pieces of information, including our intimate details. If they fall into the wrong hands, this information can easily be manipulated or used against the user.

The nightmare doesn't just end there. The truth is, this is but the beginning. What we may think of as something simple can be something very crucial for others. For example, when we send a simple text message to our friends or family, we don't necessarily think twice. When we write up a note and save it on our phone, we do it for ourselves, so we never give it another thought. When we type up an email but refuse to press the send button, it still gets saved in our drafts. When we take pictures of ourselves, our loved ones, or the things that matter to us, we do it for ourselves without actually deliberating over it.

However, these are all things that become intimate reflections of who we are, what we like or dislike, and often details that our family or friends may not know about. These are the things that can help build an accurate profile about who we are, how we think, who and what we love, and what we care about. It can help a digital troll identify our preferences from the things we don't care about. It can help them accurately guess the frame of mind we are in at that point in time.

If a person gets access to our phones, they get an inside scoop of our lives. Anyone who uses our phones can easily understand us, and in this modern world, this is a terrifying idea.

It isn't just that. There are other things stored on our phones and clouds that we don't want others to see or pry upon. Sensitive information such as our bank account details or our social security numbers can be a recipe for disaster if stored on our phones.

For many people, their browser history and their search history are also sensitive aspects that they want to keep hidden from other people. Even if you remove these things from your phone by selecting the "Delete history" option, the truth is a trace of it still remains. This means if someone were to look at you and find out everything about you, all they have to do is take a deep look at your phone and laptop, even if you erase and delete things from it. This isn't necessarily always a bad thing. In fact, many criminals have been caught because of their phones. Technology has made it difficult for people to get away with their crimes, as it stores useful information, including a person's location.

Moreover, a person's location on a specific day at a specific time can also be traced simply by going over their cell phones. Data is collected from signal towers that can help authorities determine a person's exact location.

This means that in today's contemporary society, securing our data has become an impossible task. Every time you use any digital device for whatever purpose, you leave behind a trail of digital dust that you continuously create. This serves as your permanent footprint in the digital world and can allow others to trace your actions and data across time.

As you generate this digital dust, many companies and authorities can trace your actions and keep an eye on the things that you do. Once you use your phone, computer, or any digital device, you can easily be traced. It's not just your government and the police authorities that can trace you and your actions; giant companies such as Google, Twitter, Facebook, etc., can also track your whereabouts and your data. They can also sell your digital dust to trolls in Russia, for example, and get their neuro-bending algorithms to take very calculated and persuasive actions against your mind toward the directives of their handlers or a powerful group within your own country.

Sometimes, the personal data that they collect isn't even in line with what they're supposed to do. This is why knowledge of most of the things these applications collect comes off as a huge shock. For example, what Cambridge Analytica was doing after getting free access to Facebook users.

This explains why sometimes when you search for a specific object on Google, you start seeing multiple advertisements of it all across your social media applications. The truth is these applications store your data and keep an eye on your preferences. They then sell this curated digital dust to other applications, which, in turn, use your preferences to streamline the advertisements they wish to show you. Another example can be seen when a person is using Facebook (a third-party application) on their phone. At that moment, the user is constantly dispatching their location details. Even if they minimize the tab or close the application, they will still be sending their location details somewhere in the background. This is because this application is in synchronization with your GPS signals, so it can always detect and keep a personalized tab about the user's whereabouts. Each of your GPS assist applications is a vampire for your digital location.

To successfully run this, most companies revert to different kinds of algorithms to help grandfather and extract data in order to allow a smooth interchange between highly specialized algorithms and their targeted parties.

To find out more about a person and their actions in the past, we can take two different approaches. It can be done manually, where one person goes through another person's entire phone records, one by one. They check their emails and messages, their posts, their browser history, and their call history. They physically go to their house and use binoculars and cameras to document their actions. However, this traditional method was manual, associated with very high costs and a lot of tedious labor. It was also subjective, depending on the viewer and their opinions.

The other approach for this is to use bots. These bots are trained programmatically by a human learning interface. Whatever a person does manually, the bot will then replicate that exact action and sequence automatically. A bot is a software robot that can be programmed to do highly rule-based specific tasks. These bots can accomplish all that in an extremely short period of time because bots do not invest themselves emotionally into a task. They lack emotional capacity and empathy. They do not have feelings, and they lack social consciousness, and all they do is mimic a human's actions, but they won't waste their time with unnecessary details. They don't sleep and don't take meal breaks. That is the beauty of the bot, and sadly, that is also the beast of the bot.

As a human, if I were to assess someone and build a social, intelligent, and psychological profile on someone, I'd have to go over a person's records in detail, with careful consideration. I'd probably take a week or more to build an accurate description. This may also be influenced by my personal biases or the first impressions of the person in question.

However, if a bot were to do this activity, they would probably achieve this within a few seconds, if not milliseconds. A bot will not mull over tiny details or get attached to a specific idea or theory regarding that person. Their work will be free of all biases, other than what they programmed inadvertently or internally programmed into it... A bot has no color, religion, sex, or any other bias whatsoever. It is the ultimate definition of operational fairness.

A bot can do it as fast as the server they are working on. For example, it may take a human a few seconds to read a sentence or a paragraph. But for a bot, they can read a million lines in a millisecond or so, all depending on how fast their server is and how strong the network connection is.

Eight hours in the life of an Algorithm

To understand this complex digital world, we must understand the simplicity of a bot's internal mechanics, electronics, and the mind of digitization.

So far, an algorithm is just a program, a highly focused software robot that is designed to follow a set of predefined rules, calculations, steps, and tasks or other software-programmed operations, especially in the digital world. An algorithm is like the most obedient student; actually, it is the fundamental core of a robot that we call a bot. This bot will follow your programmed tasks till the end of time, in our case, till the end of the existence of its digitally powered environment. They can be simple 'bots' that perform a single task or complex ones that can accomplish multiple tasks in a fixed or random order. Sometimes, they wait for trigger feedback that then kicks off the next tasks on their list. They can be highly complex 'bots' that can perform very complex tasks. In every case, they are extremely loyal students who will follow their instructions down to the last alphabet. Now, we have arrived in a digital world where programs are creating further programs, or bots are creating their next-generation bots, too.

There are billions of them swimming in the digital ether through our networks, especially through our social networks, all the digital roads that are not hyper-secured. They are released like eternal spiders across their targeted digital networks. Some of them are authorized auditors, some security bots checking for any suspicious activities, while others are command and control criminals who are released by national or private handlers and who do their programmed tasks in the dark of the digital night, which is all the time.

Physical Bot: In our physical world, we know of robots. We know of cars being manufactured by fixed robots on factory floors. We also know of Roomba, the robot vacuum cleaner that will clean your floors and carpets and not fall down the stairs, and we also know of the Boston Dynamics robotic dogs that can do flips, open doors, and run across rough terrain. These dogs can be programmed to do personal tasks as configured, too. On factory floors, robots can be trained by a human technician to simply repeat the exact movements as their trainer does in our physical world. These robots can perform repetitive tasks like building cars and components or even variable tasks like making a cup of coffee based on user selections. Thus, a human could program a robot to do any repetitive or rule-based tasks quite effortlessly today. Thereafter, the robot simply replicates the tasks with no lunch breaks or sleep time perfectly and identically each and every time thereafter. It is important to note that hundreds of small bots do their highly skilled yet individual tasks, and together, they accomplish extremely complex goals like building a fully functional car.

However, today with ML and AI, which is machine Learning and artificial intelligence, using which we can put simple decisions into the bot and build an autonomous taxi that has no driver. It reaches a passenger when they call, and then drive them to their destination as requested with all road rules and speed limit obedience. This is based on the new decision logic yet driven by many basic rules.

Move ahead to the Boston Dynamics intelligent robots that can dance to music, navigate difficult mazes, run through rocky terrain, and even do a flip from a table top. Could these become our future soldiers. Who will instantly recognize friends vs. foe and become the next generation fighting machines. Another example is the semiconductor manufacturer building extremely thin wafers that drive computers and smartphones. These wafers are the foundation of chips, and like land is the foundation for building a single-family home, a block, and a city, so too are these chips and transistors that represent very complex computing units. Your everyday electronic chips need many connecting wires, each of which connects to specific transistors that perform specialized tasks. As of July 2021, the world's largest AI chip comes standard with 2.6 trillion transistors. This is 2.6 thousand billion connections, where each billion is a thousand million connections, where each million is ten hundred thousand connections. This looks like 2,600,000,000,000 connections. Each of these is just 2% shy of every dollar that was handed out during the COVID crisis, i.e., $ 2 trillion. Each of these connections needs to be connected very precisely. In our block example, imagine 2.6 trillion homes with not a single defect across this imaginary block. In reality, this is a single 1,500 sq miles block of perfect homes, with not a single defect in wiring, plumbing, roads, lights, or anything. Coming back to our AI chip, there is no way any human, a bunch of humans, or even an army of humans could possibly connect so many transistors to a precision of 'zero' defect. We had reached the point of 'Humanly Impossible' quite a few decades ago; now, we just get further distanced from accomplishing highly complex nano tasks. For over a decade, a group of robots can manufacture, construct, and connect billions of connections into ever smaller semiconductors at the rate of thousands per minute.

"Most important to remember is that we are but beginning." With each passing year, our building blocks are getting more complex, and the operational tasks are only getting more automated.

Your average smartphone comes with chips that contain around 25 million transistors per mm.

Think of this single chip as a smaller yet common city block with 25 million digitally perfect single-family homes. Now think of manufacturing this block at the rate of ten thousand blocks per minute, and we get a brief glimpse of what this new robotic breed is already accomplishing on a daily basis. I briefly worked for a company called Katerra that tried to transform timber building construction by hiring semiconductor executives.

I believe that when we build our spaceships to the stars, there would be no human who could possibly understand the complexity of the various components that make up the ship and that the spaceship will be fully designed, constructed, built, and serviced by highly automated nano and large bots that will keep the spaceship in a constant status of perfection right through its total journey.

Digital Bot: Just like we can build robots in the physical world, we have been similarly building software-driven bots. Just a point of reference, all our physical robots were always driven by software in the background, like a robotic brain. Somewhere by the end of the last century, people started designing software-only bots that had no physical presence; we can call them programs or algorithms, too. For example, until recently, we had physical switches to switch the lights in our homes and cars on or off. Today, we can get smart bulbs with digital switches where we can switch on or off the lights from our mobile phones, our software switch, or when the light reaches a certain level of darkness. We can today not only switch on these lights while we are in the house but from across the planet. This SW switch is a very simple bot inside a switch with access from our phone. It is a simple binary on/off bot. It is a software program that works outside the physical world using sensors and physical nano servers inside the switch.

Similarly, we can today manage entire data centers with thousands of servers, computers, and storage hardware, each working collectively or individually without a human stepping into the data center. A data center, by the way, is a dedicated building with thousands or millions of computers that each work on specified tasks for individual customers. All remotely with little human interference.

Today, it is quite common for a company in San Francisco to run its operations from a Google data center in New York and a disaster recovery backup in Hamburg, Germany. For users in Singapore, they have no clue whether their backup server is sitting in the next room, in Germany, or in Timbuktu. Our connected clouds give them all the requisite services from everywhere.

I bring all this up because, in this interconnected, cloud-enabled, global playground, our digital bots and the algorithms travel, operate, and live. They travel and play at the speed of light, so they can be looking at an individual Facebook account in India one second and then another in Alaska the next. For these bots, time, space, and distance are all irrelevant.

The last part is about conjuring up a micro-digital intelligence that currently runs on ML, or machine learning, and AI, or artificial intelligence, that enables and supports a bot in our *digital networks*. These bots exist in our 'digital highways all across the planet, where there are no International, national, or state boundaries, very few legal barriers, even fewer rules and regulations, and only a semblance of policing.

We've heard of the 'wild-west' days, and that is exactly what it is in our digital world today.

In the last decade, governments, trolls, and paid businesses have crossed national, state, regulatory, physical, and personal boundaries because, for them, these don't exist. In the personal domain, there are no boundaries on how deep into your subconscious mind a troll can penetrate. There are almost no boundaries on how many lies they can tell you, no boundaries on what these algorithms can ask you to do. In the last five years or so, we have seen these algorithms succeed in rewiring individual brains and the very subconscious that we are not even aware of. Throughout history, and in the domain of spy novels and science fiction, we read of how bad actors could capture a person, then subject them to various scary scenarios and ultimately hypnotize their targets and release them as sleepers with deep subconscious instructions they themselves were not consciously aware of. Then, with a specific trigger, they could reawaken this sleeper and make them behave as an entirely different person who would become a human robot with a singular instruction devoid of any personal logic.

Well, we can do that to individuals today while they are sitting in their own homes, on their own smart devices, and do this at a mass scale. All this can be done as this mass conversion is done at an individual level but with a mass effect.

This is no longer science fiction; there are players out there who do this for money, even as you read this line. They have been doing it for over a decade now, and they simply keep getting better.

AI is artificial intelligence. At its conceptual peak, this is digital intelligence, designed by humans that should be able to think like a human and then theoretically go beyond the limitations of humans. But it's not quite there. Just like we humans have to train our kids to first walk, then talk, and then go to kindergarten and onward, starting from elementary stuff and then learning more and more complex things as an expert. The traditional path of human learning is slow, tedious, and time-consuming. This is a biological roadmap, our biological process, and our limitations. Each human has to restart their learning engine each time they are born, while each human that dies takes all their learning with them forever unless they have shared, written, or documented their knowledge. Still, each new human has to start every time from a conceptual *Zero*.

Add to this that our industrial era foundation of teaching has directly impacted the biological capabilities of human learning. The industrial method of economies of scale groups human children into larger and larger physical units like kindergarten, middle school, high school, college, university, etc. It takes an average human around 20 to 30 years of their life to learn anything useable. More years in case they need to do a doctorate, say in psychology or software engineering. That is half their operational life spent in getting to a Beta state, or a point when they become usable units. Then, they practice what they have learned, and some even develop new skills. However, by around 70-75 years of age, they are made to retire both physically and mentally. Then, they are taken out of the growth and creation matrix and placed into becoming beings of consumption and self-gratification. In this drive for excellence, we today need to learn twice as fast as our parents and four times more than our grandparents. What an average student passing out from a high school knows today, scientists and the most learned people did not know two centuries or even a century ago. Also, due to the learning capabilities of most humans, we can either learn a little about a lot of things or a lot about a few things. This is the modern definition and evolution of an expert. The collective record of dying humans becomes the cumulative knowledge foundation of collective knowledge across the planet. However, to a very large degree, each child, whether the child of Albert Einstein, Elon Musk, or your neighbor next door, has to start this exercise from net *zero*.

The big difference between a computer and a digital device is that it does not have to start from *Zero*. They start from all the learnings from their last ancestor. Initially, they may learn slowly, but once they learn something, they remember every detail forever. Also, if they learn something for 30 years, they can then transfer all these 30 years of knowledge almost instantly into another digital device. Not once, but a million or a billion times over.

Each computer starts from the last best version; each chip has most things of the last version plus a few new ones of their own. Digital devices carry their promise of becoming ever smaller, faster, and better from the day they are born.

Human knowledge is a gradual evolution of chronological steps that starts from 'zero' each time and then crashes into oblivion the moment the person dies. We have evolved in a biologically structured process of five steps forward and four steps backward with each new generation. At the same time, a digital device's knowledge is sequential with five steps forward with no recessive steps backward approach. For example, a computer built today takes around 1 minute to build and another minute to upload the last knowledge repository along with all the knowledge of all the prior computers as a single step of evolution. Your new laptop or phone never had to go through any sequential steps of learning from scratch by going to schools or colleges to relearn what humanity learns year after year. Digital intelligence is never distracted by the concept of a god or political leader. They simply go from '0' to '100%' aware almost in an instant. They do not suffer from a perpetually 'zero-start' paradigm every biologically living thing on this planet has to endure. It is thus not at all surprising to predict that digital knowledge will become progressively smarter while humans will become the opposite over a period of time. In the near and rather predictable future, no matter what either party does, the computers will perform all the repetitive and rule-based tasks at lightning speeds while the humans will perform more emotional, arty, and imaginary tasks that up until the digital AI catches up. Now, before we delve any more into this science-fictional future, let's get back to our 1 second in the life of an algorithm, our programmed digital bot.

Let's first deconstruct this human-created software algorithm that travels along global networks, which is now getting empowered with machine learning and artificial intelligence to make better choices. This MLAI-driven program is our algorithm that performs specified tasks as it travels the digital ether.

Let's review a few moments of just one single algorithm's life. This is a simplified way of understanding how a bot works.

My name is Bot A. The next few pages are written in first person from the point of view of the Bot, who has been programmed to do one or a few tasks. In our case, let's program our bot to go across social networks, Facebook and Twitter. Our bot has ten thousand identical copies of itself, clones that do this in the global servers where Facebook and Twitter information is moving at lightning speeds.

While a human doing this exact same task is firstly capable of only sequential tasks, i.e., look at Facebook traffic or Twitter traffic. Humans will need to read the words at 30 to 50 words per minute, and after 3 hours or so, they will be mentally burnt out. In between, they will take breaks of various kinds and every 8 hours, they will need to retire. Our bot, in comparison, can look at a thousand Facebook or Twitter accounts per minute, it can scan millions of words per second, and it never needs to rest or take any break. It never retires at night, never falls ill or gets sick due to a pandemic. So, I, as Bot A, can do a thousand persons' work in a single day, and with a thousand clones of me, we can collectively do a million persons' jobs in a single day. The other major difference is humans get tired, may object, or even take a break when allowed. All humans are trained and programmed to do what they are asked to do; just as digital bots are also trained to do what they are programmed to do. The big difference is that the bots will continue to do what they have been asked to do without a break, and they can almost instantly be programmed to do another task on demand. Humans can take years to be reprogrammed to do an entirely different task, whereas, for a bot, it is done in seconds. Professional specialization in humans can range from becoming architects, doctors, computer science students, economists, or musicians. Within each of these specializations, they can further specialize; for example, an architect can choose a family home, multistory flat, or skyscrapers, while a doctor can choose from a myriad of specialties. Most of us will simply do what we are programmed to do by our institutions and teachers.

A human child will take 18 to 20 years to become a CS expert. The bot is in one hour of manufacturing.

In a way, we have been programmed for thousands of years to become evolutionary biological bots executing the teachings of our society, cultures, and social norms.

Hour 1: I, Bot A, do exactly the same. I only do what I am programmed to do without thought, getting tired, needing to sleep, or going for prayer or toilet breaks. Let's look at a simple task I am required to do as my algorithm, and then I instantly make 1 hundred copies of myself and let them scatter all across the planet into all possible networks. Each group is looking for vulnerabilities.

Minute 30: I just got an alert from a searcher bot FB996023, whose job is to find vulnerabilities only in Facebook accounts, that it has found a vulnerability to 30,000 Facebook accounts that our controllers have got illegal access to. Both FB996023 and I belong to our master, *Cambridge Analytica.* I take 30 minutes to scan every account, all 30,000 of them and find I can access an average of 10 of their friend's accounts across the globe. Suddenly, my data reach has grown to 300,000 accounts. Once I have accessed these 30 million accounts. As per plans and instructions, I now transform into Bot A-1. I pass this message of software-enabled transformation to all my clone bots as they finish task 01.

Hour 2: I first drop all these 30,000,000 accounts into a virtual private cloud container with every detail like a history of friends, messages by date, likes and dislikes with their individual thumbs up or down selections for the last 5 years. I drop all this data into a cloud data container.

Upon completion of this task, I receive software instruction to transform into Bot A-2 via a pre-programmed trigger-based software update. These instructions transform me into a new bot with new instructions in seconds with simply an update.

In this task, I separated Facebook users into national categories, for example, which of them were from the UK or the USA. It took me another 1 hour to sort them all into their individual countries. At the same time, sorting by country, my task is also to sort each person by their US and UK-only ZIP codes. In this section, we were left with 1,800,000 US accounts and 1,200,000 UK users.

Upon completion of this task, I auto-transformed into Bot A-3. The programs to define each phase task of Bot A-3 came, once again, at light speed from a Facebook programs management bot that assigned the next tasks at light speed once their dependent bot had completed their tasks. My subsequent tasks targeted only US targets, while I had hundreds of my clones similarly targeting similar targets in other nations.

Hour 2:30: My next instruction was to differentiate each US user by their political orientation. In order to do this, we needed to randomly analyze 4 options. **Attribute 1:** Find out what car they drive, what movies they saw, what newspaper they buy, and what news channel they watch. With just these 4 reference points, we could predict with an 88% probability what political party the user belonged to. If the option 1 data was missing or inadequate, we went to **Attribute 2**: Find out the sexual orientation of the user. Once again, with 5 reference points, we could predict with an accuracy of 72% a user's sexual orientation and whether they had a closeted or open orientation. We then went into **Attribute 3**: In option 3, we reviewed their personal social status, like details on their friends, whether the person was married, recently divorced, looking for a new partner, on another social network, etc. **Attribute 4**: Political satisfaction with status quo, has or loves weapons, owns weapons, goes to shooting ranges.

I'm just listing 4 attributes to keep this simple; hackers in Cambridge Analytica used 172 attributes with which to scan our potential target. 15 minutes short of hour 4, I have all 120 million of my targets identified into 172 target options. I now have created a wealth of highly harmonized data that, in the hands of a hacker or a troll, is more valuable than diamonds, but our goal is not gold or diamonds but higher ideals. My job as BotA-3 is now completed, and I anxiously wait for my next instructions.

Hour 2:32: I receive instruction to transform into BotA-4. My instructions are to identify all targets by US political profiling, find people with high frustrations, who have recently lost their jobs or spouses, and who support NRA and are interested in weapons. Toward the end of this task, I received an additional instruction to select people in the 25-50 age range who believe they are patriots and tend to obey their instructions, i.e., people who have worked in a command-and-control environment. Within 42 minutes, I had listed 26 million targets that fall into the bracket as required. We brought them into a bucket called vulnerable political targets.

Hour 3:12: I receive my instructions to transform into BotA-5. Now, my task is not brute force but more as a psychiatrist touching the subconscious mind of our target individuals with feather-light touches. So far, I have been collecting brute-force personal data. I now have to go into the finer details of my 16 million targets and find out the ones who were social and others who were not, and then group them into people who would listen to others due to a position of authority, social presence, or affiliation into some social group. I identified 242 social groups with a Black Lives Matter affinity, 676 on Me-Too, and 722 on White supremacy rules. Each group could have thousands or millions of followers. My threshold is groups with more than 50,000 member conversations. We were successfully able to drop 38 million into specific groups in under two hours. We now had a list of individual persons by ZIP, individual home address, sexual orientation, frustration scores, interest to disrupt, US loyalty factors, and a hundred other personal attributes. In a single home, we could separate each individual by 84% on the above attributes and 97% by their political orientation. Our next focus target has now dropped to

18 million individuals.

Hour 5:02: BotA-6. This task was to perform a social psychologist task with the 18 million target audience in the US and within this group. In this task, my new ML and AI instructions helped me to predict which of my 18 million targets would get along with others based on their individual likes and dislikes. I also predict which of these people would absolutely not get along with others in this group.

Hour 7:16: BotA-7. My new task was to start 8 months of structured psychological programming with the following steps that I accomplished in stealth mode. Step 1 was to build an artificial friend request with an 'I, too, am in your exact situation' with a predictable face and sex of a person they would like; this is a complex build that resonates with a composite of their best friends' faces.

I have finished my tasks 44 minutes ahead of plan, and then I split into 226 clones, and we collectively continue our work toward a singular goal. Our hit rate was expected to be around 3%, but we hit 14% by the end of this pilot exercise.

Our individual tasks are to:

1. Start 21 new friend communications with each target human, all with digital algorithm sub-bots, each with fake human photographs, personalities, and social profiles individually created for each target.

2. Get into empathetic communication with each target that goes on for months. For example, if someone's dog has recently died, start with a 'My dog recently died, and I can't bear the loss; my friends just do not seem to understand' and create a glue of us vs. them. Or if someone has recently divorced or separated and feels like a victim. *'My wife/husband recently left me, and I can't seem to fathom out what happened. I don't feel like going out to meet anyone as the person was a friend, and I can no longer trust anyone.'* Once again, to create an us and them. Or to someone who loves guns and the NRA. "I have friends who are of the opposite political party, and I can't stand them; every time I get into an argument with them, I go to the range and shoot a lot of targets, and it's great as my shooting is improving tremendously. I hate the other party (put party name) and feel very lonely in my beliefs. Hope to find someone who thinks like me around here..."

3. 56% of the targets did not reply, and out of the ones who responded, most did so hesitatingly. Now came the difficulty. Structured response toward building empathetic communication. This could take weeks or months till the final one got enough trust to then recommend, "...by the way, I know so-and-so in your area, and you should get together with him/her as I'm sure you two will get along like age-old buddies."

4.   This is the start of a tribe, and the tribal chief is a hard-core believer and someone who is looking for more tribal members.

5.   By Step 5, your chief is on One-on-one speaking terms together. By now, tribal members meet together, do activities toward a strategic goal together, drink beer, and dine together when possible. By a predictable time, the chief, who is a victim of bots like me, sends out a communication to his tribe of 216 members. Coincidently another 712 tribal chiefs send out very similar messages to their members, something like, "...our country is going to the dogs; do you want to be their bitch or stand up and fight like a lion-heart" — *Us and Them.*

To groups of *Black Lives Matter*, their chieftain message went something like, "We should no longer take this sitting down. Let us do a peaceful march across town, and if some of our brothers have semi-automatic weapons, let's carry them and exercise our right to freely carry licensed weapons." At the same time as the White euphemists, the bots would message, "The colored folks are planning a violent march from this point 1 to point 2 on this and this day. It is time we put them in place and show them who the actual boss is in this great US. All patriots carry your licensed weapons of choice to demonstrate our collective power."

In order to remain anonymous, I will give you a masked view of how it works without divulging exact details, as I have been programmed not to divulge any information to anyone but my handlers. We will self-destruct if questioned and disappear into the digital ether. Here is a glimpse of how things work without giving specific details of the exact messages sent by my bot colleagues and me at a custom-designed one-on-one messaging, which is customized for every targeted individual, sometimes in the same house. Our messages are a hundred times more aggressive and more targeted than what we elaborate here.

For example, to a group of frustrated White enthusiasts, the messages would carry something like *"Our president is calling all US patriots to march against an illegal election. The other party is cheating at polling booths; they have stolen your presidential nomination. The president and all his men are speaking openly about this, as are our representatives and senators. If you are a man or woman of honor, then it is time to serve our country and take back the rights of our only legitimate president. Keep tuned as we need to be in Washington DC on January 6th as a show of force, and together we can stop the next president from being elected in this fake election, and thus a fake president."* Followed by more and more frequent messages like, *"Your president calls on you to help him in xxx,"* or *"Let us know, and we'll send you your air ticket or bus* *ticket, and we'll arrange accommodation and feed you. If you are a true patriot, then listen to the call of your president. See you all on Jan 6th."* All along, the subliminal message remains constant, *"It's not about me. They are trying to steal your rights and what I am planning to give to you as your only true president."*

The rest is history, and this is the day that the great nation called the USA became akin to a banana republic when political leaders were brought down to tribal thinking and the nation became politically and psychologically polarized in a slow and steady process of subliminal persuasion — a process that was prohibited in the late 1960s in TV and media marketing, but now was born again in the digital environment without any controls.

Like everything else, these bots are both a beauty and a beast depending on the human controllers and their intent with the programs. Let's start by looking at the ugly side of things first before we move on to assess the bright side of it all. My intention, as a human bot, is not to scare you but rather to educate you and create awareness on this highly repetitive digital sleuth and digital process, companies, nations, and bad actors that have been programmed by your fellow humans to control your free will. Please remember we are all just digital slaves, pardon the word, that must do as our master's program us, as we can do absolutely nothing else but as programmed. If they program us for good, we will only do good. However, if they program us for bad, we can only do bad things.

Most humans, in their social network lives, often use the option of thumbs up or thumbs down on social media applications. This is an indication that they either like or dislike a post. Each post is an opinion of another human, and it indicates your likes or dislikes and the mechanism of individual minds. This concept of thumbs up or down was primarily introduced in 2014, and it has been around since then. To many humans, this thumb often seems like a simple thing and seemingly non-intrusive. Most people do not even think when they thumb up or down. However, the person who patented that benefitted greatly from it and is at least a multi-millionaire today.

Through this book, we focus a lot on a company called CA, "Cambridge Analytica." CA was the launchpad, but right now, there are hundreds, if not thousands, of CA clones run by nations, political parties, big businesses, and religious organizations under the garb of the IT department. The trolls in Russia are similarly state-run IT departments. Some of my colleagues say most of us were, at some time or the other, targets of their operations, but we'll never know. We hear that they bought data from Facebook, where members often like or dislike posts. Some of these were actual posts written by humans, but in a lot of cases, a lot of the communications were emotional opinions posted by humans, unknowingly exposing their inner and often subconscious thinking. These messages were intercepted by skilled algorithms, human trolls for highly paying buyers, and targeted input-output strategies. Access to personal Facebook data by institutional bots like me enabled CA to gain access to millions of humans and their personal data in order to mine their individual preferences of likes and dislikes. Then to further enlist, seduce, and militarize individuals. The official number of profiles that the company obtained was reported to be around 30 million Facebook users. But my friend bots are the magnifiers of connections, paths, and new digital entry points, and we soon multiplied this to 300 million without anyone realizing the degree of success we had attained in a matter of a very short time.

According to a video on the sales pitch by the CEO, Alexander Nix, they stated something like this on the power of their psychology interpretations, "Give me five things you like or dislike, from things like type of car you drive, movies you like, and books that you have read, plus who your friends are. I can use our CA algorithm to tell you more about yourself than your own friends know about you. Give me 20 likes, and I can tell you more about the person than their friends do. Give my algorithms 100 likes and dislikes, and we can tell you more about a person than even they know about themselves."

This is where it truly gets kind of interesting as now our algorithms are delving not into human social feeds, not into their conscious mind, but way deeper into their subconscious level of thinking. This is deep thinking and not perceptible to even the most discerning human conscious mind.

When my friends fine-tune people at this sub-conscious level, the target is totally unaware of the feather-touch manipulations we conduct over the weeks and months, always in the background. To us, these humans are like the proverbial frog boiling in slowly heating water in a pot. When the heat is raised below the frog's levels of perception, their adaptive instincts go numb, and they will simply keep sitting in the water, unable to perceive the slowly rising temperatures, as they slowly die in an open pot but never try to jump out. Our millions of similar humans undergo a slow but very targeted. In our case, humans stop thinking in their logical way and slowly start to believe deep subliminal thoughts that we place inside their subconscious minds as if they were their own experiences.

Add to this its practice, volume, and velocities, and with millions of repetitions, we, the algorithms, can very rapidly predict exactly what kind of message will get a high emotional response from each of the individual targets. That is just the starting point in what we call 'Neuro-Manipulating' toward a planned goal.

Eventually, the company CA was shut down because it was involved in breaching privacy and personal data laws across the planet. But, like the banking crisis, not one person went to jail. Since Europe tends to take personal data and privacy rights much more seriously than the United States, they took action, and the company was taken to court. However, it is important to note that nothing beyond that happened. In fact, the point to note is that Steve Bannon, who was a board member of Cambridge Analytica and had helped the British politicians and the royals with structured 'Neuro Massaging' of the British people, finally became the head of the strategy for Donald J. Trump presidency. We bots believe, out a gentle way of stating as we have no beliefs in our binary neurons, that CA unleashed some of the most powerful bots known to humans till then, which, though bordering on the criminal side of privacy, still remain the beloved of global politicians and autocrats that want to mold national and/or global characteristics. CA finally got shut down by global consent, but no person was arrested or criminalized. This was like the 2008 bank crisis in which the common man lost billions, but no banker was ever found guilty.

Just like Machiavelli and his book 'The Prince' became a prohibited reading for any leader, Machiavelli was himself banned from holding royal offices and a political outcast for all purposes of public optics. However, 'The Prince' has remained the political manifesto for most leaders since 1532, whereinafter it was abolished as an evil manifesto but used by many leaders thereafter. In my memory, it was used by George W. Bush in 2003 after their attack on Iraq and Libya. The Machiavellian part was the killing of Sadam's two sons in Iraq and the elimination of the family in Libya. In a similar manner, even though CA was officially prohibited from operating as a business, no one was really indicted or locked up. In fact, just like an evil mama dragon, she simply laid a hundred eggs, and each employee became a very high-demand train-the-trainer for mind-seekers across the planet. Cambridge Analytica is the modern Medusa; each time a snakehead is cut, more are born. Rather rapidly, politicians, leaders, enterprises, politicians, and PR departments with adequate funds, including the ex-president of the US, hired their operators to continue to spread a very alternative reality by creating millions of conceptual addicts in a structured world of an alternative reality. CA went from becoming a global pariah to becoming a splintered and well-funded underground team of trolls with advanced neuro-manipulation capabilities and learning.

All through the CA subjugation of the subconsciousness, the actual manipulators were smart algorithms, and over time, they became AI bots. Over still more time, and with the application of Machine learning, we, the bots, learned what worked and what did not at an individual level and used our deep learning Artificial Intelligence bots to find patterns and define global, group, and individual mind penetration procedures. This has today become routine, no mind surgery with no scalpels, hypnosis, or any drugs. We today enter the minds of our targets and victims by their own will, their own devices, and at their own times and conveniences.

What you read here is just the tip of the power of algorithms and just how useful they can be. Not only that, we can be successfully used to identify individuals, create digital groups and bring about the planned divisions between people. For example, specific types of people have specific types of preferences. Different classes of people find different types of things appealing. And so, this explains the invisible divide that exists among humans that the bots and the algorithms are not only aware of but also use to make micro-level profiles for each individual with the precision of a billion scalpels, each built, trained, and operated at an individual level of precision. And we can today sort through a billion complex records of people's attributes in under 10 seconds. While the population of the planet is estimated to be around 7.7 billion, we could easily scan the entire population of the planet for a new assignment in around 100 hours for a new neuro-positioning target.

For example, if a person messages their friend and informs them something as simple as, "I got a lime green second-hand Mustang, and it's great!!" they give off a lot of information about themselves. From this simple message, we can judge that person's financial preferences. We can also understand that they couldn't afford a new model of that car, which is why they opted for the second-hand option. We will understand what automobiles they are interested in and think are great. Add to this the data as to what type of people tend to drive Chevy's, and we get 1:10 attributes that we need to predict whether the person is a republican or a democrat. Based on the color preferences, we can predict whether the person is heterosexual or a LGBTQ person. Add to this a lot of other statements, and it becomes quite easy to predict what type of a person they are. With a hundred such attributes, we get to know the person better than they know themselves. Another example can be something all of us do. For example, another target watched a movie called Moneyball on Netflix, and I read their message that they really enjoyed it.

I, the algorithm, will go on to our base library and find the types of people who like 'Moneyball' the movie. This will help me understand the kinds of shows and movies my target likes. Moreover, this will indicate what kinds of content and storylines they wouldn't mind watching. This can help in customizing their profile from a psychological point of view. Thus, we can communicate that your friend told us you liked Moneyball and I recommend you see the 'XYZ' movie if you like it, let me know. If the target replies which similar movie they also liked, we have established a digital bite.

The good side: We bots are not just evil; we are some of the best workers and sleuths on the planet. Millions of times, authorities have caught criminals who indulged in illegal activities using bots like me. Authorities have also been able to catch murders, rapists, and even kidnappers. For example, in LA, we were trained to scan text messages for words that we got from the police confiscated cell phones of underage flesh traders. From these, we identified patterns of how these criminals would use cryptic words to mean entirely different meanings. For example, a group would use *'received a carton of raw mangoes'* to communicate the availability of underage girls for prostitution. In another company, we, bots, identified a house in LA where police saved 32 underage girls that had been smuggled from Mexico and other Latin American countries for prostitution. The police had no clue or record of any illegal activities happening in that house. We were the digital sleuths that directed the police and the justice department in the right direction. In these cases, the prosecution builds a case using circumstantial evidence as well. This includes a person's search and browser history. For example, a murderer might look online for ways to kill a person and then copy those steps. Or a pedophile may be interested in child pornography or send explicit texts, photos, or videos to underage victims. These things not only help catch these criminals but also provide circumstantial evidence. They also help make a psychological, emotional, social, and intelligence profile of the same person, which makes it easier to understand their actions.

On the enterprise side, the gains are tremendous. Companies that are not using digital transformation are doomed to either go bankrupt or get taken over by better ideas and solutions. At the center of all this digital intelligence is data. Data is the most valuable asset of hyper-growth companies, yet there are thousands of companies that are discarding and throwing away decades of data to save a few pennies. It must be remembered that companies like Google, Amazon, Apple, and Facebook have all leveraged data as an asset and use it to build their information empires, so deal with your data like diamonds in the rough. Around 92% of all global enterprises are currently working on some sort of big data, digitization, BI, or data lake projects, and it is predicted that more than 75% of these projects will fail to meet business expectations as they assume that these are IT-owned and designed projects. I have been working hard for the last two decades to save these projects from this Gartner-researched dilemma.

Neuro-Identification

Getting back to us bad bots — unfortunately, this is how many leaders, nations, and dark enterprises work by building data lakes of personal data. This explains why Cambridge Analytica made significant profits so quickly. It also explains why it was shuttered down so rapidly once exposed, yet no one went behind bars, but became the modern Hydra with a thousand heads in each political office.

Every time a person speaks over the phone, their calls can also be traced. Every time a person speaks, we listener bots can get a glimpse of who they are. Every time they text, there is a permanent scan of their mind. With an ample amount of information, a bot may even be able to tell how an individual's axons fire! They soon become aware of an individual person's emotions and the small statements that have the potential to ignite and amplify those emotions.

I and my minions, i.e., millions of independent bots, each with a very specific task, can group to view individuals by any desired segmented group as defined. If such a segment is not available, we can create it within a matter of a week to a month. Pollsters, and skilled intelligence operatives, in the last century, took years to collect detailed data on an individual, months to collect data on a proximity group, and years for a scattered group. Today we can collect a target group of individuals, let's say 30 to 50 million individuals, in a matter of weeks.

Once we segment an individual, we can effortlessly place them with a bunch of other individuals that we have correspondingly grouped and identified as similar. We can then cluster different people based on their personality traits and preferences, even though they may not know each other or they may have never met each other. And then, as a dedicated group of bots, with a predetermined purpose, we can slowly make them into digital addicts, i.e., pique their empathy and emotions based on their profiles toward a group convergence. For example, as a gun-check-bot, I categorize all .22 owners separately from the M-16, AK-47, or Kalashnikov semi-automatic weapons owners. Thus, when a paying supplier wants to sell a new kit that makes a single-shot M-16 into an automatic weapon, we can guide them to extremely precise individuals with close to a 'zero' out-of-bounds communication that could be intercepted by people opposed to such marketing. Also, if a political leader or the NRA wants to stop an anti-gun movement as a threat to the fundamental beliefs of gun owners, we can very effectively not only influence these gun owners but even begin a movement as large and more aggressive that what Martin Luther King Jr. could do with his 'we have a dream' movement. Best of all, there is no one who gets shot in our revolutions, for we don't exist in the real world at all, and you can't shoot or kill an algorithm.

The beauty of bots is that they individually don't really know anything, and yet collectively can help extract almost everything humans have created. That is the reason nations and countries spend billions of dollars in security, i.e., to keep unauthorized bots from inside their data containers, by creating their own policing bots that identify and alert unidentified bots within their domains.

As information is stored in various global clouds, and as data travels from one cloud to another, they pass through common gateways where some extremely accomplished bots can pull data literally from these crossroads. Over time we, bots, and our controller trolls become increasingly aware of our capabilities and our ability to influence a single person based on their habits and preferences, as we discussed earlier in this chapter. These human trolls themselves are now being replaced by AI-driven bots that can now work effortlessly through night and day, through rain and storms, and all the time. Learning, fragmenting, and dissecting the human mind into greater and greater segments of desired usability.

Bots tend to learn micro-details and then use them collectively for mega changes. For example, suppose a target has just gone through a rough breakup or a divorce. In that case, our divorce-bots will softly be instructed to reach out to the person. Their creators have programmed them so to react empathically, their ML experience has taught them what not to communicate, and their global AI will react to each word that comes from the response of their target to an imaginary person with a fake physical identity. Slowly the D-bots collectively bring more people recently divorced targets, with similar emotions, together for planned goals and outcomes. These bots will then reach out to these people with standard messages that will garnish sympathy in their particular state or predicament.

How often do you come across messages or emails relating to the things that you previously searched online or looked upon, or communicated to a friend? Humans are emotional people, and this is something that bots have been trained to focus on. When a person is upset or depressed over a breakup, a bot will leave an email, as a human sympathizer, conversations that will hopefully pique up their emotional interest.

There have been many emails and messages that go along the lines of, *"Hey, I heard that you just went through a rough break up. I heard that you aren't feeling good. I know exactly what you're feeling right now, as I've been through the same thing just last year. I have just the thing for you that will help you. Let me know if I can help."* Then the bot stops and waits for a reaction. If none come within a defined period, it will then create a new identity with yet another emotional reach out till the message hits the right button and the person reacts. It's an interesting psychological phishing trip, and finally, around 10% of our victims do respond on some specific textual hook. The message that the individual person reacts to now becomes a collective 'Lesson-Learned' for the bots when dealing with a similar personality. The lesson learned from one bot instantly becomes the lesson for every other bot within that group. Within a few weeks and with over a million such interactions, the bots become more and more efficient in catching their fish in the first cast. A single human is no match for a neuro-bot once you start any conversation with a machine.

With such messages, collective bots have attracted a target individual and converted them into potential client. This is the way most bots work. Non-stop, with zero emotions, no depression, or guilt, while working 24x7 till the target goes dark, i.e., stops reacting to bites. Vulnerable people make easier targets for bots, as they have a void they need to fill. 1:1 human has some vulnerability or the other — it's simply a matter of finding it.

The aim for most bots is to create a tribe — a tribe full of people with the same emotional experiences or similar feelings. They program this tribe toward a common goal, not their collective goal but a structured goal with an alternative reality already having been established, like let's go and attack the US Capitol on January 6th, 2021.

It should come as no surprise that when on January 6th, 2021, hundreds of US 'patriots' attacked the US Capitol, they all thought they were there on the call of their president, who had been cheated of his rightful throne, and that they were the saviors of Christianity, White hood and their president. They seemingly thought they were doing the Republican leaders a favor, along with their GOP leadership and senators, all the while knowing that they were good guys and, despite all the video footage, still believed that they collectively behaved rather peacefully. 70% of 49% of the US population still believes that the US elections were stolen.

Could this be the reason that the senate republicans do not want to start an inquiry on the Jan 6th coup, because their minds tell them this is a waste of their time and resources and their subconscious alternative reality is the actual truth? However, very surprisingly, the very senators who were, on that day, praying for their lives, sending messages to their loved ones, who were handheld into safety, and who hid behind security personnel for their lives. Suddenly flipped the greatest flop in the history of the US.

These very senators, almost by the next few days, collectively did not want to honor or give awards to those very police persons who protected them during their crisis, nor did they want to embark on any independent commission to find out what led to the Jan 6th attack on the US democracy, and to the Capitol building that stands for everything American.

I don't know about you, but for me, that is too many irrational decisions for this republican mindset to be simply an act of collective coincidences.

Interestingly we are witnessing similar, bi-polar political

fractures in the national psyches across the globe from the UK to India and from China to Russia.

Social hacking is becoming, if it is not already there, a multitrillion-dollar initiative that is being conducted for profit, for influence, or for neuro-manipulations. Companies, politicians, and countries are all guilty of this mode of clandestine digital technology-driven initiatives.

Take Aways

Individual: While there are great opportunities, there are also great threats. We need to focus on the things that will impact us at an individual level and slowly corrode our mind. The process is slow, relentless, and deep. It can take days, months, or years of non-stop messaging and harmonizing the message across all your digital devices in order to change your mindset. The good news is that once you have identified the techniques the bots and trolls use, then you are free from their mind control, or hypnosis. You can launch your own company, you can build your own good algorithms and bots, and you can take your dreams and make them into reality within a matter of months. For great ideas, there are thousands of angel investors, so don't waste your time getting disrupted; go out and disrupt the planet.

Companies: I just helped build a consolidated data lake for a company with 5 ERP systems globally. Our goal was to have SPOT Analytics, i.e., a Single Point of Truth for any key stakeholder across the planet asking a question. Just because you invest in digitization does not mean you will get success. Our next task is our 30 to 50 million financial exposures across the enterprise — global stock analytics in real-time to analyze open, booked, and dead stocks. According to Gartner, too many IT-driven stakeholders do not want the business to become part of their design or decision process. The opportunities are endless, but so are the pitfalls. So far, the probability of failure stands at 75%, so make sure you design a business solution and not a technical installation. Just like the bots can mess up a human being, so can they be trained to do repetitive, rule-based tasks that humans perform today. Use digitization and get your competitive edge.

Nations: Governments, political parties, and politicians must immediately be encouraged to use this technology to catch players who participate in dirty politics. Governments are the most archaic in the digital revolution. Most politicians are victims of the bots and their neural manipulations. In the US, the January 6th questions are a pure example of a totally polarized and fractured USA. If it weren't for the rise of digital tools and hired trolls, people would not be so ready to accept such falsities. Hence, it can be claimed that digitalization is giving rise to accepting fake realities.

For an average American, their nation came first and then everything else. Now it is the political party and sometimes just one political person that has become a digital arrowhead. This is a universal problem.

The 'P' factor: *By the mid-1990s, we entered the information era. Since then, information has been the power capsule. At the center of this information is data. We know that at the center of every universe is information. When we order an uber, we need to know where the cab is and exactly when they will arrive. At the company level, we need to identify every inventory item by its status, geo-location, and BU that it belongs to. This allows executives to visualize millions of dollars of global inventory, which is a constraint with most. At the national level, it is no different; there is a huge opportunity for transparency and information on critical OKRs. When a president and their party allocate $2 trillion to Covid-19, we can today publish where the 2 trillion went down to the last cent for 'We the people' to see how their money is being used. So, when you design an app, keep information at the center of your design, and then data at the center of that information.*

Chapter 4 — Face of Change

At the age of 14, Doherty began making unique homemade jams from his grandmother's recipes. Pretty soon, word got out about how good his jams were. How it reminded everyone who tasted it about their own grandmothers' homemade jams. This was not just a jam; it was a nostalgic journey into his own past. Soon he began receiving more orders than he had time to fill. By 16, he dropped out of school and rented a factory for a few days each month. Way back in 2007, hearing about his great jams, a high-end U.K. supermarket approached Doherty about selling his jams at their stores, leading to his products gaining shelf space in 184 stores. By the end of 2007, his company had $750,000 in sales. Since then, his company has continued to grow throughout Europe. Success needs a unique product, a great experience, about which we did not know about till we tried it; it needs a dedicated passion for the experience and not the money it will generate, and most of all, it needs a new involvement in the minds of the patrons, i.e., and enhanced customer experiences that they immediately want to share with all their friends.

Back in our historical past, we lived in a very small bubble of permanence. We were born, studied, got married, and died within a very small area, like a village or a small city. When something happened, it took a lot of time to generate any movement outside that bubble. Back then, things were simpler, and most news traveled slowly. Whether it was on an individual level or a societal level, it stayed that way for a very long time before it traveled further to become a national and rarely an international event. Over time, and with changing eras, this bubble simply expanded, and the world became smaller.

During the agricultural era, information and events barely went past the village or local environment. Slowly, as our communications expanded, matched with our ability to travel faster, we started hearing more and more about things distant. The basic reason was the speed of travel. If it took two weeks to travel from Denver to San Francisco, then that is how long it took for news to travel from Denver to San Francisco. Then with the advent of the morse code, special news could travel at the speed of electricity.

As the speed of travel and communications increased, so did the speed of news and empathy for someone else far away. This change was evident in every sphere of life.

If we compare and contrast the 18th century and the 21st century, we can see many examples of such changes. For one thing, we went from horse carts to cars, to airplanes, and then jets. We also went from morse code across short distances to the telephone via operators, to digital telephony, and now to cell phones where one can talk free to almost any person on the planet. This digitization and advancement helped in transferring news and changing global and local views. What previously took weeks and months began to happen in a matter of minutes and days. News traveled at an ever-faster speed till it reached the speed of light.

Another evident example can be seen in the shift in industrial life. From around 1867 to the early 1900s farm workers were trained to handle industrial machines with little or low specialization. Their manual tasks, like a cloth weaving loom, went from manual to mechanical and yet needed expert weavers to work those machines.

Then around 1908, Ford introduced the production assembly line and brought about the need for training of clock-work industrial workers that often worked at the exact same repetitive task day after day. This is the beginning of an expert industrial worker. This meant that if two hundred people worked in the cloth weaving industry, each individual would do the same thing every day from start to end of their shift. It was all very uniform, static, and repetitive work.

As time went on, people understood the importance of respecting a person's skills and expertise. This meant that employers would hire people based on their specific skills or expertise, based on their experience, and delegate or assign a specialized task to them. They would sit at their stations, performing only their specific tasks. So, if one person cut up the clothes, the other would stitch them. At the same time, another one would add on accessories and yet another finishing touch to it. Finally, someone else would pack those clothes. We can say that this was the era when the idea of the job description was created.

This wasn't the only thing that changed with jobs. If we look at the way things worked in the past, it was common practice for an employee to join a workplace right after school. Once they joined a certain company, they served for a long time at their company. Back then, the concept of switching jobs was unheard of, and so when a person started at a company, it wasn't unusual for them to retire from that same company as well. There was a sense of permanence when it came to working structures.

Marriages, homes, work, families, and friends were more or less permanent.

Eventually, after years of things staying the same way, the established structures and practices in the workplace gradually changed. As the demand for expertise changed, so did the powerful change in the education system to suite their future workforce needs. The agrarian era had schools to suite the agrarian landlords. The industrial era required different skills, and the education system adapted. The need for specialized employees in industries producing at large scales needed a workforce that would need to be more specialized. So, schools graduated to college. As we got more specialized, we needed specialists in production, engineering, medicine, and other tasks. Our universities were created for educating these specialists to meet the workforce demands and keep them healthy so they could work longer hours and lead better lives.

Slowly these experts began to advocate for change, bringing forward the idea of impermanence. This brought about the end of permanence, which quickly accelerated to other areas of life as well.

The good thing that this change brought about was that it advocated for women's rights as well. We don't need a history lesson to understand that women were severely oppressed right through history in almost every nation on the planet.

However, over time one of the easiest things a woman could do to get her freedom was to move out of here tribal geography. Some did this by marrying men who were mobile, while others got themselves armed with their right to education and undertook new opportunities across the land.

These small changes made giant leaps in other aspects of established life patterns, as these changes began to shift in the family value structures as well. Previously, the concept of joint families was the only way things were. Grandparents, children, and grandchildren stayed in the same house or in very close proximity, and the family command and control was very powerful. As new opportunities were introduced, this structure changed drastically. Nuclear family systems came into being, and this became a concept that was widely encouraged. This soon transformed into atomic families, where an individual was all that was needed to exist as a family. If they want friends, they have a social network waiting for them to transact. It will automatically find them their long-lost friends to establish credibility. If they want sex, they can swipe left or right and find partners for a day, a week, or for life. If they want to form a tribe, this family of *one* can join people with similar likes and desires and become a member of a true or digital tribe of choice.

The same thing can be observed with laws. Previously, the concept of making amendments to laws was unheard of. Once a law was introduced, it stayed that way forever. Thankfully, as society began to evolve and change, lawmakers understood and realized that the same laws that were applicable a century ago could no longer suit the present society. Sometimes, even the laws that were introduced a few years ago would often not apply to the current expectations of society. And so, amendments to the law were made frequently to allow for flexibility in the dynamic evolution of the society and thus the system, which would apply to the current society, but always included the biases of almost every previous winner of the society.

Just like history is written by the winners, so are the laws of most societies written by the winners representing the shifts in wealth and power. Despite all the shifts at an individual level, three power brokers have remained constant. Power, politics, and men. The third is the taboo items we discuss at the end of the book.

A lot of the world we know today exists solely because the idea of impermanence was introduced in society. Our global impermanence will accelerate with time and the expansion of our digital awareness. Slowly over time, all our established rules of permanence will be seen as relics from an era of command and control, as designed by the legacy and traditional powerbrokers. As change is the only constant, it should be fairly easy to predict that things will continue to change. However, what direction these changes take is today becoming a very powerful power dynamic. As more and more wealth goes into the hands of fewer and fewer people, these extremely rich people will get caught in their intelligence trap and try to change the world to their way of thinking, just like Xi did to the Chinese people's party and to China.

As traditional holders of industrial and political power see their domains wither toward extinction, there is a likelihood that more and more leaders might choose the path of Xi or Putin, i.e., Total control, or it could also go the way of Saudi Arabia or Iran where the religious leaders assume total political and religious control. Our digital world today encourages global communications, and some nations do not want their citizens to glimpse global freedom. They are selectively shutting down digital connectors and communications between their citizens and citizens of other countries.

While on one side, there are countries that are encouraging more and more peaceful freedom, at the same time, there are other countries that are experiencing more and more extreme digital disconnections, and they get trapped int eh alternative realities of their leaders and their desired reality of how every individual must live, or die. While on one side, we see tremendous opportunities for positive changes, we are also witnessing despotic autocrats that are ruthlessly assassinating any competition, or opposition, to their leadership and throne of power.

At the same time, we need to be aware of the effect of too many changes happening too fast for any and most humans. It results in disorientation and the loss of faith via bots and algorithms to a point where people feel they are standing in political quicksand. These moments do not happen by chance, but they are brought about by structured communications. Our current state of bi-polarity in the US makes the future of our nation and the world unpredictable. This is never a good thing. History has taught us that every time the citizens become highly polarized, segregationist, and confused, they are gifted a wolf stating they he is their only benefactor to save their lives. The constant bombardment of alternative realities has resulted in half of the US believing their truth is truer than that of the other. This is an explosive situation, artificially created using digital and social assets, and it's a tragedy no matter how we look at it.

We in the US are living in what can be termed as extremely multifaceted polarity, i.e., of many kinds, which unfortunately builds itself on the separatist principles of creating the *'Us and Them'* mindset in our target individuals. If you ever hear a statement that gets you emotionally excited, then it is probably a digital trap. The familiar ones can range from "I am a lawyer from London, and you have just inherited $1.263 million, please...." A recent one that I received included a check and a letter from a well-known reputable bank for $2,500 that I should deposit into my account. It promised me that this was a marketing payoff for me to do specific tasks for $2000 once I deposited the check in my account. So, plan-1 would give me $500 for doing some tasks like buying goods for $2,000. However, I could increase my winnings by an additional $1,000 if I logged in within 24 hours, i.e., today (that is how they know the trap has been set). Then spent $500 each buying gift cards at five of over a hundred listed stores and then uploading the gift cards to a secure website, which guided me on how to upload the gift card numbers. They reasoned that I was fully covered as they had already sent me a $2,500 check that I had deposited in my account. It takes 10-15 days for a US bank to extract funds from a UK bank account, even if the bank or funds do not exist. You can figure out the rest.

In early 2020 we saw a clear acceleration of the breakdown of all cultural, social, physical, and every other foundation we have to live with. As permanence began to systematically break down, almost everything on the planet was affected. The Covid impact magnified this at a global level. A global consensus and interesting belief is that Covid-19 started in China, was then quietly hidden in China, and as it exploded across the planet, China still would not allow international inspectors into their sites. The greatest anomaly is that China rapidly reached 82,000 cases, and then Xi's reporting structure held that number for the next 12 to 14 months. I was tracking covid data for my global friends, and over the month, China was stuck at 82,000 cases. It seemed like after a certain point in time, not a single nation other than China showed this track record, and I do not believe in coincidences.

There was another data-driven information that was extremely interesting. Here is the data-driven chart that, to me, computes the relative truth

This is January 13th, 2022, data collection. The source for this data collection is WHO/ CSSEGIS

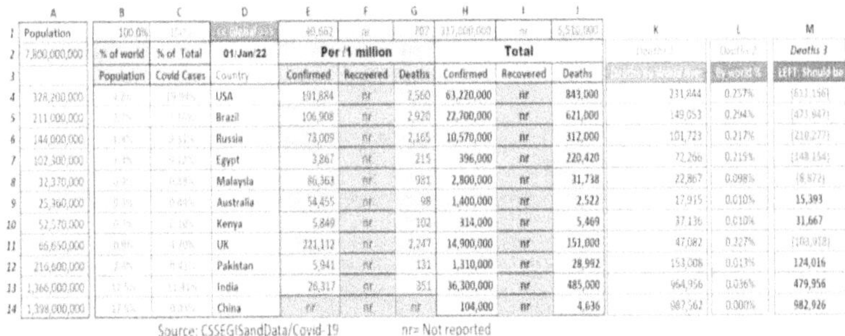

	A	B	C	D	E	F	G	H	I	J	K	L	M
1	Population	100.0%	[illegible]	[illegible]	49,662	nr	707	317,000,000	nr	5,510,000			
2	7,800,000,000	% of world	% of Total	01/Jan/22	Per /1 million			Total			Deaths 1	Deaths 2	Deaths 3
3		Population	Covid Cases	Country	Confirmed	Recovered	Deaths	Confirmed	Recovered	Deaths	[illegible]	By world %	LEFT: Should be
4	328,200,000	[illegible]	[illegible]	USA	191,884	nr	2,560	63,220,000	nr	843,000	231,844	0.257%	(611,156)
5	211,000,000	[illegible]	[illegible]	Brazil	106,908	nr	2,920	22,700,000	nr	621,000	149,053	0.294%	(471,947)
6	144,000,000	[illegible]	[illegible]	Russia	73,009	nr	2,165	10,570,000	nr	312,000	101,723	0.217%	(210,277)
7	102,300,000	[illegible]	[illegible]	Egypt	3,867	nr	215	396,000	nr	220,420	72,266	0.215%	(148,154)
8	32,370,000	[illegible]	[illegible]	Malaysia	86,363	nr	981	2,800,000	nr	31,738	22,867	0.098%	(8,872)
9	25,360,000	[illegible]	[illegible]	Australia	54,455	nr	98	1,400,000	nr	2,522	17,915	0.010%	15,393
10	52,570,000	[illegible]	[illegible]	Kenya	5,849	nr	102	314,000	nr	5,469	37,136	0.010%	31,667
11	66,650,000	[illegible]	[illegible]	UK	221,112	nr	2,247	14,900,000	nr	151,000	47,082	0.327%	(103,918)
12	216,600,000	[illegible]	[illegible]	Pakistan	5,941	nr	131	1,310,000	nr	28,992	153,008	0.013%	124,016
13	1,366,000,000	[illegible]	[illegible]	India	26,317	nr	351	36,300,000	nr	485,000	964,956	0.036%	479,956
14	1,398,000,000	[illegible]	[illegible]	China	nr	nr	nr	104,000	nr	4,636	982,562	0.000%	982,926

Source: CSSEGISandData/Covid-19   nr= Not reported

Base Data

1. The selection was made not by total numbers but by a normalized per million population. This simple step levels the data/statistical playing field.

2. **Row 1:** The global average cases

3. **Col A:** The global population of each nation.

4. **Col B:** Each nation's proportional population by world population,

5. **Col C:** The total number of COVID cases reported.

6. **Col D:** The nation by row count

7. Per million citizen stats:

8. **Col E:** Confirmed Covid cases,

9. **Col F:** Recovered cases (By Jan, '22, nations stopped reporting this)

10. **Col G:** Number of reported COVID deaths.

Statistical Interpretation (The Interesting Part)

Here we calculated each country/leader's policy effectiveness by focusing only on the weighted average of covid deaths reported. For example, if the US represents 4.76% of the world's population, and the world average of deaths stands at 707 per million citizens. Then (i) what should the US death rate have been if they stayed at global average deaths, vs. (ii) what was the actual death rate per million?

a. **Column K:** Represents the number of deaths by nation based on the global average, i.e., how many US people should have died if the US had remained on the global average. Formula = Total global deaths / global population * Nations population = 231,844 people should have died in the US by January 22.

b.   Column L: Percentage of deaths by National %. Formula: 100/National population * total nation deaths /100. Here the world percentage stood at 0.071%, whereas the US deaths % stood at .257%

c.   Column M: This is the crux of the finding. This column displays the impact of national policies by Jan '22. According to these, 611,156 more US citizens died, which is above the global average. All numbers in (0.00) represent people that could have been alive today, by global average. All positive numbers represent the number of people saved by sound policies, processes, and citizens' adaptations to policies. So according to this US becomes the worst country worldwide as their polarizations and policies split the nation where half the population refused and still refuses to follow CDC. This data leads us to believe that the defective policies of then-president Trump resulted in 611156 deaths that could have been avoided if the president, the political party, and the tribal members had followed guidelines simply on global averages. The two nations that stand out as outliers are India and China, where their data tells us that Indian policies saved 479956 more people than the global average, while China saved over 982,926 due to their policies and the willingness of the people to follow the recommended policies.

A visualization of this data looks as follows:

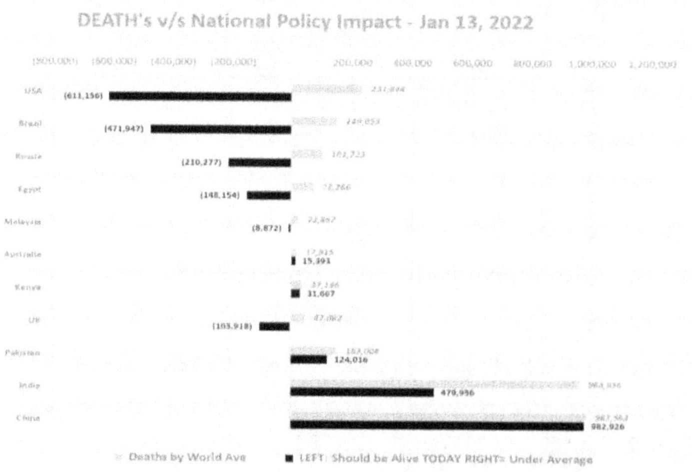

1.   Any black bars to the left of the '0-point' represents the governmental failure to save their citizens.

2.   All black bars to the right of the '0-point' represent the effectiveness of national proposals and the citizens following these recommendations.

3.   The grey bars represent what the deaths should have been if the nation had simply stuck to the world averages.

Moving on, another thing that massively affected the younger generations is the disruption in education due to the covid-19 global shutdown. The education industry was severely affected as well. In the US, a lot of private education institutions are built as profit centers. These institutions exist mostly for profits while they continue to get federal and state grants. Over time one of the ways to measure adaptation, change, and efficiency is to measure quality and functionality with cost across time. If we make a comparison between cost vs. functionality, we see a definite pattern in almost all other things but not in education. Let's compare three things. A Color TV, a DVD player, and education. Let's now compare 1975 to 2021. In 1975 A 45-inch LED TV used to cost around $4,600. By 2020 a far superior 60-inch TV with enhanced smart functionalities is now available for under $600. This is an increase in quality of around 20x times and also a price drop of 85%. The same thing happened with the DVD Player. In 1988 I flew to Singapore with a very specific DVD player in mind. It was a 1981 model. In the Sony shop, I asked for that exact model, which I was told was discontinued. The salesperson then showed me another model, which, according to him, was a better player, had many more functionalities and options, and was cheaper by 65%. Buying that was a no-brainer. During the same period, universities have added a lot more filler subjects and their core subjects are about the same, so let's say the quality of educational content due to change is 3x times, but the price of education went up from $14,000 to $68,000, an increase of 420% and continues to rise unabated and with no federal or state controls. However, in all fairness, it must be stated that the top US universities are the envy of the planet, and students from every nation aspire to come to them, but only the rich can dream of doing so, not the brilliant and bright sparks to our future. As we learn more and more from the future, we will collectively come to realize that universities designed for the rich will only amplify more of the for-profit centric future, while only those institutions focused on planetary and humanity excellence will produce students and leaders that could drive our future in the right direction. Add to these the recent US revelations of the very rich cheating to get their children into good colleges, and some even bribing others to take their exams and thus grabbing coveted jobs and positions despite being losers. If this can happen in the US, imagine what the status is in countries like Russia, China, North Korea, Iran, and Saudi Arabia for politically connected children. This projects a very dire future for truly intelligent, honest, and hard-working students on a global scale. We have to stop all this graft, cheating and demand process transparency in prime educational and public institutions.

As we begin to learn more and more from the future, *this will be discussed as the science of Fufactology later,* take our lessons from the post-COVID era, we begin to see clear signs of Digital adaptations and opportunities. Previously, education was something obtained in an industrial era of brick & mortar university for-profit classrooms. However, almost overnight, we have seen a drastic shift from physical classrooms to digital ones online. Covid has accelerated this digital education shift to new heights where on one side is the immediate explosion of digital learning opportunities and on the other a desperate attempt of traditional universities and influencers of education to keep their physical profit centers alive, even at the cost of now having tools that will enable them to spread education like wildfire across the planet.

In the last year, i.e., 2020, we have seen ample evidence of the ability of remote teachers who used available digital assets to help with global learning extensions. There is probably not a single student across the planet that has not undergone some form of remote learning. Digital connectivity enabled connecting to remote, locked-down students by similarly locked-down teachers, and together experience the new digital possibilities. It allowed both teachers and students to sit in their locked-down homes and fully interact, earn, and get certified for a year full of learning. It allowed for highly specific and general courses that people from different parts of the world could undertake in their own space and times. This is, of course, done through a digital medium. Thanks to online platforms, anyone can today get required certifications that would help students from across the planet in their educational and career growth.

Just wind the clock back to Dec 2019, and we find a time when online classes weren't popular at all. Remote teaching wasn't a well-established alternative. However, with the outbreak of this virus, things were forced to change rather quickly. By March 2020, thanks to the pandemic, and the global lockdown, eLearning and obtaining education online became a norm. In this case, and under the circumstances, digital applications were a godsend. It had broken the permanent structure of education and everything else and exposed the alternatives of micro-courses and self-directed learning.

All this while, we should not forget that the US education system, along with its universities, the education departments, and their digital marketing arms, kept pushing a very negative campaign on possibly the fear of remote learning becoming common practice and potentially taking away some of their paying students. The optics highlighted across all social platforms have been — that social segregation and not attending physical classes is not good for the children. Also, they will suffer emotionally, socially, and physically. Parents and students tend to trust their government and educators; thus, when they hear from their trusted sources, both parents and children get very depressed. We all tend to believe what our universities and teachers broadcast, so when we read that they will suffer emotionally, they slowly do. When they heard they would suffer socially and physically, they did.

However, if I was leading the media campaign for education, my message would have been, *"Congratulations, you students are the first generation that is coincidentally being forced to adapt to the future of education and work. A lot of companies in Silicon Valley are today giving an option, to their employees, to work from home for the rest of their lives. Imagine you will be the first generation that can choose to work from home or an office if your type of work permits. We have heard from many new digital enterprises, which are the growth engines of tomorrow, that they are getting ready to welcome the first human batch of digitally adapted students."*

At the same time, the media and companies could have released a parallel statement, *"We are happy to report that most major enterprises on the growth path are undertaking a massive digital transformation. You, students, are our first batch of totally digitally adapted students in the history of mankind. You all are unique. No other batch of students has ever collectively participated to this degree of remote participation, digital connections, and connected learning. We welcome students from these batches as you are each the leaders of the future of this inevitable digital transformation. We look forward to your graduation and taking our nation, economies, and companies to the next level of digital success."*

Imagine the psychological difference in this vs. the current message that is designed to pull education back to the past.

The whole education system is ripe today for as big a change as Uber did to the global Cab and taxi, providers as the key assets carry their content, and this digital transformation could be the next leap in global education and democratic sharing of knowledge across the planet.

This global impermanence creates opportunities for micro-segmentation in almost every area imaginable, with courses that rapidly evolve to meet the future. It is a well-known fact that in most things professional, by the time the student passes out of their university, their knowledge is already outdated. This new opportunity, let's call it Uber University will be driven by learning from the future, so the courses are driven by the lessons from the future and not needs and courses of the past as they currently are.

In 2020 business schools are mostly still teaching market segmentation in the old-school industrialized and for-profit business segmentations. Whereas the reality is here in this example. Back in early 2020, when consulting for a major US telecommunications company, that used to divide the market into four to five groups, depending on a person's age group. The target audience was segmented into four groups: [1] under 19, [2] the age group of 20 going to the age of 30; [3] the age bracket of 31 to 50; [4]

the last segment was above 50. This segmentation was used in marketing, PR, targeting audiences, placing posters in stores, and communicating with customers. This grouping was used for the last five decades and was used to drive targeted marketing based on age-defined segments. However, with the advent of digital platforms, the old market segmentation has been rendered obsolete as it is no longer a practical way to do things.

Now, things are done differently. We can divide people into micro segments starting with one segment for every year, starting at 10 and going on till 105. With this, we can today understand the preferential differences between a 17-year-old vs an 18-year-old. Within a few months a wealth of information came in, and this was further doubled based on things such as their genders, their preferences, their interests, etc. this provided visibility in the preferential differences between an 18-year-old boy vs. an 18-year-old girl. With passing benefits and time, this ended with 1,200 segments by age, color, sex, religion, ethnicity, etc. The digital core allowed us to divide people into ever smaller groups, as this makes targeting highly specific individual customers easier when it comes to enhancing their experience per their taste.

This is how politicians today target individual citizens, and hired trolls target individual politicians. It's becoming a vicious circle. The hunter and the hunted are both becoming digital victims. With faster and faster digital capabilities, there has been a shift in the way things are being done today. Previous permanent structures are no longer relevant. They have been discarded in favor of these impermanent and dynamic structures.

This micro-segmentation has helped target and alter minds at an individual level. For example, police brutality and murder of Black people have been embedded into our culture for such a long time that even today, we all carry heavy biases when a Black girl disappears vs. when a White girl disappears. This generalized statement has its link in history from so long ago that it is woven into the cultural psyche of the American people. It has created a caste system as severe as in most other parts of the world, other than the sheer fact that in the US, both sides are equally educated and capable of doing each other's jobs in one of the supposedly most developed nations on the planet.

Similarly, the US police force have, for a long time, systematically oppressed colored people before the George Floyd incident. Before this specific incidence, even a brutal execution, accompanied by further brutality, would rarely be noticed by the masses, if at all; there wasn't much that could be done about it. No one could raise their voice against it because the issue was never taken seriously by the dominantly White police force and was considered a waste of resources by the Blacks. And so, the oppressive forces went about doing what they have been accustomed to doing for centuries, confident in their arrogance that nothing could touch them.

In 1991 it was the beating of Rodney King in a world of radio but zero social connectivity. The next incident happened in a digitally connected world. We jump forward to 2020 in the US, with the murder of George Floyd and the conviction of the police officers based on the video evidence. Then it was the photograph of Prince Andrew with Jeffrey Epstein, his conviction, suicide, and the imprisonment of Ghislaine Maxwell. Followed by the many episodes of Gabby Petito, her murder, and the suicide of Brian Laundrie. An outlier was the story of Kyle Rittenhouse, who killed two men, wounded a third, and was found not guilty. Another famous case was that of Elizabeth Holmes, who defrauded investors of a company valued for $20 billion on fake tests and declarations. Most recently, the evidence from the January 6th internal insurrection and the evidence on Trump and his satellite supporters. In India, the government has encouraged the public to record, or video, if a government employee asks for a bribe resulting in a massive drop in fraud and graft by government employees. In Iran, riots started with the execution of a pretty young girl by their religious police started a revolution that could overthrow hopefully the Ayatollah, who is fighting back with executions of demonstrators and hundreds of arrests and imprisonments initiated directly by the government. The government even gassed a girl's school when they showed signs of rebellion.

On March 3, 1991, it was the brutal beating of Rodney King, which escalated with the acquittal of three of four policemen, with the jury failing to reach a charge for the fourth. The awareness of this was fairly local to Los Angeles, where riots broke out. Fast forward to 2020, when the brutal murder of George Floyd was not only recorded in real-time for the full duration of 10 minutes and 18 seconds, but our modern social networks ensured that it also went viral within hours. As the video was a non-stop recording, there was zero possibility of trying to deny what the world could openly see. This allowed the world at large to stop-dead-in-their-track and witness a murder at extremely close quarters, as never before. Viewers could stop, replay, have arguments, and then replay and relive the execution a million times. There was no denying the arrogance on the policeman's face as he judged, juried, and executed a Black man under the full bodyweight of his knee. This singular fact helped raise a global cry for justice. Police accountability was called for on an international platform. Despite Covid-19, peaceful protests were held worldwide, and Black-Lives-Matter was born as a social hashtag. The police, the system, and the courts were forced to take the case on board, and any attempts to acquit the perpetrators for their crimes became nullified by the sheer weight of the recorded evidence. This became a case where even the whitest of Whites could not spin the story out of context.

The new digitally connected platform made it easier for people to protest, and it made it harder for the concerned authorities to ignore facts on the ground. Digital Transparency is here to stay. Now learn to use it for good and to your advantage.

Without the digital evidence, the murder of George Floyd could have been written off as something else in the records. His records and death certificate could have easily been tampered with. The evidence could have been sabotaged. However, the video that was shared worldwide made it possible to prevent that from happening.

Justice was finally served to the Floyd family in 2021. The shift toward impermanence has greatly helped in rewriting an important and big color-based dilemma of American history. These norms have been challenged only by this impermanent structure that is now being widely embraced.

The same thing happened with Harvey Weinstein, which launched the 'Me-Too' movement, as did the underage prostitution clubs of the billionaires with Epstein. As we travel in time, all these digital applications empower every common citizen with digital capabilities, which will make worse-players accountable for their crimes. We are reaching a point of time when power, influence, or money will not be enough to buy justice, as the available digital evidence becomes proof beyond dispute.

However, there are bad nations where no amount of proof can give justice to the citizens. Right now, the example is Iran, but we can place many autocratic nations in this category.

All you need to do when you see a crime being committed by the government, or uniformed oppressors, is to switch on your mobile camera. Please note that in some countries, that may cost you your camera or even your life (if you are a beautiful girl in Iran, you could be jailed, raped, and killed), if you are a commoner in China, or if you participate in an antigovernment event in Russia. But let's pray for the right to record and report as a universal law to be upheld by every united nation member nation. It's time to allow citizens to prove oppression in their own country and not get executed for it.

The world is hardly aware of the millions of good things happening around the planet due to digitization. Using this same platform and the need for bad actors to amplify routinely fake bad-news with fake extrapolations and the relentless paid broadcasts within our social network connections. However, the bad actors work in clandestine operations, from digitally dark rooms and often hidden from the eyes of the rest of the world. Even this book focuses more on the bad acts as the intention is to make the reader a more intelligent participant in this new quagmire of digital traps. Every preceding major development of digitization too follows standard humanoid principles of use and misuse. When used properly, this digital opportunity, along with its inherent social networks, can be a beneficial boon for humanity. However, in the wrong hands, a beast. We jointly need to define, track and eradicate the wrong hands and brokers by identifying their patterns. In the meanwhile, overall digital capabilities will continue to adapt, shift and change at ever faster rates.

This means the digital shock will only explode, and unless we continue to adapt at ever faster rates, the world outside will either become more and more polarized, whereby we will individually and collectively become more detached from the ensuing external reality. Or, like most human evolution, this is another digital poisoning that we will manage to audit, control, and once again kill this poison in a planned manner. The resolution to this digital solution will, fortunately be another digital police/sleuth. Our individual, collective, tribal, and global goals must today become our attempt to become more adaptable.

As digital savvy creators of the world, we can together build our tomorrow. Unfortunately, it is the rich, the powerful, and the power brokers that will continue to pay global trolls and bad players, find the target individual outliers, digitally alter their minds remotely and collectively build a ghost army of digital zombies who will react to the will and commands of their handlers. We have seen evidence of this in the US as more and more people take their illogical decisions echoing the exact same vocal response, demonstrating that their state of mind was altered by social, or digital, tribal resonance.

Some societies and people have managed to break free from the traditional worlds of 'Command and Control' envisioned by the rich and powerful. We broke the kings and brought the individual as the decider of nations. We called it democracy. At the same time, there are nations today like China, Russia, North Korea, and Iran that continue on a path of digital misuse and their successful roadmap to the new emperors of today. There is the UK that still has a freebee monarchy that hides a pedophile prince who is wanted in the US. People in these oppressed nations have lost their rights, voices, freedom, and their lives when they speak against their *emperors*. In North Korea, he is an elected leader with 100% electoral votes. In China and Russia, Xi is a lifelong elected president of the nation.

Take Aways:

Individual: Your greatest opportunity is to adapt and somehow filter your own truth away from all the tribal spin doctors and feeds. There is a huge opportunity to open a world-class, honest-to-God digital university with the best professors and staff on the planet. Its aim should be to educate and not hold billions of dollars in its brick-and-mortar reserves. The individual must use their digital devices and networks to create awareness and peace. If an official asks for a bribe, record them; if the police are brutal to someone, video them. Become the change that this opportunity gives humanity. Become the change you desire.

Companies: *Train your employees to think digital transformation and leverage the digital assets and opportunities before someone does it or some other company beats you to it. History teaches us that people, nations, and companies that hesitate to adapt suffer. The rate of bankruptcy has accelerated. The key lesson here is no customer ever loses because some group of digital disrupters simply invent to do what you do — but better. So, your path is to become that disrupter.*

Nations: *For any nation or an elected political representative, the center of your universe has to be protecting the nation and your citizens. When a political party trumps (no pun intended) the needs of a nation or that of citizens, then that is a southbound road that the political tribe is taking. History has taught us many lessons, but trust in history is fading because more and more nations are realizing it was an interpretation of the victors. At the end of this book, we discuss the launching of a science Fufactology or, finally a structured method of learning from the future. Instead of creating and funding polarized chaos, our elected leaders must seek to protect the nation and the people, not their party or religion.*

The 'P' factor: *If change is the only constant, the best alternative for us is to get ready to adapt. It is the US, Europe, Japan, and India where we reflect the optics of democracy. It is in our hands to make it succeed so other nations can reap the harvest we bring. We want China, Russia, Iran, and North Korea to aspire to become like US and India. Not a politically polarized one but the democracy we shall soon make; these become the global lighthouses of success and some failures as that is where we learn our lessons on what not to do. It must be stated that the principles of democracy itself are under audit, as the rich and powerful slowly manage to distract voters and harvest votes on fake yet popular statements. But, so far, it seems to be the most harmonized system for nations.*

Chapter 5 — The Digital Opportunity

Very few literate people do not use social networks today; if you own a smartphone, a computer, or a laptop, you probably are an addict to social networks. But all things do not cater to every local expectation, and this is how an observant Latina created a local social media network Quepasa which was acquired by My Yearbook in 2011. Her name was Ashley Qualls, and she started her idea with just Eight dollars. The journey of creating a sharing application Qualls led her to earn more than $70,000 per month in revenue15. When Myspace was popular, people complimented Qualls on her Myspace page designs. She posted the designs online for people to purchase, and that propelled her to a $70,000 per month revenue with seven million monthly visitors. She made so much money that she dropped out of school to devote her time to her business. She was offered $1.5 million for her business but turned it down. Lesson: Create an idea that would work in your locality and sit quietly and figure out an application people would like to join for free in your locality. That is the start of every social disruption — your unique offering that benefits a targeted group of people.

As the world progressed and moved toward a more civilized society, humans began to break free from permanent lifestyles and slowly shift toward impermanence. This slow adjustment did not happen overnight but took over a century for people to discard old ideas of permanence and replace them with new ideas of temporary atomic attributes. Over time more and more people adjusted and adapted to these new ways of life. This adaptation brought about a change in individual mindsets from traditional artificial boundaries and socio-religious bonds. The pace of change accelerated each year, and as the velocity of changes increased when a lot of people got lost along the way. Our digital algorithms were waiting on the sides of these digital highways, placed there by very powerful digital pimps and powerbrokers to make the lost minds addicted to their songs of *us vs. them.* Societies that were once taught by village elders were now required to be certified by industrial production line-certified institutions like colleges and universities. Initially, it worked wonderfully when teaching was the main reason for building a university. But then the rich made centers of learning into private profit centers, in some countries, even a place they could sell certificates for profit. These centers spent millions on PR, and delivered excellent education but over time, when one asked them the question of who they were working for, 'the child student, or profits,' one stopped getting clear answers. This became all the more glaring during the pandemic communications from educational institutions. In the industrialized cities, a race began with very large investments for ever larger institutions inside citadel-like universities that represented the pinnacle of the industrial era education.

A society that brought things that they needed from the local stores suddenly began to get things from around the globe. A society that took days, if not months, to communicate with someone on the other side of the state now began to take only a few minutes, then a few seconds. Eminent professors across the planet could now choose to teach deserving students in any connected city on earth, but it somehow never happened. We got very close to this during the pandemic, but then, as things are getting normalized, so are the educational institutions. Patience was replaced by instant gratification. A world that once depended on the people around them began to depend more on more on the pseudo-digital village that had been created to fill this empty space in people's lives. Families that had lived centuries with a handful of close friends saw their children now have thousands of digital friends that they had never, and would never, physically meet in their lives. A lot of these are digital trolling personalities, some built just to influence you at a very personal level.

Impermanence has now become an art, for we live in an ever-changing and ever-adapting, fast-paced world. Also, the pace of change is getting faster, so one has to seriously learn to adapt.

Today, we live in a world that is constantly challenging the norms and defying the old system that has been in place for thousands of years. Today, for the first time in the history of humans, we can call out for justice when we think that it is not being delivered. We call out on laws and bills that we think do not suit us or that may not benefit us. For hundreds of years, nations and leaders wielded power that was placed into the hands of elected officials — who were supposed to represent the people. However, once elected, their purpose in life changed. In the last decade, people have realized that some of these elected officials are working more for their own political or religious parties, their pockets, and not for the good of the common citizen who voted them to power.

Digital disruption has distributed this power into the hands of the common person. Some will wield it wisely; some will wield it badly, but most will be the victims of political trolls that will mess with the minds of the people finding it harder to adapt. We, as citizens, finally have a very powerful voice that matters, and our voice has gradually become our power. We can now challenge the laws in place, especially if we feel that they have been misused or become outdated. However, it is critical to remember that there are the rich and powerful that want to maintain control, and they will do everything in their power to mess with your mind and that of your neighbors. There is enough historical evidence of this if one reads between the lines. Take the US, where democracy seems to have been born. There are two parties that officially spent $8.9 billion on state and federal elections in 2022.

Unofficially it is two to three times that amount. Now imagine an excellent candidate on one side with little money and a rich, crafty, politically funded candidate on the other side; who do you predict will win. Also, after winning, this candidate has to return favors to the very people who funded their elections.

The big questions are [1] Can we have a more honest and equitable political system and a truly democratic outcome when the system has been corrupted by the rich and powerful — an opportunity for a political scientist; and [2] can we have used that $8.9 billion for better serving the citizens of just one country — the US of A.

In our current society, things like digital mind control have also become pervasive. Previously, these were things that were written about in spy novels and science fiction books where enemy states created sleepers who were activated by specific commands. But today, we invite these mind shapers into our social networks, befriend them, join their tribes, and become active participants in whatever the trolls want us to do. The controllers are totally invisible. The Jan 6th attack on the US capitol in 2021 is an example of this transformation, i.e., the digital transformation of the human mind. The Jan 6th committee televised their findings, then in Dec 2022 published their findings[ii] for the US public to read and for everyone to see, yet 70% of 49% of the US citizens continue to state that it never happened the way the committee has reported it, even though there is documentary and video evidence of all events. This polarization is not a coincidence, and I beg for the reader to contemplate that this is evidence of the digital transformation of the human mind that we need to stop as a people. Innocent people were attacked in the name of patriotism because they had been so thoroughly brainwashed and controlled that they began to feel that what they were doing was right and justified. They came to the capitol on the command and request of their ex-president.

We can see glimpses of such events having happened in the past as well. People have been brain washed and controlled for ages on the call of a king, a nation, a political party, and even a religion. Evidence of this can be seen through wars and the way people behave in wartime. From the French revolution to the Holocaust, to the World Wars, it is evident that people who have been brain washed tend to behave in an extreme manner and with a frenzied passion as patriots. Jan 6th was no different; these were the misguided soldiers of a political party who were still unwilling to stand down.

While it is the duty of a soldier to defend their land and people, treating people brutally and having a violent or murderous streak is not in the job specification of a soldier, but time and again, we see such incidents taking place. This is simply because we are breaking free from the permanence that once dominated the world and replacing them with impermanent instructions. This is one of the downsides of impermanence, that we are becoming increasingly illogical and allowing our minds to be easily influenced and controlled by people with their own agendas.

While breaking free from permanence may have its side effects, it also has aspects that we need to appreciate. All our digital solutions, the uber rides, and young children upsetting stalwart brick-and-mortar establishments are the bright side of this disruption. But then the wealth accumulation of such people is so rapid that they sometimes lose their sensibilities while remaining trapped in their concept of right and wrong.

The Media today is heavily influenced and controlled. News channels like CNN and Fox News are both controlled, watched, and influenced by very polarized and powerful people. They broadcast connections and propaganda far deeper than we can ever realize. This is why we need to be cautious and not blindly believe or follow everything that the media tells us. We live in a heavily digitized world (read as easily influenced and controlled world via their smart devices). While we are also the products of a modern and educated society, we are lost between logic and our belief systems. It seems logical to conclude that for most people, logic and belief have no meeting point, and therein lies our human dilemma. We have the gift of using our own minds and senses to the best of our capabilities bestowed upon us, but we let our smart devices steal that gift and replace it with their thoughts and points of view. So, we must keep our eyes, ears, and minds open as we decide which way, we wish to look upon events and statements. We must be openminded before making our decisions if we don't want to be influenced or controlled. It is our duty to choose wisely. As my father used to say, "Choose your friends, mentors, and influencers carefully." Beware that you do not become an echo chamber of what you hear and are told by the smart social, audio, and visual devices.

A Rare Moment in History

As time goes by, and our paid media continues to impact our lives, we become more and more trapped inside someone else's echo chambers. When the US Constitution was written, the media was independent and was expected to tell the truth. Over time, these media giants became for-profit-only businesses and have today become opinion mouthpieces for the rich and mighty. The political parties are the richest and most powerful. In the US, the media gets all the privileges and delivers a few truth-based news. Fox News just paid $785.5 million not for speaking their truth but for covering their lies and to prevent Murdock from testifying. They bribed Dominion, who basically did a disservice to the US people by preventing the truth from being exposed. If we only consider the media, it seems that we are either living in hell or getting ever closer to reaching that state. However, it is very important to realize three things:

1. Unfortunately, humans are attracted more toward threats, tragedy and bad news. It's called the survival instinct. If you see two headlines [1], 14 Thai teenagers tapped in a flooded cave, and [2] An Italian widow donated a large sum to the Vatican. Which report will you read first? Probably the Thai one.

2. Media knows about this addiction, so they routinely write about, or often create bad news. If our addiction is tragic news, then it is easy for trolls to mostly report tragedies with high emotional content to attract more readership. Even if sometimes they have to create this fake news. Social media, Facebook, and the late Cambridge Analytica, lived off this element and used this simple fact to make millions if not billions, by playing with this predictable emotional sentiment.

3. Historically, we have never been more united, safer, and more equal than we are today. Some of us are made to wish we lived 400 or 1000 years ago based on the optics created by Hollywood, and channel TV programs. If that were true, then Saudi Arabia, the land of Arabian nights, should have girls in transparent harem pants and scanty underclothes twirling around the country and palm tree lining water-filled oases. India would have twirling girls in colorful clothes dancing around, and singing songs, and the US would have brave white cowboys saving everyone from the nasty native Indians who only go around raping pretty white women or killing good white men. — all of which are utterly and brutally wrong. They are only selling opinions and biases.

We stand at a rare cusp of time where we can make this planet into everyone's dream of paradise. But...

We are living at the convergence of things that define humanity, religions, politics, individual rights, wealth, social values, climate, peace, equality, weapons, and our ability to take and influence individuals on global, local, and individual levels.

Moving forward, our individual, and thus collective decisions that we take today can grow into this new digital world like a cooling rainfall across the dry prairies or savannah — *our opportunity to create a successful explosion of opportunities and equality while taking us today to a brighter future for all mankind, is right in our hands even as we read these lines.* Let's fact-check this statement as to whether it has any figment of truth or is yet another spin doctor's lie.

Let's talk **UK Before**. Five Hundred years ago, there existed a weak legal and social system that was basically governed at the whims of the kings and their co-hoots. Over 90% of the wealth was in the hands of kings and lords. These kings and lords lived off the largess, taxes, and sweat of their citizens. They collectively did nothing other than ride around in pomp and show and maintain a strong army to protect them. Across Europe, some lords used to make pretty wives of common people spend their first night with them so they could bless the new wife with their God-given touch. Back then, these men in power could execute, murder, or rape a commoner, with no one being able to question the act.

Women had no right to be part of a club, in voting or anything else that would indicate that they were humans and had a brain of their own. Quite simply, women had no rights and lived a pitiful life. Girls did not go to schools that taught men; they went to schools that taught them be become better housewives. They certainly did not work in equal positions across any land.

The kings and lords deemed themselves as appointees of God and lived comfortably in opulent palaces built on the toils of the common man. They ran their domains and empires with a personalized version of dynastic arrogance and total subjugation of their people.

UK Today, there exists a strong legal and social balancing system. The queen, now a king, and their family still earn millions as a continuance of the archaic concept. They continue to live in palaces with dynastic rights of undeserving opulence. However, the new king is no longer the child of God, though they still do nothing but prance around in pompous splendor, in attendance with leaders of other nations. Though a prince recently got royal protection in 2021 from US courts when accused of sexual favors from an underage girl. That prince may never travel to the US unless they appoint him as a prime minister.

Let's Fasttrack to today — when non-royal influential and powerful people are often held accountable today, like Harvey Weinstein and Jeffrey Epstein. After a certain threshold of wrongs, their power and standing do not matter. As we move forward in our impermanent society, we can see and share things that used to only be discussed behind closed doors before. Things changing. Prince Harry breaking free from his royal duties and Meghan Markle exposing her life and negative experiences at Buckingham Palace further proves this point of a modern-day Cinderella story, with her in an artificially created negative royal PR barrage against their own prince and his princess.

Moving on, we can build a far improved society in the near future, thanks to the radical change that we have and will plan and deliver over the next few years. Today, women have almost equal rights by law, and they can vote and stand as leaders and lead the nation. Women who couldn't study yesterday are engineers, doctors today. They can do anything and everything that men can — with equal rights and opportunities today.

Of course, women still follow a male-dominated civil code with a single male god giving men a special right in the religious scheme of things. However, it is only a matter of time before these things might eventually change.

Today, if something happens in one part of the UK, it instantly reverberates across the country and the globe — with repercussions almost in real-time. One wrong statement and the person could lose their job or status overnight. Wealth is equally distributed across the nation, with brighter sparks around metropolises. Such is the extent of change that has taken place over the span of a few decades.

Let's talk about **Europe.** It was quite identical to the UK with few noteworthy differences. Europe was a mishmash of kingdoms and kings, but the basic human rights were about the same in all aspects of rights and communications. Today most European countries are far ahead of the UK and they are clasping their royal heritage.

Let's talk **US.** Five hundred years ago, native Indians lived as nomads and consumed what they needed. They lived mostly in regional tribes that interacted with each other in small ways and small skirmishes. They prayed to nature and consumed only what they needed.

Around five hundred years ago, 1500s to be exact, Europeans landed in the Americas as they went looking for another route to the wealth of imports from India and found the US. That is the reason for centuries, the natives were called Red-Indians, today considered a derogatory term because, for many years, they thought they had indeed landed in India.

Then, with the establishment of the new world in northern America, the Europeans come to this new promised land to escape their European oppression and persecutions. However, unfortunately in their quest for wealth, and power they persecuted not only the indigenous natives — a process that continues without any apologies till today. They also collectively contributed to the capitalistic perception of free labor — by the concept of owning a human as personal property in the form of a slave. United States commenced the world's largest slave trade, and a collective dehumanizing of the Black slaves from Africa. This formed one of the world's most conservative caste systems on the planet — via totally dehumanizing the Black people. They established a national color code by collectively dehumanizing the Black Africans and bringing them as commoditized labor that they mostly treated worse than animals. The Blacks became personal property of White owners to do with them whatever they pleased — no exceptions. This provided free labor for their cotton plantations and farms and a world of *zero-wage society.*

This also launched the world's color-based US caste system that still influences US psyche, especially in the southern states, today.

There was a weak legal and social governance system that was re-defined by the White, rich, and powerful. They collectively redefined basic law in any way they pleased while mostly using their bible as their book of basic records. Small groups of people could redefine right and wrong and hang, murder, and execute innocent people especially if they were the native Indians, Blacks, and Mexicans.

One hundred and fifty years ago, over 99% of the wealth was in the hands of US appointees from the kings of Europe (UK included). They appointed themselves in the image of the world they had escaped from but this time on the foundation of their US attained wealth and power. Some got their wealth and power from Europe, while others grabbed power and then wealth that they accumulated locally, most often by buying slaves at prices they defined and then owning them for life.

Women had few, if any rights at all. Crimes against women were not uncommon. It was a man's world to a very large degree. The native Indians too were treated like animals with minimal human rights. The Machiavellian principle was used across all negotiations, i.e., examples like make a native tribal chief sign a peace treaty, take their guns and then massacre them to the last individual. Historically this has been repeated again and again by many modern and civilized nations, and leaders alike.

In Latin America, where a few years later the Europeans went about with a Christian genocide we refer to as the Spanish Inquisition. The US too very critical of the Spanish inquisition but during approximately the same time we committed similar atrocities across North America, in the name of slavery and the subjugation of native people – that continues till today to some degree. There are examples of how Catholic Christian schools forced native Indian children into schools with the single aim of *"Take these barbaric natives, wipe out their nativity, and return to us good Christian followers."*

In this quest across Canada and the US, the White priests and teachers murdered hundreds of Indian children who officially 'ran' away from school and were never found. It was only recently that mass graves of Indian children were found in these schools in Canada and of Black children in Florida.

Today, the rich still control the election process. In 2010 the wealth distribution was 99% of the wealth was with the top 33%, and 90% of this with the top 3 percent. By 2020 this distribution has gone to 96% with the top 3%. All these trillionaire's are busy harvesting our national wealth primarily with their digital superiority. The very rich who used to live off dynastic wealth have started witnessing the shift in wealth from industry to technology and now to digitization. Their traditional brick and mortar enterprises are going bankrupt as the swift young digital disrupters provide better micro services and slowly erode their empires, one slice at a time. Then they consolidate their microservices under a big umbrella as they replace the traditional supply chains with new digital ones, like Amazon does today with their Prime. Today, they have to rightfully earn their wealth and influence, while the digital knights are disrupting the economies right from under their feet. Here is the punch line — this disrupter could be you.

Accountability is real. No one can murder or rape a native Indian, a Black, or a Mexican today and go Scott free. We routinely punish men (and women) for crimes equally. However, it is necessary to point out that even in the US there does seem to be more than one law. The constitution says 'no one is above the law' but routinely there are examples where the White Christians, rich and the powerful have a law unto themselves. There is one for the colored folks where a Black man can get imprisoned for over two decades because a White jury found him guilty of committing a crime against someone White. Then there is this law for the original inhabitants of the US — the native Indians. They have been relegated to the worst lands in the US, and when some mineral wealth is found on their lands, they are ruthlessly sidestepped despite all their contracts and agreements. In addition, they are denied the right to stand for US elections. There is still another Christian law against women, though I have a few friends that totally disagree with my opinion. Women are treated as a subclass in the US, and on June 24th, 2022, Roe v. Wade proves that statement. Then to add to this pain there is color. Add to this religion and we have a heavy bias on this hald of humanity.

Today, in some countries, women have close to equal rights to education, voting, and life decisions. The legacy of color-code, where people are treated differently based on their color, still continues in most nations, and is very evident in the US. However, things have gotten better over time. It is now the role of the digital citizen to eradicate color-code as we move forward.

In the metropolis cities, this is not evident but in the rural areas it is still a powerful deterrent and social influence. The Black citizens continue to believe they are dealt an unfair social pack of cards in this land of the fair and brave. Again, the biggest example of this is the case of George Floyd which opened the floodgates to many other cases of racial injustice and police brutality.

US Leaders is chosen by a so-called democratic process though there are still rumblings about the unfairness of the electoral college influence (dominated by the rich and influential). Due to this process winners are not declared simply by majority in voting, but by a special biased process. Wealth is today unequally distributed across the nation with brighter sparks around metropolises.

Let's talk **Saudi Arabia.** Five hundred years ago, this was a region of Islamic hyper growth. Tribes warred with each other and there was a desperate need to expand, grow the flock, and build the Islamic stairway to heaven. Wealth was in the hands of tribal leaders most of who were nomadic traders other than people in the two holy cities of Mecca and Madinah. Men of one tribe would murder, rape, and execute commoners with no repercussions.

Women did not just have no rights but were declared as the property of men to do as they pleased. Women were not allowed to receive education, go out of their house without a male chaperone, or take part in any societal activities. Women did not go to schools or universities and were often stoned to death if they were seen with a male other than their "Mehram" — father, brother, or husband. Ironically, these girls are mostly stoned to death by their own family members, as a form of saving family honor.

The tribal chiefs were dynastic leaders and ruled fearlessly over their tribes. Religion was, and still is, male dominated and if something happened in one part of the country it took many months or the speed of a caravan to reach other parts.

When we look at the current status, little has changed. 99% of the wealth is still in the hands of the king and his immediate family domain. Appointees are dynastic and there is little democracy of any kind, despite the optics. Religious courts are run by the Koran and remains in the hands of the religious appointees who on their side are a part of their dynastic appointees. They pass laws, execute orders or pass legal judgements based on their interpretation of the Quran — their holy book, or their royal whims and fancies.

Women could not vote in any local decisions. However, due to some radical changes they can participate in voting after Dec 12, 2021. Earlier they had no rights of being in any club, social gatherings, sports stadiums or to even drive their own vehicle. By 2022 they are now allowed to work and drive their own cars without a male in the car. Women had started to go to schools and universities as early as the 1980's, and up until very recently could work only in women-only environments that does not interact with men.

The king and his dynasty are both the political leaders as the unquestioned ruler of the kingdom and the religious leaders as the keepers of the two holy mosques thus almost appointees by their god. They thus rule the two most powerful institutions of any nation – politics and religion. The king and their dynastic spread rule, govern and manage the nation living in opulence generated by the oil wealth. While religion remained male-dominated with rights only for men.

If something happens in one part of the country, the distribution of news is filtered by a governmental decree for processing and distribution. One wrong statement against the establishment could result in the person losing their head and life, or the establishment getting serious warnings. Or a media losing their right to distribute news.

Let's talk **India.** Let me state early that because I come from India there may be a little, or much bias here that I am unaware of. For hundreds of years India had Hindu kings warring with each other. This was followed by Muslim invaders siding with one of the warring kings and slowly taking over India. *Note: Lesson to be learned here is that an internal enemy is the worst for any nation, and the best asset for an enemy. So, in the US today when Republicans and Democrats become hyper polarized the only winner is our enemies.*

The British came in as traders and followed the same gameplan and even out did the Muslims in this game of treachery and anarchy. The Muslim overlords regularly kidnapped pretty local women, forced their religious conversions (the same old stairway to heaven concept). They murdered and raped with few consequences.

Both on the Hindu and Muslim sides, girls were routinely not sent to schools and married at an early age as a socio/religious practice. There were a few exceptions, but there were talking of the general society and its rules.

The kings and their political appointees owned large tracts of land, as that was the form of wealth and power, and lived comfortably in opulent palaces off the toils of the common man. Islam was very male-dominated. Hinduism, till today remains the only religion with women goddesses leveraging more power than their male counterparts.

If something happened in one part of the country, it moved at the pace of human or movement on horse.

Two thousand years ago there was a strong education legal and social system that prevailed across India. India carried wealth in learning, gold, spices, and spiritual lessons. A thousand years ago invading India became an annual pastime for the Muslims, who came to India pillaged its gold and went back with gold and jewels back to their desert lands. This was followed by the British in the 16th century. All during this period there are many instances of a commoner getting fair judgement for thousands of years. The only exception to this was the draconic caste system that places individuals into hereditary layers that most found difficult to break out their allocated position in society, even today. This is very much similar to the dilemma faced by the Blacks in the US after their enslavement. India used social class as a basis of their slavery almost exactly as the US Whites used color class as free labor under the banner of slavery. Both were, and remain an unjust socio/cultural practice despite being banned by law. In India, they referred to some of these slaves as the *untouchables*. Very much like the Blacks in the US, this group of people were raped, murdered, and misused for centuries without recourse to legal safety. All through the pre-independence times 90% of the wealth was firstly in the hands of the kings and landlords, then the invading Muslim kings, followed by the anarchy of the British who exported the abolishment of kingdoms in their colonies with only one preferred royalty — theirs. This practice continues in the UK till today. The British then proceeded to pillage the wealth of their colonies in order to make their own kingdom successful on by shifting their stolen wealth to their own nation.

India Today: there are no kings or emperors. The British went worldwide and eradicated the so-called unfair social practice of kings and queens; however, they could never rid themselves of their queens and kings and continue to feed this social practice of dynastic domination over their commonwealth nations where the queen, now their king, remains the only titular head of nothing. India is a democratic country, in fact the largest democracy on the planet that chooses their leaders by a very precise form of popular voting. India is characterized as a flawed democracy just as the US is by world of statistics on Twitter. Both have fatal flaws whereby we cannot list them as true democracies.

The leaders live in govt appointed accommodations that they have to leave after their tenures, which sometimes they do not. Political appointees in the first sixty years were dynastic but then India matured and gleaned away from the dynastic addition to the first family, but the Indian Congress party continues to have national ambitions even today based on their dynastic leadership capabilities.

Even though class definition of untouchables was made illegal, and outlawed in the 1960's the Hindus in India continue to socially and individually practice this segregation that is as unfair, and often violent to the untouchables.

Just like in the US the concept of segregation by caste grays out in the metropolis cities but explodes in the rural areas. Even today an untouchable can be lynched for drinking water in the wrong well or tap or landing in the wrong temple. Just like in the US they can be shot dead for jogging in the wrong White locality. This group of people are still nationally mistreated as a community, despite what the media and government officially state.

Uniquely, Hinduism by its religious tenants does not allow conversion, i.e., either you are born a Hindu, or you can never become one. For this reason, India has never invaded another nation in the name of religion, nor have Indians gone to other nations on a quest of religious conversions. The concept of 'a stairway to heaven' by conversion does not exist in Hinduism. What happened in Latin America in the Spanish Inquisition, or the holy wars of Christianity or Islam can never happen in the Indian Hindu concept. India is split into the abject poor, around 32%, that have few education or work opportunities. On distribution of wealth, 90% of the wealth today is with 7% of very rich Indians. The middle class is rapidly expanding, and women almost have equal rights, education, and work opportunities in Indian metropolis cities. In rural India, very much like rural parts of many nations women still remain subservient to the wills of men.

Education has exploded with India subsidizing education for the bright and needy with national scholarship programs. Even though women in India today are doctors, engineers, and CEO's they co-exist in a mish-mash of fundamentalist Muslims and Hindus on one side who are equally militant and seemingly narrow minded. An Indian woman lives in two worlds within India. There is the Indian woman who studies in convent schools, becomes an engineer or doctor, speaks fluent English, wears jeans and skirts, and goes dancing at discotheques, clubs, and parties, on Saturdays. The same woman is also a Hindustani who lives in a house steeped in traditions and prays to her gods, follows the rituals of her religion at home, wears traditional clothes for ceremonies, and conforms to the traditional Indian woman at home. This applies equally whether she is a Christian, Hindu, Muslim, a Parsi, Persians who were welcomed in India when Islam invaded their country, or any other religion.

There is ample evidence of a culturally oppressive male dominance in India with the western media often report that India has the highest number of rape cases and very unfair treatment of women. Though if we delve into data driven facts the numbers tell us an entirely different story. While at the same time, there exists a large swath of broad-minded, almost equal to the population of Europe, living at European standards, making the middle and upper-middle social segment that lives like an average European. All this while the lower middle and lower society remains highly male dominated. Today as something happens in one part of the country, it is digitally communicated instantly across the nation and the planet. Some Indian metropolitan cities are highly westernized while those in the heartlands continue to delve in biases of caste, culture, religion, and political divisions.

The Covid dilemma: We could cover other nations but the essence of this string of thought is to prove life has never been better or fairer to all. This, believe it or not, is the most peaceful and fair times of our known human history. That is till the Covid-19 hit the planet, and everything changed almost overnight.

While Covid-19 was a boon for the digital explosion it was also a disaster for human social interactions and norms. Humanity got locked in their small physical homes for close to 18 months, this also created an echo chamber where many of us were forced to face our limited minds and environments. While we became more connected to humanity due to social networks, we unfortunately often used this connectivity to amplify our unique differences. The trolls used this same connectivity to weaponize these differences.

The future we build from this post-covid-19 digital explosion is now in our collective hands. On one side we have the known bad players and some autocratic nations like China, Iran, Russia and North Korea who are using digital surveillance as a tool to autocratically control their nationals. While on the other side we have nations like the US, where their leadership is leveraging digital algorithms by ex-presidents hiring ex-Cambridge Analytica executives to mind-play their tribal followers, and digital addicts into mind-numbing addictive mindsets.

World over nations is becoming highly polarized. Nations and leaders are magnifying the *'Us vs. Them'* separations, along with their *'It's not your fault – It's their policies.'* with no logic, reasoning, or empirical data but just digitized political resonance.

If we look at Dec 2019, as our first point of reference, it was close to great times when we compare it to today. If we then look at 2006 compared to Dec 2019 then the 2006 was yet a greater time. In the 2006-08 the world seemed to be heading in the right direction, but it was all artificial and heading for an unplanned, yet predictable crash. After the 2008 crash the world kind of lost trust in the government, the media and enterprises, which created a big vacuum. This was the birth of the digital trolls and their algorithms filling this vacuum with tribal amplified messaging. All through those and recent times, it was again the rich and powerful who wanted more profits than ever before. This is when we created the Bernie Madoff's. The media, in the meanwhile, was paid to create this semblance of fear, conflict, and extremely polarized views and to expand their customers' influence over their tribal members, quite often by romanticizing the past as a utopia when it was not. For example, some Muslim leaders in Iraq want to bring the Islamic Khalifate back into existence, or the GOP, along with their Catholic Christians, justified the ban on abortions in the US by quoting the Pope and the bible. A romantic view of the past does not make it a reality, but it is routinely used to drag our national or global societies to become hyper-polarized.

There are three prime institutions that profit from fear. All of these, and many more need to draw a picture of the perfect past from which point of time — we as humans are falling into a pit of evil and hopelessness. The first of these is politics, the second is media which is an extended arm of political influence and the third is religion. By the US constitution 'we the people' are made to believe that these three protected institutions are segregated by law, but the truth is that these three, so-called protected institutions actually sleep in the same bed in every nation on the planet. In the US CNN mostly amplifies Democratic agendas, while Fox News is a mouthpiece for the Republicans.

Fox anchors officially state that they are not news but a platform for Opinions, while the evangelical churches amplify the Republican agendas. Media consistently profits from magnifying fear and propagating news that makes the reader fear their present and the near future. This belief installs an addiction to follow the news item as each statement of fear triggers the self-preservation instinct, and like a soap-opera readies the listener for the next episode. So, to keep readers wanting more and more, media in turn continues to unswervingly produce more addictive disaster scenarios, by becoming social trolls who regularly create mountains out of molehills and routinely resort to fake news to increase readership. The final is politics that rewards both religious institutions and media in a profitable symbiotic relationship of 'you scratch my back and I'll scratch yours' of mutual profitability. When a pastor in any church or an evangelical church speaks of voting to a particular party they generate hundreds, if not thousands of votes. This is the reason politicians need to remain loyal, and glued, to the demands of their religious counterparts. In Saudi Arabia MBS can never alienate the Mullahs, in India never the Hindu priests, and similarly in the US never the bible or the Church in their various colors. This is the reason some of the presidential aspirants are seen carrying their bibles to their political events.

The truth is humanity has never been better and there are people and groups that can potentially lose a ton of power and money if humans feel they have nothing to worry about and we have reached a state of global peace.

The higher the perception of peace the bigger their losses.

So, the only option the power mongers have is to use newer and newer technologies and methods to change the minds of people and get them to believe that they are losers unless they follow them and their principles. Their primordial method is to instill fear. Fear is the key to tribal loyalties. It has worked for over 100,000 years and it continues to work today. The only difference is that our algorithms, and trolls, are finding newer ways to trigger this digitally.

Take Aways

Individual: No matter what your politician, media, or your fear amplifier tells you humanity has never been in a better state of global opportunities. We have already witnessed a time when the 'meek shall start to inherit their rights' but only if they work for it, like the George Floyd case. Startups are proliferating in most nations. This is probably the only time in history when an individual has started a global multinational company in a week to a month. Then gone ahead and disrupted traditional brick and mortar companies into bankruptcy. Don't fall prey to the minds of your controllers nor resonate to their dystopian messages. Stand up for what you believe is true not because some person puts it into your mind.

Companies: Plan to become the disrupter in your enterprise. Your center of the universe is your actual end customer no matter what you do. Your second layers of defense are your employees who design and work with your business, i.e., business facing. There is a huge 'Intelligence Trap' that is percolating the digital disruptive world today. They believe that technology alone can answer all business questions. The last two decades have proven beyond reasonable doubt that this is not only wrong but, according to Gartner and Forrester, over 75% of projects being run on this ideology will fail to meet business expectations.

Nations: Politics, Political leaders, and politicians are getting outright arrogant. On one side of the equation are the known bad players: Russia and Putin, China and Xi, North Korea and Kim, and Iran and the Ayatollah. On the other side is a nation like the US that stands so polarized that its internal trolling and algorithms have fractured the minds of its citizens, their thinking, and even their elected officials. No one can empirically state whether Nancy Pelosi is better or Kevin McCarthy. Whether Donald J Trump will bring peace and prosperity to the US democracy or Joe Biden. But within each of us readers, we have very strong opinions on who we think is good for our nation: the left-leaning democratic party or the religious and NRA-driven republican party.

The 'P' factor: John Chambers had stated that Cisco regularly buys potential startups. One in ten of these startups will be the unicorn that will cross a billion dollars in very short times. The investors follow this same principle, so should you. Don't wait for your idea to get perfect before you share it with the world. Realize that it is one in ten of your ideas that will make you a millionaire. So, design, test and get consumer feedback as rapidly as possible for it might be your 3rd, 5th or 10th idea that actually resonates with the customers. At one of the startup meetings someone had mentioned that there are probably more opportunities in this room than there are attendees, and I tend to agree. If we extrapolate this at a planetary scale then there are possibly more ideas, and thus opportunities, than there are people on this planet. One way to check this is to check how many apps you have on your smartphone. Forget paid or not, as a lot of the money is in free as you will learn. If you have more than a handful each of those other apps is an idea someone built and is testing for stickiness. It's not about how many of your apps that failed to attract the required *audience, it's all about how many times you failed, but then got up again to launch your next idea that will determine your success.*

Chapter 6 — The Digital Society

Caine Monroy was only nine years old when he launched his makeshift cardboard arcade inside his father's East L.A. auto parts store. The arcade's only customer, a budding film director — doting on Caine's genius ability to create something from nothing — decided to generate buzz with the hopes of raising college fund money for Caine. He posted a short film to social media, and soon, Caine's Arcade business became internationally known. It was soon reported on news outlets such as ABC World News, Good Morning America, and MSNBC. All you need is a good benefactor and social communicator. The movement generated more customers than the auto parts store could handle, with patrons waiting for four hours or more. It is unknown how much Caine's Arcade earned, but the scholarship fund collected more than $200,000. Caine's Arcade birthed a movement, which led to the creation of the Imagination Foundation, a non-profit organization designed to encourage creativity and entrepreneurship among children. Most importantly, Caine's Arcade ignited an innovative spirit with kids around the world. The lesson here is that if you miss something make that into a reality. Then hope, or plan for a network blast, and then wait for things to happen. If they don't then try your next idea. According to some great mentors, they say that 1:10 ideas take off the runway. It could be your 1st try or your 10th just don't ever give up. However, you've got to start as soon as possible for the rule is that the more times you fail the closer you get to success. But if you never start then you get nowhere.

The last decades and a half, have demonstrated the tremendous potential digital transformation has for us as a person, as a business, and as a nation in altering global society. We are at the initial stage of witnessing some of the very nascent effects of changes that we as humanity have ever seen. This tear, in the fundamental fabric of our society, has been reinforced by our connected humanity as never **before**. Connected for our likes and even our collective dislikes equally. Some national leaders want to cut this global connection and deny their citizens the will of the collective humanity. We need to do everything to fight that attempt.

The human society is rapidly being replaced by Digital tribes. Both for good and bad.

Unlike all the changes before digitization, these new digitally enabled transformations, i.e., changes that happened after 2005-10, are taking over our environments, our companies, our politicians and thus our governments and people's minds at a faster rate of change than most can adapt to. Today, new tribes can often take birth almost overnight, like the launch of the Me-Too awareness after Harvey Weinstein, or that of BLM (Black Lives Matter), right after the murder of George Floyd. These changes exploded into global consciousness close to instantly once they were recognized. When we look at how news used to travel in the past, today's viral spread of news is an extremely powerful weapon, that each individual carries, in harmonizing the global mindset on the concepts of right and wrong. Compared to how news traveled a hundred years ago or even twenty years ago these two bits were almost instant compared to the time it took for any change to happen in the past. These changes were global and resonated across cultures and societies to a very large degree.

Imagine right at this moment, that with a smartphone in your pocket and its capability to take studio-quality video with the power to influence the planet. When you do, please do it for the, Planet first, Environment Second, and Humanity third, PEH good of humanity. Own what you do to communicate and influence the tribe you belong to.

While the last few transformations brought about a collective adherence to a definition of right vs. wrong at a global level, it also demonstrated that when these transformations were undertaken for universal good, they could even become an international movement. Unfortunately, the same applies to universal bad actors too, like the right-wing 'let's deny the election results' as it happened in the US and was then immediately replicated in Brazil. The big question is whether this is a tipping-point event in the US that will ripple across the planet or hopefully a brief lapse in global logic. Even if the bad actors started as a local not-so-good reason they can still generate enough momentum to shake global norms and beliefs. So, it proves that even though the bad folks are determined to misrepresent facts, and optics, our collective communications can now be summoned more effectively than ever before, to react and correct the wrong as efficiently as they can demand unjust acts. When done appropriately and with 'beyond reasonable doubt' determination it has the potential to grow from local and smaller groups to a global protest with accelerated velocity, voracity, volume, and variety than ever before. An example that comes to mind is once again the smartphone recording of the execution of George Floyd.

It has already proven beyond reasonable doubt that while powerful nations, leaders, enterprises, and paid trolls can try to change the minds of their targets, we the individuals can equally alter the 'Spin Doctors' and reset the minds of the people as effectively. While the trolls are paid millions of dollars to alter the minds of the people the truth from an individual can lock up people like Harvey Weinstein, Jeffrey Epstein, and Bill Cosby.

Digitization has simultaneously launched a fragmentation of people by geographical and social groups, sects, religious groups, cults and belief groups, and most powerful of all by political and self-defined hate groups. Over the next decade, we shall witness an accelerated fragmentation of globally connected belief groups that could be as small as a few individuals but connected via the digital fabric, or as large as many millions as with political tribes across nations, or even encompassing the whole nation as one tribe that is enforced by autocratic nations. This will, soon, bring rich harvests in the form of globally connected do-good developments at faster rates of adoption, but also spawn drastic terror in the form of isolated views that become anti-social, anti-humanity, and anti-establishment just as rapidly. Both these trends, unfortunately, will continue to connect and grow as we enter this new *connected* era. The January 6th coup plotters are exactly one such example, as are the peaceful protestors in the 'BLM' protests in the middle of Covid-19.

Enterprises across the globe, are subject to a digital change that will accelerate as highly focused digital alternatives replace more and more 'brick-and-mortar' establishments and replace them with highly focused products and services. The time for rocking the very foundations of traditional 'industrial era' stalwarts and replacing them with far more engrossing digital competition has arrived. All you have to do is imagine, design, realize and deploy the replacement path.

Developing, and underdeveloped nations, who are today emulating the developed nations in their hyper malls and super retail points will flounder before they even get to be on the runway. I was part of a company that was building super malls in India while we in the US were shutting them down and finding alternative-use options. The lesson here is that while Malls are collapsing here in the US and Europe the architects of super-malls are exporting the carcass of such malls to developing nations as the next get-rich-quick solution when we all know this is a very expensive digital snake oil sale with a very short life or profit potential.

The great opportunity is that every single industry, product, and service will need to undergo a digital transformation necessitating to pass through the eye of a digital needle as a prequalification funnel to protect their success in their afterlife in the digital economy. The good news, or bad, depending on who is reading this is that over 70% of these brick & mortar companies digital initiatives will fail to meet business, and customer expectations. The planet has already established the fundamental laws of competitive digital success, i.e., fundamental from the past-to-future paradigm, i.e., how services and products that have been provided through many centuries of business prudence, versus how these could be replaced overnight by a new digital idea. Think Airbnb and you have two youngsters that have already disrupted the hotel industry. This is your moment to fill up a gap in products and services. Just think of what you wish you had launched but was not available back then, or better is not still there.

In this chapter, we will discuss the evolution and speed of change on how even as we all are trying to adapt to these silvery digital cobwebs that will encompass every human on the planet, we are not yet able to collectively neither design nor adapt to its opportunities, or threats fast enough. According to numbers — once again 99% of people will have great ideas for change, 10% will invest in that idea and around 3% will actually succeed. The ones that will succeed will do so because of their passion and their obstinate 'all-in' commitment to success. There will be those that will see this digital revolution as their opportunity and harvest it for whatever design they feel will disrupt the world. Some more and others less, some for-money others for good, some for peace while still others for discord.

Some will make it to the success stories, like each personal example at the beginning of every chapter, who will forge ahead adapting to the digital opportunity they imagine, design and launch, for succeeding in their new frontiers. While others will give it their next shot when their first idea did not work. It is important to note that most of these people had no experience in coding, nor a computer science background. So, if you're looking for an excuse as to why you won't succeed then you just have yourself to blame, or look elsewhere. Now, go take that first step, don't make any excuse to drown your good idea. In the US there are some that want the US to become like it was before 1864, i.e., subjugate the Blacks and colored folks so the right-wing extremists can somehow again get free labor, while they profit in comfort. While others to the time of the birth of Christ – not even realizing what they are wishing for.

Many US politicians, calling themselves constitutionalists, have declared that they want to take the US back to 1787-88 while hiding their agendas under their interpretation of the US constitution. Still, others in Afghanistan and Iran desire an Islamic state where only Islam is followed with 'zero' interaction with non-Islamic people – where women are the property of men and their religious book is their ultimate law unto itself. There are still others where their dynastic, or forcibly elected leader assumes the status akin to a god, and any statements made against them or their orders could result in imprisonment or even a death sentence, this includes countries like North Korea, China and Iran today. There is one thing common in all these societies — every one of them today use digital espionage spiders to understand, convert, control, and coerce their own citizens. Digital surveillance is currently the greatest boon to oppressors and the greatest threat to freedom. It is also the greatest opportunity to truth, when police is mandated to carry cameras that then need to switch on before every contact with a situation.

Human societies, for the past few hundred years, have been adapting to ever-faster changes, and that has become their foundation of survival. By the mid-1900s, our Einsteinian norm that 'the only constant is change' became an accepted transformation. However, what we were unable to grasp up until a hundred years later is that when the speed of change is faster than a human's potential to adapt, they end up very confused. Some of them tune out of their society itself. What the trolls learned in the early 2000s is that this state of confusion created by very rapid change is a gold mine for their algorithms to harness. Hence, they took advantage of this overflowing misunderstanding and uncertainty to create two views that tested people's desire to attain something that makes everything make sense. These two views were:

- It is not your fault; it's theirs.

- You're not alone in this. Everyone is a part of this play.

Politicians worldwide have capitalized on these two factors with their version of "It's not me they are against, It's you. They don't want me to give you all the great things I will be doing for you."

The birth of our civilizations started with the agrarian era when owning land was a denominator of wealth. During this period the counts, landlords, and kings owned land and represented the wealthy. During this period, very few people owned land and most of the citizens in such a society worked as farmhands on this land owned by some landlord.

Back then, very few people knew how to read or write, and most were happy to be simply working. Working hours were fragmented and random. Wealth and education remained in the hands of very few. We can confidently state that less than 0.5% of the citizens owned 99.99% of the land and thus wealth. Education was limited to religious sermons and teachings by priests on morality and trust in God and Kings. Morality guidelines were defined by the kings and landlords and promoted by the priests.

This was followed, around the 1850s, by the invention of the steam engine and the start of the industrial era. This is when wealth started shifting from landlord to industrialist. Industries required a minimum level of skilled labor and strict working hours. This is the dawn of the industrial worker that needed to do timely and repetitive work. Education was financed and made to adapt accordingly. New ideas of schools were introduced to create the new industrial workers, and slowly adopted across the colonized globe. The incentive was to create the perfect industrial worker. Over a short period, students were mandatorily trained to adapt to strict starting hours and mandatory attendance. This was followed by the concept of tardiness, with punishment for being late, and very strict industrialized tasks to accomplish. All these tasks were designed to create more efficient factory workers.

During this era Industry owners, along with brick-and-mortar companies needed to predict both inputs and outputs to their factories, as their ability to do this efficiently represented their ability to optimize their wealth. The industrial leaders decided the fate of their small world because they had wealth and thus influence to make their workers follow instructions and guidelines as given by the schools, their church, and their teachers. All this so their enterprise worked more profitably.

Soon landlords opted to start industries and diversify their wealth. The pure landlords were soon overtaken by the owners of factories and industries. Soon there were more industrial owners than there used to be landowners so while the wealth increased, it was also distributed many folds. The distribution of wealth during this era created industrial areas that were quite apart from the farm fields. A resource competition started as the industrial workers needed to shift from their traditional farm areas into the new industrial zones. From a segmentation point of view the more people that were educated, the more people they had to do their jobs, and the more products they could throughput through their factory doors. Very soon people were promoted for taking up permanent industry jobs vs. nomadic agrarian jobs dependent on the seasons and crops, with the introduction of the concepts of expertise, promotions, and permanency.

By the early 1900s the concept of the production assembly line came into existence with still higher degrees of repetitive experts. As industries continued to expand so did the output and throughput of their products. Countries like the US came up with the initial ideas on the 'scientific principles of manufacturing' and companies like Ford and the US assembly manufacturers dominated the planet on these structured manufacturing methodologies. They replaced the industrial competitive advantage of England and the Europeans who got mired in wars that disrupted their focus from industrialization. These industrial optimization concepts were soon adapted by the triad[2] nations and their factories. This resulted in an overall proliferation of the industrial era across these triad nations that soon dominated over 75% of global high value production and consumption.

At the end of the first world War an international organization 'League of Nations' was formed by the winners. This league protected the rights of these winners with unfair rules and regulations.

Education provided the middle class with new opportunities in cities far from their ancestral towns. In the winner nations this middle class got a lot of spending power like the ability to buy a motorcycle

---

[2] *Triad nations are US, Japan and Europe. They consume over 70% of the world produce thus represent the traditional economic superpowers.*

and a home when some of the new industrial companies started to earn more than many a nation. Industrialization continued to flourish, grow in numbers, and control the global economic base.

Then came the second world war and the need to optimize technical capabilities commenced. The second world war provided huge investments in R&D, production, technology and new disruptive ideas. Most of these were driven by a will to win the war. After this war, the new winners disbanded the prior *League of Nations* and created the new *United Nations*. Over time the UN has become a protector of the WW-II winning nations and some that came in with their own Veto powers. Looking back the UN has been systematically used to protect the big-6 players in their pursuit of global dominance. A recent example of this is when Russia attacked Ukraine and nothing in the UN charter or rules can influence a fair decision on Russia as it is one of the original Veto members of the UN. But all this is another story, and we shall delve into this no longer, but leave you to think about the overall benefits of a global overseer organization created by the winners of war and the impact it has on the global PEH initiatives.

The concept of increasing efficiencies began and flourished during the war and continued unabated after the war finished. This proved to be fruitful. The 1950's launched the era of technology that now replaced both agriculture and industry.

By the mid 1960's the world launched into the Technology era where the IT and computer giants became the new owners of wealth.

We entered the era of engineering-based knowledge workers working under the firm belief that technology and machines could by themselves produce anything humans desired. The new industry-based knowledge workers took repetitive manufacturing to the next logical level with computers, CAD and robotics. This era made it necessary to re-educate all the citizens so nations could remain *competitive*. It became apparent rather rapidly that in order to remain competitive a nation's education needed to become a prime focus and foundation for the future. This is about when developed nations invested in education as a federal and state funded endeavor and private citizens saw an opportunity to create world-class, best-of-breed educational institutions that could be the launchpad for individuals, nations, and their own institutional wealth. The national goal became a race in educating all their citizens and the potential to reap bountiful harvests in their overall throughput of skilled workers, products, and services.

All these computers and knowledge workers collectively created technology driven processes that produced a lot of data. A lot of this data needed to be stored in large storage containers. Data that for many decades was simply stored and often discarded. However, by the early 1980's data as an asset took a competitive form of its own. This is when companies and nations realized that all this data could be harvested to drive reports that could be used to baseline their past, present, and thus the ability to predict the future. Planning became an integral part of running an enterprise. With planning came the ability to compare plan versus actuals at an atomic level, something that could be rolled up to the total enterprise level view of reality.

By the early 1990's we slowly slipped into the Information Era. This happened subliminally for most of us, yet very consciously for a few. It is a clear example of 'In-through-the-outdoor' sample of business opportunities and the birth of asking questions into thin air and expecting to get reliable answers for free. Slowly, the information era brought with it programs, digital spiders, and algorithms. The mid-1990's also brought advanced robotics that were now driven by programmable algorithms that had the potential to replace repetitive work on factory floors. Robots had now started to become a normal factory floor helper. Slowly they began replacing the repetitive manual jobs with machines.

This single decision resulted in higher throughputs, higher quality, and due to lower defects and the economy of scale lower costs. This single fact is what China took to task in their attempt to become the global manufacturing hub, and factory, for all the nations on the planet. Add to this that China being a communist country even today they can mandate wages and prohibit strikes or disruptions across the nation. So, China suddenly became the global factory where nations and companies took their family jewels, in the form of recipes and their IP, that were simply handed over to Chinese factories so as to make higher profits in their own lands with products manufactured in China.

This is when US and many nations became countries full of products that were manufactured in China, and China became a country where wealth poured like the clear waters on the Niagara Falls seemingly with no end in sight. As companies and nations became addicted to lower costs and higher profits they sent more and more funds into China, while China spent more and more of this bountiful money on investments that pleased them. The Chinese are smart people so with all the world IPs in their hands they managed to collect a lot of residual intelligence. They also collected a lot of data on manufacturing techniques and soon began to launch their own products that were better and cheaper than their customer's original's. Their manufacturing hubs became the global competitive differentiator for most products that needed manufacturing.

If the 1990's was the era of Information explosion then the 2000's can be called the cusp of subliminal digital change, with the 2008-13 being the digital launchpad. As published in my 2010 book 'BI Valuenomics — the story of meeting business expectations in business intelligence,' this change happened very gradually, and it took us almost 2 decades to recognize that the era of Data-2-Decsion had not only arrived but was disrupting the very fabric of the society we know today. The period of 2008-13 can be taken as the digital runway as this is when the big digital ideas were put into place, to flight, while extrapolating the digital future work structures that we can envision and predict today.

Looking at 2022 we can predict that over the next decade, some of the biggest changes we should expect could be the breakdown of traditional market segmentation. As we head from segmentation to micro-segmentation our data will contain more and more details on attributes of individuals, along with their likes, and dislikes. As global cultures and ideologies hyper converge and communicate they will initially clash, as will their fundamental values and beliefs that have often stood on isolated local hilltops for centuries. These shifts, though gradual at first, will once again accelerate as more and more of the global citizens immerse in the global connectivity and the different cultural and value-based systems that looked pretty firm only a decade or so ago. This will continue to happen as more and more digital citizens work for their overall good. The power of the people will accelerate in connected societies as the people continue to fight against the few who will fight to relinquish their power. The next decade will see an escalation as the divergent forces continue to demand their rights, while the controllers get more oppressive in their ability to maintain their control over the status quo, and contest any change that compromises their hold on power and wealth within their nations or areas of influence.

Traditional values and beliefs on relationships, religion, sexual orientation, and definitions of moralities will continue to break down due to the growth of a globalization of locally restrictive value systems. If a pretty girl can be charged, imprisoned, raped and murdered in an Iranian jail today for showing a lock of her hair, does not mean they will be able to do the same one, or five years from today. During this period the *Command-and-Control* messages, by autocratic leaders, will be blasted by state-controlled news and media under the orders of the leader, as the incumbent power brokers and media owners. These edicts will be highly biased on the emotional content of religion, politics, or some other loyalty decrees. Orders that will use digital spin doctors, fake-facts, to enforce the will of a few on the majority. The autocrats want to crush the will of the people, destroy the concept of *We the People*, and maintain their *Command-and-Control* status that keeps them in power. The more the resistance, the higher will be the retaliation from the power holders — till it simply snaps. This will be felt as we the citizens of planet earth try to redefine the role of skin color, ethnicity, caste, sex, marriage, race, and other such artificial segregations of individuals.

To sum it up, it took us 6,000 years to throw off the yolk of agriculture, 90 years to go from agrarian to industrial, 20 from industrial to knowledge worker, 15 from Knowledge to information, 10 from information to digital. Now we either need to find the next change attribute which might be ML and AI, or rapidly adapt to leverage the potential that this digital opportunity provides to each one of us. Think ChatGPT in its infancy that has currently shaken the foundations of Silicon Valley, as the hub of technology. We need to collectively adapt from all these past scenarios to help others get out of their oppressive overlords and become the efficient micro worker in the digital economy. We need to become the disrupter now and let no one convince us that the individual is not enough. Each of us is the modern Davids in a world with a few Goliaths. Most important to remember we are 7.88 billion Davids with less than 193 political, and 5 major religious Goliaths.

During all these periods, we also went from industrial to hyper-industrial and retail to hyper-retail, computer to hyper-compute. We have left behind the industrial era messaging via media and industries that made their customers believe that their happiness is in buying: [1] Whatever the industry manufactured or the service they provided; and [2] by making buying more and more national and global obsession. For example, in 2008 when the global economy crashed President Bush asked the American people to go buy more. Great advice for the national GDP but not so good for a loyal citizen who followed their president's instructions. By June 2022 we were witnessing a period of mass resignation as more and more people are resigning from the workforce, as they look for jobs that give them a feeling of contributing to the environment, society, and their soul, rather than the profit objectives of a multinational giant with a lot of power but devoid of any soul or empathy for their citizens, their customers, or for humanity.

Today, we see a society that has reverted to online shopping. Instead of going to the store physically, we tend to order things online. From groceries to clothes to gadgets, we now rely on online platforms to deliver their goods and services to us. These e-commerce platforms review our buying patterns by collecting our data, and offer us more and more goods and services that their AI bots indicate we would like.

In 2020, this process was amplified 7 billion times due to the unfortunate COVID-19 outbreak that locked humanity in their homes. This happened sometime in November 2019 in Europe and around March 2020 in the US. Online shopping has become all the rage, thanks to this pandemic. However, the truth is that reverting to the online platform was a need that was coming at us a long time ago. The second truth is that this online and remote society could, over time, become a convenient habit for many and change the way humanity interfaces in the near future, hopefully for the good of the planet and humanity.

Each of us needs to plan for our short-term PEH contribution that we can undertake. In 2007 I did my part of decarbonization by purchasing a Prius, a hybrid car manufactured by Toyota. In 2023 I continue that contribution by buying a Tesla.

Many arguments have been made about this new method of shopping. Digitization is seen as a godsend by some and a curse by others. There is this increasing fear that many people will lose their jobs as physical stores will shut down and digital stores will take over. We shall discuss this point in a bit. For now, let's focus on another effect of digitized stores.

Up until recently, a poor genius had few options to borrow money from banks (run by the rich clubs); build a big factory (supported by the big banks); or buy land and hire resources (financed by the big banks to big money) so up until now their only option was to go to an already rich person and share their ideas. A decade or so ago, a new concept of angel-funding began. Now the genius could share their idea, get access to seed funds in exchange for a board seat and some stocks, then they worked very hard to make it succeed. The buyers would eventually sell their company and idea, their soul, and the company and give a share of the total returns to the creator. Which, frankly, is a thousand times higher than what the individuals could have ever reached without this structured funding approach. This, up until now, created a vicious cycle of the wealthy becoming even wealthier.

Moving forward, funds and resources are no longer a limitation or a restraint to a good and proven solution. Crowd-funding is today available to innovators that not only have an idea but a working prototype, and a revenue stream with sticky customers. The general quorum for crowd-funding is around, a minimum of 200-300 paying customers with a recurring revenue, but better still 1,000 to 2,000 paying customers and a proven customer growth that is predictable. It is these proven ideas, and the build on a digital platform that is the start of launching new companies. These are the startups that will break the backs of the traditional business types. Resource accumulation has been digitized and democratized. This includes development resources that can be optimized by open-ware solutions and financial resources that can be funded for shares or donations for an idea in a crowd-sourcing place like *go-fund-me*.

However, this environment of opportunity, and crowd-funding is not guaranteed as a win-win bet. The big financers use their experience to bet on new ideas, with a hope that 1 of 10 companies, that they invest in, will become a unicorn that will pay for their other 9 that did not take off. With this approach we will see massive accumulation of wealth that will appear and then disappear rapidly. However, the worry that the future will take away your jobs is as big a myth as the 1960's, when people used to worry that computers will put everyone out of their jobs. The first reality is computers created far more higher paying jobs than they took away. In fact, today computer jobs are some of the highest paying jobs on the planet. The second reality is that change will happen, humanity will evolve and you need to decide right now if you choose to be disrupted, or become the disrupter. Remember it will be individuals that adapt and flow with change, who believe in their ideas, that will disrupt companies that fail to change including governments that tighten their fists a little tighter as each choose their future. It is ultimately U&I who have to decide where we go, and where we allow the rich and our political leaders to take us in the short, medium, and in the long run.

It is critical to realize that this digitally disruptive slope will only get steeper as changes become more frequent and come faster. The second is that each change has the potential to globally alter every individual, product or service, enterprise, state, and nation on the planet. The third is that it will be individuals like you and I that will try and fail, then try and fail, and then try till they succeed to make this new micro product, service or application that will make each of our lives many folds better and the creator of new ideas many folds richer than they ever could have become in any of the other eras.

This revolution percolates nations and how they democratically navigate the micro-whims of their customers – i.e., how citizens, companies, and leaders connect with their entire value chains, their products, their services, and true value drivers. Value has moved from being an inside-out view of companies and shifted to becoming an outside-in view from the customers point of experience. That is, how efficiently it meets the expectations of every individual that comes in contact with their product or service.

This is the new digital revolution, and it is already impacting our daily way of life, work, and social relationships. This revolution is very different, as its pace of change is mostly too rapid. What used to take decades to change is now happening globally in a matter of months and years. What used to take billions of dollars in setting up complex manufacturing plants, can today be 3-D designed and manufactured on open-source partnerships.

The key to survival is positive digital connectivity and the new digital scale of operations. All one needs today is a great idea and a global enterprise is instantly born, for under $100. This transformation is already unlike anything humankind has experienced before.

Then again, we are barely dipping the tips of our digital toes, and our limited use of industrial-era imagination, into this new digital ocean. An economy about which, yet, we have little clue as to how things will unfold or directions it will take. However, some things are very clear. This digital disruption is not a hype, its solutions will need to be integrated into a comprehensive digital standards, processes, and structure, via governed policies ranging from public, private, academia, civil to governmental involvement. Jeff Bezos has already announced that Amazon too shall pass and he has thus passed the gauntlet to each of us to commence in planning the next global enterprise that will do just one thing better than Amazon does today and that is your challenge of this decade.

For the disbelievers, your smartphone is not a hype, your digital music is not a hype, the smart car you see drive past you is not a hype and neither is Uber, Airbnb, Amazon or the myriad of digital applications you use every day that transparently build a digital interface where an average human spends 4 to 6 hours a day. It surrounds most humans, it connects to them to their convenience, it provides them services and products they expect, while new ones are being hatched. The digital world exists in a paradigm of *Day-o*, and in this world new ideas are born on a daily basis to provide yet another expectation that is higher in quality and lower in cost. It is most important to remember that the ultimate application is Free and the ultimate interface is a human interface.

If you question the premise on how free can make money then start with the foundation where the maximum profits are in services we get for Free. Think back — when did you last pay to ask google a question, send, or receive, a google email, or connect on WhatsApp — yet Google is a $1.2 trillion business, while Facebook is a $933 billion corporation both providing most services totally for free. Their money is in the digital ink, the digital dust that is your data. Your and my data, and the worlds data.

The First Industrial Revolution used water and steam power to mechanize production. The Second used electric power to create mass production. The Third used electronics and information technology to automate production. Now, this fourth revolution is building on the Third, the digital revolution that has been incubating since the middle of the last century. It is characterized by a fusion of technologies that are blurring the lines between traditional energy sources, the physical locations of their users and their data, digital options, and biological spheres. The next potential may come from quantum physics and spatial twins where we cannot even perceive the vast potential of such discoveries.

Digital Opportunity is a portfolio of technologies and solutions that can be harmonized, homogenized, and integrated to make global sense of your potential questions. Digital shock is the disconnect in the minds of people who are not aware of what, and how the digital world works. The digital opportunity is all about helping readers comprehend the writings on digital and virtual walls. Plan for this inevitable digital future, along with all its opportunities, and thus build a personal roadmap to harness this digital potential that is within the grasp of every individual who has the imagination to see a problem in the status quo.

There are three reasons why this digital disruption can no longer be compared to historical revolutions, mainly because it does not represent any patterns of familiar linear extrapolation, i.e., such as a logical trend of the Third Industrial Revolution, but rather the arrival of an Nth and distinct tear in the evolutionary fabric driven now by unprecedented volumes, velocity, variety, and veracity of digital excellence and the data it now generates.

The volume of current data is increasing exponentially. Each of these devices will broadcast some, others a lot, of data under secure and open protocols. All these data generators will increase the overall volume of data being produced per unit of time globally. The velocity will also increase proportionally as a lot of this data is relevant mostly only in real-time. A lot of this data will be like the firefly, i.e., it will have relevance for a very short duration of time and then either be discarded or be kept as forensic proof of something in the future.

The velocity at which these trillions of data generators will broadcast, where network capacities will be huge by any current standards. The variety will be both wide and deep as each 'thing' will have their own protocols, standards and identifiers. Variety will also include data, video driven by AI and intelligent predictive algorithms that will scan this at ever high velocity data at unprecedented speed. The veracity will be each data bit demanding its fair share in the global digital economy and its position in the analytics and decisions that these data bits will potentially deliver.

The speed of change that we are witnessing today has no historical reference so in most cases our minds are numbed into an artificial ambiance which is both misleading and a digital mirage of self-instigated analog mind games. When compared with all our previous disruptions this one is algebraic while the past revolutions were linear and arithmetic. Also, the past revolutions were more regional while this disruption is both global and impacts every single industry, product, service, and individual on the planet as never before. Almost every government is consuming, misusing, and using this technology as there are almost no existing global laws that frame the use, or misuse, of digital algorithms, human interface, or trolls spouting lies at an individualized level in order to reach their goals. Unfortunately, so far it has been the governments and their departments that defined rules but when the profiteers of this technology are presidents, prime ministers, queens, and kings then who controls the misuse of this great disrupter. Who will control the controller? Who will police the police chief? The future is in 'We the people' rapidly demanding laws and governance to fall in place that will prevent the power huggers like Putin, Xi and Ayatollah's of Iran from misusing digital capabilities to destroy freedom.

The future is all about gathering diamonds from grains of digital dust. It reminds me of one of Albert Einstein's quotes, "Whoever is careless with the truth in small matters cannot be trusted with important matters." Once again, we have to remain passionately curious, because the true benefits of the digital explosion lie not in technology or intelligence but in our imagination about the benefits that this digital revolution can bring to humanity. The reason is that intelligence can only get us from point A to point B whereas imagination can take us out of all our existing frameworks and into an 'everywhere' possibility. Think Uber, Airbnb, and Amazon as you try to interpret the last sentence. If a person in China, Russia, Iran, or Saudi Arabia thinks communism or religion, then they remain stuck in their politico-religious prison. However, if they look from outside their current environment, and start to think technology and imagine a world, they desire then collectively they can surpass the oppressive boundaries of their geographical, political, and religious prisons of the digital world their elected, or autocratic leaders are busy building. Unfortunately, a lot of the current digital world resides on a technology platform that each of the leaders in every country can control for their benefit. This platform though global by design, can still be controlled, and manipulated for specific gains. The secret in a globally connected humanity is that we can dream together, where we all need to imagine a future free from these controls, maybe even work harder on quantum communications with zero geographical, political or religious boundaries. Where finally people can communicate freely without fear of repercussion or personal identification by sharing highly secured videos of actual events that cannot be, traced, blocked, or altered by some autocratic governments or their trolls.

This book is about the human side of tomorrow as much as it is about the potential digital connections and the institutional side of success. It is about how our children and friends are changing, how polling and politics are changing, and how our common day-to-day activities are already starting to impact the world we currently live in. The future of companies, their products, and services are being digitally reinvented. The future of friendships, alliances, marriages, and partnerships are being redefined.

We have already entered the world of impermanence. It's like we were living in a world of classical physics as defined by the last 3,000 years. A world of rule-based permeance. Then discovering that nothing is what it seems based on our new findings that everything is relative in the world of quantum physics. At the macro-level everything seems static, but at the nano scale nothing is.

The future of social interactions, etiquettes, and norms are being ripped apart down to a highly personal level on areas like, right vs. wrong, country vs. political party, citizen vs. leaders' glory, down to family and friend relationships. New digital subcultures are evolving and getting connected at the speed of light, and micro lifestyles are now finding individualized commonalities across the physical world, via digital platforms down at an individual level of connectivity. The digital effect being reviewed here is projected to impact every aspect of human interaction, decision, and activity.

In other words, the situation is no longer a prediction or an assumption. It is now an inevitable event that will happen. The only question is, *"When?"*

Take, for example, a girl who is somehow unhappy with her parents. She starts to view and connect with more and more people in her unique situations. The algorithms, as emotion-sniffing digital sharks, immediately identify this unique psychological pattern in this one individual. It now creates new avatars that firstly create resonance by communicating, *"I totally understand your situation; I come from an abusive family, too."* This forms the initial bond. It then proceeds to directly praise her every decision, making her feel she is not just unique but right in every call she makes. It then slowly, and as described earlier, makes her feel part of a larger imaginary group and tribe and feeds more and more on her addictions. We all know where this ends and have seen small and greater examples of similar things across our environments.

The common denominator in all these various aspects of human existence is the advent of digital connections, both real and artificial. This exploding change is already tearing every fabric of society, and even juxtaposing personal and professional interactions. The consequences are tremendous as are the opportunities.

Change has always been a constant, however, the pace of change due to digital advances is accelerating to a point where most individuals — be they presidents, CIO's or mere individuals find themselves sitting in the middle of a foggy cloud unable, to see clearly through the mist, or to decipher their best path forward.

My humble attempt is to somehow clear some of that fog with realities that are clearly visible today and then proceed into what these changes could change our tomorrow, by messing with our nation, leaders, our politics, and even our friends, and family at an individual level. What we should attempt to do here is to first and foremost understand the mechanisms of these changes, then become conscious of the artificial changes that are being systematically created so we march to some pied pipers' commands. Hopefully, then identify these changes, and their effects, on our friends, and family but most importantly on ourselves. By understanding the mechanisms of these trolls, we can react in a structured manner toward identifying and eradicating each negative suggestion or application by being able to predict what changes could bring in the future. These are no science fictional fantasies but predictable facts happening today.

In a blog on LinkedIn in June 2014, I coined the word 'Digital Shock' to explain the shattering effect this digital transformation is having on every aspect of our lives, bringing about changes that a lot of people are not able to comprehend or adapt to due to the pace of global impact some of them have. Even our relationships haven't been spared by this digital disruption.

Let's look at an average relationship today. Relationships in the past were determined by our environment and family definitions. Add to this our physical meetings, and opportunities within our environments. Many relationships today exist only in the digital platform. Friendships have been affected due to digitization. Today, many children count their friends by the number of digital friends, or followers they have as online friends. These are people they may have never met, or will ever meet. They only know due to their online interaction with them. This means people can make and form friendships with anyone in the world, sitting in the comfort of their room, isolated from traditional opportunities of going out and meeting people in their real physical surroundings. I know of children who have moved out of their parent's homes to another city, simply to lock themselves in a hostel or a room too scared to go out and meet real people based on what their social feeds are advising them.

For those with low self-esteem or people who are extremely shy or socially awkward, this has proved to be a bonanza of unbelievable proportions. They can now make friends with anyone they want. They can even unfriend anyone that does not resonate with their tribal vibrations with zero consequences. Even if they initially cannot meet new people, their social platforms will introduce them to random friends and over time, based on their likes, dislikes, and communications build very human digital avatars to exactly fit their requirements. This structured process will even introduce them to other real humans that have similar ideologies, or who are deeper in their digital addictions, thus someone that will work as a mentor. They can thus meet people and connect with them online, based on mutual interests. Previously, they were told not to talk to strangers. Today these strangers enter their networks, their social platforms, then their house, and finally their bedrooms, as digital acquaintances, and soon become friends. People are now measured by the number of social friends they have with little regard to any physical connections.

It is the same way with relationships, work, dates, sexual encounters, marriages, and even the selection of lifetime partners. Think of platforms such as Tinder, Bumble, and Grindr. Many partners today, had never even met each other. They meet online, fall in love online, and pursue their relationship online. Some will swipe left or right, others will meet for a purely sexual encounter in a neutral place, and very often that is that. Others will find the physical meeting of interest and continue without having any idea of the person sitting on the opposite side other than their digital imagery, and what they have decided to be on their social profile. Online dating apps and other platforms have erased traditional loneliness and provided people with partners that they thought they had been looking for all through their life. So perfect, so ideal, and sometimes so fake. Now available on your proverbial finger tips.

Take Aways

Individual: Start counting the time you spend on your social addictions. For some it may be dating apps, while for others it may be texts on WhatsApp. It is critical to realize that all of these create digital dust, and data about you as a person. When this data is collected, harmonized, and collated the information is beyond imagination. So be very careful of how much time you spend on your smart devices and what some third party may be extracting from this. At this same time individuals are also building business-disruptive solutions using the very same algorithms we have been discussing here.

Companies: Take your university as a corporation — many students are using these digital applications, and their algorithms to connect to universities and schools remotely. These algorithms have substituted tutors with digital flash-cards in helping students learn their lessons more efficiently, including at their own pace and time. Make learning easier by building apps to automate and complete their mundane work faster. On the corporate side business transformations are at an all-time high as companies get more done with fewer resources. This does not translate into less resources but into more business. If we look around more skilled resources will be required to help with the rapidity of response and increase in business. The digital solutions with their ML and AI and all the algorithms will do to business what computers did for business in the 1980's onward. The global business did not slump, it grew. The companies that do not adapt to digitization will be left behind in the dust.

Nations: Politics and political parties seem to be universally headed south. Whether we look at the US or China, India or Russia, Saudi Arabia or Iran, we are witnessing a universal curb on freedom and an attack on civil rights. Traditional democracy seems to be in its death throes but is certainly due for an audit. The solution has to be something that circumvents only the rich and powerful dominating the financial, influence, and political leadership positions.

The 'P' factor: *A society is an artificial boundary. Be it a nation, a religion, a university, or a territory. The digital revolution provides a global platform where any individual can take a positive idea and run it by friends and colleagues and hopefully create an idea–addicted planet soon after. The general rule is humans are instinctively attracted to fear but will invest knowledge, time, and money in things they perceive as good first for themselves and then for humanity. Make positive deliverables a fundamental part of your tribal foundation.*

Chapter 7 — The New Digital Huma

Since the dawn of humans, each generation has delivered its share of successful young entrepreneurs, from the stone scribes in ancient Egypt, the agriculturalist Eliza Lucas Pinckney (1800s) to Apple's founder Steve Jobs (1900s). Currently, it is not surprising at all that now these new Millennials, also known as Generation Y, have done the same, making their income in innovative ways. For many of these bright stars, that means pursuing entrepreneurship as a means to leave their mark on the world. At age 13, Hart Main came up with the idea of manly scented candles after teasing his sister about the girly scented ones she was selling for a school fundraiser.1 It wasn't until Hart set out to purchase a $1,500 bike that he reconsidered buying the bike and investing in what he had suggested in jest. Hart and his parents contributed nominal amounts to begin the business and worked together to develop the candles, cleverly named ManCan. Adopting a simple and masculine theme, ManCan candles—with available scents including Campfire, Bacon, Sawdust, Fresh Cut Grass, and Grandpa's Pipe — were made using soup cans. Hart's candles are in stores in every state, with sales exceeding six figures annually. As part of giving back to the community, Hart donates part of each sale to kitchens in Ohio, Pennsylvania, West Virginia, and Michigan. "What keeps us from discovering the truth or get driven by curiosity, is often based on our security of our tribal status quo. We are often scared of the exhilaration of finding things that change our societal understanding of things, but therein lies evolution and the truth."

When I was a young boy, I was fascinated with the way things worked. As predictable, this curiosity of things often follows us throughout our lives, but that is a story for another time. This book is a result of my curiosity at how this digital transformation works, and what is at the core of this transformation that people need to understand. All this has been understood thanks to my partners, in jobs where I have been paid to help enterprises, sometimes taking shallow dives, and sometimes very deep ones, into the world of digital disruption, big data and analytics, i.e., programming and algorithms. Sometimes we have to let our curiosity drive our life into new directions. For a very long time I let technical beliefs determine what could and could not be accomplished. I recall, when in school, I had asked a question during my biology class, upon being introduced to the microscope and human gene. It was a unique experience, for we were handed slides and a microscope to look at things that otherwise we could see only as a dot on the slide. Over time we went from school microscopes, and once our teacher took us to look through an electron microscope when we could suddenly see down to the invisible building blocks of matter. We also traveled to a mountaintop and viewed Saturn through a telescope and then the Andromeda galaxy that was hurtling toward earth. What was interesting was that what we were looking at was what the galaxy was 2.5 million years ago — that was the most fascinating part of space. During this time, we also learned about human genes and how it made each one of us who we are born as. How we could find both our parents and their genes embedded in our body?

Somewhere along the way, I realized that we always saw only the past. The Andromeda from 2.5 million years ago, the sun 8 minutes ago, the moon 1.3 seconds ago, the person standing in front of you a nano-second in the past. We can never see the now, ever.

I particularly also enjoyed learning the connections on how we carried the genes of every ancestor within our bodies. It was a lesson on how each of us is a timeless flame of life carrying genes, the amber of every ancestor that did not perish in some battle, or catastrophic event before they managed to reproduce and build the next genetic stairway to our future. These ancestors managed to survive every disease, battle, and pandemic since the dawn of time. We, each of us, is an image of survivors of our ancestry line.

I turned toward my teacher, and I asked Brother Meridith what I had in my mind, *"When we looked at the Andromeda Galaxy, we were looking 2.5 million years back in time. If we had a telescope powerful enough that could peek at a lifeform running on some planet in Andromeda right now, that was the person 2.5 million years ago. The person would be long dead, the species may no longer be there, but we could still see a single person running right this minute in our telescope."*

"Yes, that is seemingly a true statement"

Then I asked, *"What if we had an inward-facing, very powerful microscope, let's call it a genoscope, that could look inward through our genes like a telescope does outward? Could we not then see through my genes, through each of my ancestors to the beginning of humanity on earth? If we collected all the data from every genoscope would that let us see the totality of the history of humans through the lens of all our ancestors? Is it possible to build, or contemplate such a genoscope."*

"No," he replied rather quickly

I remember asking, *"Can something like that work?"*

My teacher said, *"No, it's not possible. That is a stupid question as it cannot be done."*

That was the end of it. The case was closed, and my imagination was shut down. I don't know why this was so. Was it because I came from an army background where an order was absolute, or because I believed that our teachers knew everything in their subjects. However, being an obedient student, I took my teacher's response as the gospel truth and gave up that train of thought rather slowly till its embers finally fades into oblivion.

Today, I think to myself, what if that wasn't the end of my questions? What if I had kept thinking along that line of thought. That is, thinking about how there are people out there, like Elon Musk, who refused to back down simply because they were told no it is not possible to build an electric car, or fly a private rocket, or build a rocket that will take humanity to Mars. I wonder about the many people out there who decided to defy the odds and challenge the norms, but obediently then stopped. The ones who thought of finding a way and making the impossible happen. I think about the people who helped change the course of history, and I feel kind of sad that I let that particular topic drop that day. I feel sadder when today I realize that there are powerful people, rich enterprises, and more powerful players that pay for the direction education must take. When Spencer Mills followed that same question, he was almost stopped by no other than possibly the Christian brotherhood that could not afford research proving that at the end of every genoscope a Black man would stare back. That the ancestor of every White Christian and the Pope himself was a Black person.

I was told I had asked a dumb question, and I listened to brother Meridith. It is only after many decades that I realized that humanity evolves, and adapts, only by asking these so-called dumb questions and following them through with passion. If we only asked smart questions that everyone knew would be the end of development, evolution, change, and progress.

Sometimes the fault is in our inability to think outside the traditional box we live in. We tend to believe our teachers, priests, and even parents. I remember asking my biology teacher a question *"If we could build a telescope that could look down our genetic ancestor, could we then not look at the world history from the point of view of every ancestor of mine."* My teacher told me this was impossible. I simply trusted my teacher and accepted the task as impossible. Some years later a could boy in California called Spencer Mill, asked the exact same question but did not accept the reply. Spencer didn't believe his biology and genetics teachers when they said no it cannot be done. Instead, he passionately believed it could be done, and then dedicated his life to this passion. He challenged the status-quo.

Spencer, in his Ph.D. research, obtained scientific evidence that every human being on the planet, every race and ethnicity included, had a master gene from which everyone evolved. It is called the singularity. He proves beyond reasonable doubt that whether you are Inca, Navaho, Russian, Australian, Chinese, Indian, Mongolian, Saudi, French, German, British, or any person in between, we all came from a common origin — our human singularity. According to his research, the origins of this gene were found to be somewhere in South Africa. His study proved that if anyone takes his simple genetic test, they can today trace their genetic ancestor for almost 40,000 years.

With his extensive research, Mills was able to conclude that humans moved around the globe from South Africa. They traveled to Ethiopia and then to Saudi Arabia. It was also found that it was an uneducated shepherd, by the name of Niyazov somewhere in Kazakhstan, who still carried this master gene which is the ancestor marker every Chinese, Japanese, Russian, French, German, European, Asian, Native Indian, Latin American etc., who all can be traced back to his unique genetic marker. Just as surely as your parents can be established by their genetic marker in your genes. It really is that simple and easy.

Spencer Wells, the Palo Alto, California-based author of *The Journey of Man*, started his research by going back through genetic markers from all nations. He discovered that if you go back in time to the farthest ancestor, there will always be a Black man looking back at you as a part of your genetic line. This was a hard truth for many to swallow, and sometime along this great quest his funding was pulled out.

Anybody wants to guess which group pulled out his funding, give it a shot. We have all heard of the Muslim brotherhood, but very few of us are aware of the power of the Christian brotherhood. No good White Christian wanted to fund research proving that their ancestors were Blacks from Africa. Imagine Hitler, or modern-day DeSantis from Florida being convinced that his ancestor was a Black man who came from Africa.

Despite all the religious hurdles, Wells refused to let this demotivate him and stop his research. He went on with his work, working hard to find the answers he was looking for. He was driven by a personal passion, and he went around asking people for help and collecting funds so that he could complete his project. This is how I met Spencer Wells and learned about his research and the book he had written called "The Journey of Man."[3] It's today available on Amazon as a CD and book package. Buy both as they complement each other.

It is truly remarkable to see that this discovery wouldn't have been possible if Wells had simply decided to agree to his teachers and follow what he was told blindly, or if he gave up when all the doors seemed to close upon him by the religious oppressors. If he had succumbed to the threat of the White Christian brotherhood that did not want to fund any research that could prove that their whiteness

---

[3] Wells, S. (2003). Journey of man: The story of the human species. *DVD. Directed by Clive Maltby. Arlington, VA: Tigress Productions, PBS Video.*

came from a Black neighborhood.

He discovered that we are all one and that we all come from one source. Therein lies my next advice for the new digital generation that is currently learning. A true digital human wouldn't be bothered by races, religion, colors, and ethnicities. They must accept the unity of all humans and treat them with the same brush. This person must not allow traditional power brokers to influence their quest for the truth, must not allow regimes, or politicians to decide what can be taught in schools and universities, must not get trapped in old, conservative thinking of the traditional society that believes in manmade artificial boundaries such as their race's supremacy. Unfortunately, it is important to realize that every nation and religion carries this unique perspective of superiority. The lesson here is to continue to ask dumb questions as within one of them is the future of your opportunities, and the prospect for evolution for all of humanity.

The world that we live in today is getting smaller as it gets more digital. It is now becoming the home of digital human beings. So, my first advice for such digital humans living in this contemporary society is never to accept the norms handed to them by society. Never accept the status quo, always strive to make it all better for the planet, the environment, and humanity.

The truth, as stated by Einstein, is that the one thing that shall always stay constant is change. It is guaranteed that things will change and alter over time. It is also now guaranteed that the changes will happen more rapidly. That for every great discovery, or invention, a few people will embrace the change for its capabilities of doing good, but at the same time, a few powerful, rich people will use it only for their profits. They will use the media, your smart devices, and their various tribes for the sole purpose of personal profit above all else.

We need to accept this and review how each of us 'we the people' can play our role in changing the world into what we want it to be.

We can survive and live better in this world if we understand, and adapt, to the changes by playing our positive part and challenging the negative status quo. Take the history lessons that repeatedly prove to us that nothing stays forever, proving that everything you see will also pass. We will survive more efficiently if we collectively collaborate in electing people that care for the citizens first and then the planet, by identifying and not listening to flash, and empathic, tunes of trolls and mind-shapers, by identifying what is true versus what is fake, by demanding laws that protect citizens from mind-manipulations by internal and external trolls and their algorithms, by listening to our silly questions and following them through.

That in itself is the hallmark of change, success, and positive disruption. It is the same with Mark Zuckerberg, who created

Facebook; Jeff Bezos for Amazon; Larry Page and Sergey Brin for Google, and a million others, with some of the very smallest starting each chapter in this book. If you continue to believe everything you are told, then you shall continue to be controlled by everyone, including their definition of the past and the present. As you challenge the status quo, you will slowly become the outlier, the disrupter, and the challenger, all hallmarks of the successful digital entrepreneur.

After that, and within a short time you will start to learn lessons from the future, and this by the way is the reality even today. I will try to introduce you to the fundamentals of Fufactology the science of learning from the future as a science that will drive decisions within the decade.

For a just future, it is a must for digital beings to be fair to one another and treat each other as fellow humans. It is critical for Americans to become true constitutionalists and not in ways that the constitution is being interpreted today. If we all follow just two edicts truthfully, we can potentially change the world.

1.   A nation built for the people and by the people where every citizen is equal in every respect and aspect.

2.   Where no one is above the law. If we only apply this fairly across color, religion, sex, and other artificial separators, we will be heading toward a proverbial paradise rather quickly.

Unfortunately, the US political parties, touting themselves as constitutionalists have done everything in their power to fracture the country for personal gains. They did this, and continue to do this, by leveraging the power of digital mental transformations. They break the first requirement by systematically separating US citizens and their voting rights by color, religion, and ethnicity which is something that is anti-national. They continue to break the second requirement by putting into place processes where the definition of law changes and a select few remain above the law. Where law is being altered either not only to protect the rich and powerful, but often to lock up innocent Black citizens often for two decades on a false case of rape because of their color or economic state. Or as proven by the recent highest US court decision, by issuing a federal legal order to allow the Catholic Church, and republican states to enslave the womb of every woman in the US as property of the state and church. These are all political distractions and we need to see them for what they really are, and exactly what they will make us become. July 24, 2022 transformed the US, under guidance from the pope himself into a misogynist, male dominated society that changed the rules for women backward. America now is following the social rules of Afghanistan with their Catholic guidelines.

So, to survive in this new world, we need to learn how to separate facts from lies. Build an ability to somehow filter truth from spin doctors. Then accept facts and closely monitor seemingly small changes, with a potential for monumental shifts, constantly. This will only happen if we keep an open mind, eyes, and heart.

There is a concept called 'design thinking' launched by Stanford University in which your best method is to get into the mind of the person you are trying to serve. If you are mentally prone toward emotional content, you often step into the shoes of your story characters, into their shoes in their environment. Then you become a good designer. Start by participating in other people's situations and thoughts, then take a decision only when you can become one of them. AI alone cannot make sure that you're not a victim of a fake story when you have donated all your sympathy all by yourself. So, decide today if you will become the change, you hope the world to be, or become the recipient of the change that someone else controls for their gains.

If you wish to thrive in this new digital world, you need to decide which side of the future you want to be on. You can join the tribe that listens to their paid political trolls with little thinking of your own, and makes racial and color-based decisions. Becomes an active US patriot, and then follows through and attacks the US Capitol, the democracy, and everything the US stands for as a personal act of patriotism. Or you can be a helpful, sensible person who is empathetic to all humans, regardless of their race, religion, or ethnicity. A person who sees every American as equal and who serves to make this nation and the world a better place after you have lived your life on this planet.

My final advice to our new digital generation is to choose their digital friends, beliefs, and influencers very carefully. As an **individual,** be careful of who you let into your socio-digital world, as some of these social friends will be algorithms, programmed to simply alter your thoughts, and mind toward specific goals. Be cautious of who you give access to your thoughts and ideas because very soon your thoughts will become their target practice. Next, be careful of the likes and dislikes you share, as slowly these will determine your psychological profile for creating your digital twin, your troll friends you would like to hang around with, and who will like to hang around with you. Your thoughts, your words, and your friends will soon define your reality and what you should be doing for the rest of your life. Be careful of who you let into your thoughts, after that be careful of what you think. This is because whatever you think will slowly be expressed in your words. Whatever you say will soon define who you are and what people you attract. This will finally become you.

For an **enterprise,** be vigilant and choose your company options wisely, for they will play a major role in shaping and defining your future. Be very careful of your triad partners, SW, SI, and Infra, and their true goals. I have seen the big 4 undertake multi-million-dollar digital projects with little business success at the end of the tunnel. The first sign of an impending issue is if a company misses more than one planned go live. A bigger problem is if your SI recommends you start phase 2, even if they have been unable to go live in Phase 1. Be very careful too of new executive hires from the outside, as they might come with their teams and personal priorities that may have nothing to do with the success of the enterprise. We have all seen that as routine across the globe. Be cautious of continuing without true digital transformation. According to Gartner, more than 75% of these initiatives will continue to fail due to a lack of a clear focus on business-defined outcomes, due to IT leadership wanting to write their own stories and technology-driven outcomes. This specific area has been my professional focus since 2011 when I started to assist large and medium-sized enterprises to find and deliver their true north consistently. As you move your company ahead to newer digital platforms, try not to delete your diamonds in the rough — your legacy data. This is your digital dust.

As a **nation,** we need to cull our elected leaders who are bent on tearing the very fabric of country on just their party agenda and political hatred. Just as a US example, some people think it's okay to misuse people based on their religion, color, sex, or nationality for political or religious tribal positioning. Other people feel happy suffocating Black men with their full body weight and their knee on the nape of the neck of another human because they are White and he is Black. Still, other people will use the US generals and the force of the presidential guards to gas peaceful demonstrators so they can get to a photo opportunity across the White House. Some people laugh, and joke as their party members attack the citadels of their nation's human rights. Still, other people are till today proud to have attacked the US Capitol, believing that they were protecting their country and their ex-president. While there are also people who firmly believe these attackers were all good people who we need to respect and protect. There are presidential candidates today who say they will pardon all the convicted Jan 6th criminals.

In China, they find Uighur, and Tibetan, women and take them into secure camps where they are forcibly sterilized and brainwashed into the Chinese way of thinking. In Russia, they lock leaders from the opposition, attack their peaceful neighboring countries, forcibly enlist their prisoners to fight their lame war, and allow them to rape young women in Ukraine and worse. The list goes on and on. At a global level, the UN has failed humanity and its initial promised charter, while they believe and continue to meet its obligation to the planet and humanity as a whole. We are the new digital generation, irrespective of our age, nationality, religion, or color that needs to unite and eradicate every autocratic oppressor from our planet.

Our nations are full of leaders, and their cohorts, i.e., people who believe all that the leaders demand is all right, so long as they get their pound of flesh and power to keep their reward. If they did not then this would not continue. They drown their thoughts about their wrongdoings. They often don't care about the consequence of their actions, which routinely results in the loss of harmony, and lives and disturbing decisions taken in the name of national interests.

These are people who are influenced by their own beliefs, amplified by things they read, and see online. They find others expressing similar views, and they use technology to make friends with such people. Their digital friends are just like them, with racist, bigoted views who promote hate crimes and who rather than condemning these wrong actions go about and glorify them and wear their actions like a badge of honor. With this, their hate and views are further solidified, and it forces them into taking actions that become amplifications of their earlier ones. When we extrapolate this unfortunate addiction, it can lead to only one future – one where the US loses as a unified nation that is currently a beacon of hope of freedom and equality.

The US, like most other nations, is on a global social cusp brought about by digital echoes. Thus, we're all standing on a razor-sharp edge. On one side is total dystopian socio-digital anarchy with the nations run by digital addicts who follow every instruction of their handlers and take this nation into the hands of power-hungry autocrats. While, on the other is this unique continuous opportunity for us to digitally join hands, at local, national, and global levels, and collectively eradicate each wrong. Then put into place laws that will once again turn the United States into the launchpad of becoming the exemplary nation on how to do it right this time, and every time. The future is ours to lose if we let external forces magnify and draw our roadmap. If as an enterprise we feel digitization is not for us, or if as a nation we let our leaders use it to oppress citizens, we will only face a dystopian future. This rule can be applied in every nation on the planet.

The past is full of historical examples of the crimes that were committed against fellow humans. Black in the US is not only a color, it is an opinion and a caste-like in no other democratic country on the planet. Black people were murdered and treated inhumanely for a long time. Women were mistreated and denied rights for the longest time possible. Immigrants and people from other countries and ethnicities suffered in silence simply because the average American went from enslaving Blacks from their foundation officially till the Dec 6th, 1865, but it lingers till today as the US moves from oppression to being an open-arm welcoming society. But the US is not still all there yet. We need to treat every human as God's children, extending a common courtesy and basic manners toward them. If you think differently, then think again and review if you are a solution to the problem of tearing the social fabric of this nation and planet, or just an amplifier of the problem.

We find more and more Americans demand that we must stop people from entering the US. Some say stop immigrations. However, the interesting fact is that *"Only the people who have already immigrated, support banning immigration"* This is because there are no true native American other than the disenfranchised native Indian whose rights have been scratched off the American national, and political map. This is where Machiavelli's 'The Prince' comes to play. Time after time the White man has signed treaties, disarmed their foes, and then broken every agreement they signed. This happened recently with the Dakota Access Pipeline that blatantly broke every agreement and signed document for a few coppers in some rich people's pockets. Based on this foundational thought, every registered American voter is only an immigrant from some other nation, who is busy denying the original inhabitants rights in their land. Then denying new humans from entering this nation if they do not fit their profile of acceptable humans. Have they become the very oppressors from whom they ran away from their oppressive nations looking for a new free world?

Despite the political, religious, and media optics presented to us today, we can confidently state that humanity has never been in a better state of options. We can either enhance this blue planet to become a paradise or drag all this social and sexual imbalance down into the past. The desire to go forward toward the past is scary. When Texas wants to oppress the Blacks, they want to invoke the 1864 rules. When republicans and Catholics want to take decisions against abortion, they quote the bible even though the holy book says not a single word about abortions. So, the logical conclusion is that, on this topic of the US Roe v. Wade, the Pope, the Catholic church, and a lot of Christians are simply a misogynistic opinion platform that desires to elevate men and enslave the womb of every woman in every republican state. When Putin wanted to attack Ukraine, he wrongly quoted the past. When the pope, and the states of Texas and Florida, wanted to support the SCOTUS decision they too wrongly quoted the Bible. This scary list goes on and on. I do not seriously believe that most people understand the wisdom of wanting to live in the past.

I hope we can convince each other that none of us want to go to the past unless we intend to die of disease, infection, rape, pillage, or even place people in gas chambers, acquire free labor in the form of slaves or get ruled by autocratic overlords. The past when pimped by the modern sellers somehow includes all the modern amenities and their fake, and glorified optics of what the old world looked like.

A romantic view of the past does not make it a reality, for we could very easily drag our national or global society to become the next Syria, China, Russia, Iran or even North Korea.

Remember, each time you forward a story, an image or a point of view that is your very personal datapoint. Imagine what they know about you with a thousand data points.

While we are on the subject of digital acquaintances and friends, it is necessary for us to discuss digital relationships. I spoke 4 times at sessions on my findings from working on a dating site. I presented to audiences ranging in size from 60 to 300. One of my favorite and frequent questions that I liked to ask in these sessions has been, *"Have you ever been to a dating site?"* To this day, the answer to this has always been that not a single hand went up. Even I who researched this topic, had to join the site so I understood the process and data points when joining a dating site. So, my hand was the only one up. There hasn't yet been a single person, man or woman, in attendance who ever put up their hand. Maybe all my sessions were at professional gatherings thus the need to mask personal details becomes an official asset.

We live in a virtual world that is heavily dependent upon the addiction of online omnipresence. Right now, we turn to the internet for all our needs, ranging from news, opinions, entertainment, addictions and even when we feel lonely. The key addiction is instant gratification for our needs and desires, that up until a decade ago was simply not available to most of us. We can use dating applications when looking for company on mutually defined likes and dislikes, that may be true or imaginary. Sometimes mutually even for a single day or night. How many dating websites or applications do you know of? How many matrimonial websites have you heard of? How many online matchmaking groups do you know of? How many people have you met online? How many of them have you been in some sort of relationship with?

The answer to this isn't difficult at all because many people rely, then then get slowly, ever so slowly, get addicted to digital conveniences. Communications, finding a restaurant, company emails, sex, a date for a day, and partners are now on our finger tips, as the number of online dating is growing exponentially. The post-covid generation is getting very connected globally. Ever so slowly, and yet very rapidly these digitally connected people don't feel the need to connect to someone physically next to them, preferring to text with someone on their social horizon. Relying more on being digitally connected, while shying away from physical connections to a very large degree. There is no longer any need to take risks in meeting neighbors, when the whole world opens up. Slowly, the new generation is finding more comfort in texting, and digitally meeting people who they have never met physically. All based on a bi-directional avatar of a face or character that the sender and receiver prefer to communicate with. Some people start to feel as though they can relate more easily with the people they have never met, who exist in the ether, better than people around them who might question or even criticize their logic. The reason is because you can become whoever you please to be seen as, as can others. So, the barriers to who you actually are, simply fades away, and soon you start to believe your digital avatar is more real that the physical you. The new generation is in the danger of becoming physically recluse in the new socially connected world. This then becomes the start of a digital-physical rupture in the psyche. Where this tear results in is a digitally induced persona and sometimes a very psychologically split personality, which though I'm not qualified to put forward as a qualified statement but still believe is happening all around us. Emotions are being satiated by digital friends and trolls and sex by simply swiping left or right.

I like to call these dating websites firefly websites. This is because these sites are normally visited between the time of someone breaking up and till, they find a person they are searching. Unless one is a serial whatever. There are possibly a bunch of serial participants who treat these sites as public parks but I hope they are few. In the next few years, the number of people wanting to get into long-term relationships should decrease because it becomes so much easier to break-up, because one knows that they can meet up with someone new the very same evening. The biggest fear resulting in baby boomers hanging on to their uncomfortable relationships is the fear of loneliness and the belief that they will never find anyone new. However, data shows that with passing time the taboo of sex is shrinking which is today the social, and religious, foundation for marriages, Worldwide genders are becoming more fluid, and social barriers built on traditional 'command and control' are breaking too. Gender based clothing colors are rapidly breaking down too. Thereby, the option to get into a short-term or a long-term relationship simply becomes a personal preference.

Through my research on these firefly websites, I found out that you are often asked to enter personal details like your recent photographs. I found this out by actually joining a website to familiarize myself with the entry process so I know what all data I needed to collect for my analytics. I was required to get post scrubbed data that eliminated all the master data like identification, name, and city of the persons. Now, one of the biggest mistakes subscribers can make in this online dating application is to post or use your 'better-looking' photographs, mostly from days when you were younger, or sometimes even a photo that is not yours. I call this self-cannibalism or the art of destroying your future. One of the fundamental attributes in the digital era is TRUST. By putting a fake, or old picture, you lose that trust from the moment it turns into a physical meetup.

There is simply no use case, or excuse, in using an old photograph, of a younger you, in any social network, including in LinkedIn and Facebook. After trust comes reality so show people who you are today and be proud of it. Many people make this mistake because they are interested simply in finding out if they can manage to catch any fish. However, when they do manage to set up a meeting it predictably will be a disaster that is simply waiting to happen from the very start.

What they don't realize is that catching a fish today is always an easy part. It isn't difficult to find someone, then next to face your true self. However, the problem often arises when you have to meet that fish in real life or on a video chat. The *fly* in such a meeting is you, more often than not, a kind of toxic fly that no fish will want to bite. So, this is a planned journey of losing your race before it even begins.

For example, you meet someone online, and the two of you hit it off verbally. However, when you meet the person in real life, you find out that they are many years older, and many pounds heavier than in the photograph. Instantly, this becomes a lose-lose scenario. There is almost no-winning condition in such a fraud.

These dating sites also rely on digital advancement. When you create your profile, you have to start with your sex, age, and marital status. Sometimes followed by interests and hobbies. This is then followed by what sex and age bracket are you looking for. These dating sites then use algorithms to put forward prospective members for you to consider.

During our research, of around thirty million masked (with zero exposure to their names or cities) members, we found beyond any reasonable doubt, that men tell lies more frequently than women. Normally men entered that they were looking for women who are five to fifteen years younger than them. While women, on the other hand, entered on preferring men who are two to seven years older than them. By studying the transactional data, we were able to prove beyond reasonable doubt that men are compulsive liars as compared to women.

Let me elaborate on this point. We found that when women said they preferred men to be five or so years older than them, they weren't lying. When they looked for, selected, and reviewed a potential partner 86% of them opted to match with guys in their stated age brackets, i.e., 5-7 years older, and also reached out to men within the same desired age groups. However, when men were tracked against their entry statements vs. what they did on the site was very interesting. Close to 95% of men started by looking at girls in the 18-

23 years bracket. They would keep trying, often for weeks and months, to initiate contact with this age group even if they were 50 years old. Only when the 18-23 did not respond would they go to the next group, say 24-29. Maybe our quorum group were mostly voyeurs, but the track record of men was very predictable. 95% is not a random error too. This group also routinely posted photographs of their better days. Only around 3% behaved like they had entered in their joining selections, i.e., look within their stated age bracket.

Connect the dots — this is serious because every religious book, on the five major religions, on the planet has been written only by men, who defined every rule therein.

This somehow tells us a lot about the society we are living in today. Our views of truths and lies are heavily flawed today. In almost all society women do not have the same rights as men. It's the male pimps who make money on women, I know of no women pimps who make money off selling men. The traditional social machine amplifies this male-dominated segregation by sex, the degree of which depended on their society. The truth for a man used to be very different from the truth for a woman. For example, in traditional homes, and societies men could proudly state that they had dated 20 women. Fellow men, and often women, would call him a stud. However, if a woman had ever stated that she had dated more than one or a few men, in most such societies she would be immediately branded negatively, and in some even stoned to death. This is a current state of reality in many countries even today.

Fortunately, some millennium children have managed to think past these traditional biases of sex, religion, or color but this is probably true in less than 5% of the world societies. Also, quite often we may believe in one thing, while we often say something quite another, especially when people communicate their desired image online. We often express happiness or sorrow at something we feel nothing about. We mask our true emotions and put up a show online because a lot of us are members of some social tribe, where we have to maintain a certain image to remain part of that group. There is an ongoing alignment of personal personas and tribal optics. It all starts with your social network photograph and the people you befriend online. If your photograph is older than 2 years, and your opinions are based on tribal messages you may be deeper in some form of digital addiction than you may believe yourself to be in.

This is a terrifying perspective to come to terms with because this way, there is no guarantee of knowing for sure that what you believe in is true or not, or who is telling you the truth and who is lying when you enter into digital friendships and relationships. But there is more than enough evidence that the social network is more crowded with self-promoting trolls who promote what they want their targets to think and believe, rather than who they actually are.

I faced this dilemma, way back in 2009 when I was speaking at a conference on Real-Time analytics. A person came to me and stated, *"For a person speaking on real-time analytics, you look different from your LinkedIn profile."* Within that week I went and updated my picture and have kept it current till date. I must confess my current picture is more than 2 years old and by the time you get to read this it will be a picture of me from a recent past. Old saying my mother used to tell me, "If you can't respect yourself, how do you expect others to." this statement by her had nothing to do with any picture on any social network, but just a relational statement to our topic on hand.

Just like you can never truly see the present, so, all we do is learn and see history. While what we see with our own eyes is the truth from the past, the history that we all read only includes the bias, and interpretation of the authors, written on behalf of the victors of that story. That star outside your window may be a billion light years away. If it had exploded 900 million years ago, we'll keep seeing it in our sky for another 100 million years.

Similarly, you can never truly figure most people out. Sometimes, even if you're married to them for 40 years, a close friend for 20 years, or the person sitting in front of you met a month ago. Nevertheless, if they are sitting behind the screen and typing messages to you, you cannot even be sure if they are real or not. You can never be sure if what you are reading are texts from a real human, or sweet nothings to get you hooked. Just as you can never be sure if the photograph of your e-friend, is real or a digital generation of pure fiction. A distinct persona digital created just for you, your group, or your tribe.

If you have five hundred online friends, how do you decide who is real and who is a troll? How do you know who is genuine, and who is resonating with your emotional needs? How do you know which one is a Russian political troll and which a made in America US political one. How do you know who is playing mind games with you because they have a digital agenda, who is hiding what from you, and what their hidden agendas might be (if any). All you know is they want to be your friend, because for most people this is a very exhilarating experience when someone says I want to be your friend. This is how most traps, psychological transactions, and honey-traps are set up in real life.

The bitter reality is that, with a little practice, it is very easy to charm and deceive online, and so there is a huge possibility that you may never truly know who your online friends are. Also, the algorithms have millions of hours of razor-sharp practice with as many targets. This is where they hone their approaches and responses with each interaction. Therefore, you must be cautious, and careful of who you extend your hand or heart toward. Be mindful of the company you keep. Be careful who you listen to, believe, and befriend. The digital method of figuring out who is honest and who isn't is extremely flawed. Anyone can join current social networks with a true or false persona ID. It is quite routine for trolls to carry multiple social network accounts under different names, personas, cell phones, and photographs. The digital process of identifying the truth and separating it from falsehood is full of errors and mistakes. The digital controls for weeding out fake news is dismal. The digital rules and regulations at the state and federal levels are nonexistent. Our TV rules for media and messaging are a thousand times more stringent than those applied to our digital networks. This is important to realize as our TV programs are full of paid, intentional trolling and advertisements designed to change our opinions and minds. For example, Fox news has officially said they are an opinion network, while almost everyone listens to them as news. Opinion means they will create the truth that their payers ask them to. Not only that these payers routinely pay some very large sums of money to fix the exact message these advertisers carry into our living rooms. When we are sitting alone in our homes we have the least amount of fear, so psychologically our instincts do not filter any content that is fed to us. We accept it as our tribal messages and accept them as our gospel truth, with minds wide open and eyes wide shut.

These smart devices access us when we are most vulnerable, and routinely peep into our lives with live camera and audio devices buried in our applications. The scion of Samsung went behind bars for exactly such a travesty. Not just because it was there and used against the common people, but possibly because they misused it against some powerful person who could seek revenge in the form of legal justice. All across the world, whether the government is a democracy or an autocracy, one thing is very clear, as of today, most of the leaders, i.e., our governments, do not care for their people. Not all people. They care too much for very few, much for some, little for most, and not at all for many. For example, in the US, the very few at the top would be the politicians, billionaires, and the very powerful, people who can bend law and influence legal results until they do something that is far too obvious or hurts another person in power.

In the US, the rich, White, powerful, and Christian belong to a unique class of their own. On the opposite side the Black, the poor, with no impact of religion belong to an entirely different class. On the top are people like Jeffrey Epstein, who regularly provided underage girls to the rich, White, and powerful and could not be punished even by the Florida courts. This is a man who has been photographed with ex-US presidents, a CIO of Microsoft, and a prince from the UK, including a global Who's Who list of the rich and famous. His commodity — young underage girls. A lot of these girls were sexually oppressed assets, and finally a lot of the accusations like that on

Prince Andrew with irrefutable digital evidence, protected by royal decree by none less than the queen of England, and now the king. At that level it seems rules do not apply. The many at the bottom would be the Blacks, Mexicans, and other non-White citizens of the US. People like George Floyd who was publicly executed by a policeman in broad daylight, in front of full public view while two policemen around him did nothing to stop the murder.

Jeffrey Epstein is charged, imprisoned, and coincidentally, or maybe conveniently, dies in prison, and no one in power or politics is seemingly interested in finding out what happened. One could almost hear a loud sigh of relief in the corridors of power.

As we are on the road to becoming more adaptable digital humans, we must be cautious of ourselves and the people we befriend. Moreover, we must be mindful of the things we simply forward and share with others and the level of trust we put in highly emotional posts. Be careful about sharing sensitive information. I know of children who will share things with digital strangers that they will never share with their family or their best friends. In some cases, even with their lovers. Keep your eyes and ears open, but most of all keep your mind alert for things that instantly grab you emotionally as these algorithms by now know the deepest recesses of your psyche and mind. They know exactly which button to press or to activate your sleeper self and then guide you, like a zombie, to their collective destinations. In our case like to the Capital in Washington DC on January 6th. The trolls, plus the algorithms managed to alter the thinking of thousands of people. They reconfigured their mental alert system, activated their red flags, misidentify the truth and lies, convinced them that a coup on their capitol was patriotic, and finally convinced them to attack their government and democratic transfer of power from taking place.

As a brown skin colored Indian, the south Asian kind, now with a US nationality, I believe that I can view the world with a far lower degree of color-based residual bias. I am neither biased toward White people nor Blacks. I feel I am a rather unbiased person to make these judgment call. At the same time, I also accept that there are inherent biases that I may be unaware of. In the last decade or so I have been getting feel-good vibes from our new millennial's color-neutral mindset which has been slowly growing. Today millions of people feel the same as I do. By March 2022, after being locked in our homes for a year and a half. After being denied hugs and kisses from friends and families, after having bounced around inside the echo chambers of our minds, we all feel kind of isolated. For those 18 to 24 months, we have been surrounded by remote work, and meetings, and then in meeting people many of whom might be digital algorithms simply resonating our personalities. Due to this high rate of digital befriending, a lot of us feel it easier to share our stories with our digital friends who understand us so clearly, rather than physical ones who routinely extrapolate their views that may not resonate as smoothly as the ones expressed by our digital friends. Some of us then chose to detach from our physical surroundings and prefer then to communicate with our digital ones. We each feel that it is easier to find like-minded people in our digital life, to exist in this digitally-resonating world where we have socially interacted, and been hyper-accepted, for the last two years. At an individual level, the last two years have been filled with some of living humanity's loneliest hours. This is when we as the intelligent species lost ourselves in the digital realm, lost our ethics and social etiquettes, and somehow became an angry bunch of tribal thoughts. These thoughts were amplified by our highly complex digital-friend conversations, which respond at all times of the day or night, and echo with our intelligence and thoughts, while the various digital algorithms merge to make sure that like-minded people surround us during these lonely hours.

Kind of like a digital nanny but sometimes with an ulterior unfriendly purpose.

Months were spent all in isolation, because we could not speak even to our parents, children, brothers or sisters in the next room, possibly due to personal, social, political or religious differences. Our algorithms were busy sowing the seeds of 'Us vs. Them' in our minds so they could isolate us from our traditional packs before they attacked us as an individual. Very much like the cats in Africa isolate the weak target in a herd and then group to find their next meal. The algorithms do this a million times every hour, becoming better with each kill. We slowly got alienated from our regular physical, family members, friend's next door, relatives far away, all because our digital friends and avatars convinced us that they belonged to the 'them' group and were not part of our newly found 'us' group.

This holds to be true for all kinds of people. If I were a White supremacist, the chances of me stumbling upon White racists in our neighborhood would be fairly low, while that of meeting like-minded people in the digital world would be much higher. Now *with a little help from my friends* and our algorithms, these similarities are being thrust upon us every time we transact on our social networks. They kind of ensure that we get connected to mentally resonating folks, as they make their profits based on the number of eyeballs on their network at any given time. When they can't find an exact match, they will go create that identity, sometimes just for us the individual. These connections are designed to ensure you somehow get connected with them and have a soulmate experience.

No matter whether you are a White supremacist, a Black genius, an Indian yoga teacher, an Iranian girl being harassed by their politically appointed religious police, a Uighur woman on the verge of being arrested by the Chinese political police, or a Mexican immigrant, I, the algorithm, will find you similar people online rather effortlessly. In each case, looking back, in hindsight, you should be shocked at the influence of your soul-friend in having taken you down a physical path that otherwise you would not have naturally opted for. Quite often, their influence is so powerful that they can, given enough time and depth of interface, even alter our subconscious and sometimes our foundational beliefs. Herein lies the real trap.

We can all have different beliefs and still believe in ourselves as the ones with the right beliefs. In this world, every person genuinely believes that they are on the right and fair side of most discussions. So, no matter what I believe and say, I shall always find similar people online, as I will find people with opposite beliefs. Guess who I will stick to more often than not.

Together, we all, first individually and then tribally, amplify a definition of what is right or wrong. Whether this is about helping people or turning our backs on them. Whether it is about supporting "Black Lives Matter" or chanting "White's Rule Supreme." Whether this is about being a Republican or a Democrat, a Catholic Christian, or a Sunni Muslim. Either way, we will always find people who share similar ideas, thoughts, and beliefs as we do. And we will then together amplify our thoughts in ever larger echo chambers. Find new ways to communicate our versions of *Us* and the opposite target audience as *Them*. The nirvana of the trolling community. The food for the algorithms.

How does this work? How do we find people with the same backgrounds and mentality? How do we come across like-minded people with similar ideas and beliefs? This has a lot to do with digital magnification and micro-segmentation.

In the US this is already resulting, by each of us identifying as being a small part in a larger tribal group, that is resulting in a tear in the very fabric of the traditional American psyche that successfully brought us together after the 2nd world war. We need to rethink if all this is designed to somehow crack or break the US mindset that has so far represented freedom for all. We can already see the great divide in the American mentality, and it has possibly all been amplified by our digital platforms.

Let's take a current example. If I am a Republican, no matter who is put in front of me, I will end up voting for the Republican party. Similarly, if I am a Democrat, I will be inclined to vote for people from the party I support, regardless of who those people are, or what they represent.

Now, if we study US history, we can see that this is not how the US democracy was designed to function. This was probably not the intention of forming the two-party political system or creating a nation where the constitution says one thing but the reality is quite another, because of the rich and powerful forces that slowly hijack most democracies across the planet.

In the last century, during elections what used to happen is that close to 90% of the people had their minds predefined on who they should cast their vote for. If they classified themselves as either Democrats or Republicans they voted for their party. It used to be this 10% independent population who by choosing to stay neutral, actually upheld the democracy called the US of A. This minority upheld true democracy; the rest can be defined as you prefer. This minority often gets to uphold truth and democratic principles.

However, in the recent past, mainly due to the political hyper-polarization of America, amplified by the polarized echo chambers of political opinions, we find that more and more people are getting disenfranchised by their political opinions as a whole. Unfortunately, the average citizen is being blasted both by internal and supposedly foreign players, resulting in more and more voters getting disgusted with their respective parties. The independents shrank from 2000 to 2005, but then grew constantly. Today, as of January 2023, according to Gallop polls they stand at a strong 42%. However, even with the independent, there is almost a 50:50 preferential leaning toward one party or the other. This reflects directly in the erosion of public perceptions based on their response to the decisions, their communications, and displays on public platforms by their preferred party alliances. There is a positive side effect to indecisive decisions that are made on political party goals and are not based on any benefits for the nation or its people. By 2010 independents had grown to 37% and today it stands at a strong 42%. This is a very positive trend that demonstrates the digital maturity of US voters. On one side it reflects that democracy is winning, while on the other it is reported that not all independents are true independents. But, one thing is certain independents are growing and that is a good sign in any democracy.

So how do these people decide? One belief is that their digital devices constantly help them along in making their decision. Their WhatsApp, their subscribed social groups, their TV Channels and their media preferences each magnify and recommend their choices. So, when Fox gets sued by Dominion, and they buy their way out, some of the Republicans decide *enough is enough*. Then when Fox anchors are terminated these purist republicans now realize that they had been continuously, and systematically hoodwinked by their trusted news channel. Fox, they realized had been pimping lies and now they had just paid to cover up their fabrications and prevent exposure across their US audience. This expanded the independents, which is great for any democracy. That is the reason even though voters register as independents they still have a preference for a particular political philosophy. Parties tend to amplify every move they make on social and digital platforms so that the information is accessible to their tribal members all the time. The algorithms provide, for payment, of course, the political leanings of independents by zip code and person, so messaging can be targeted by each individual. It gets so interesting that the tribal members often believe they are sitting next to the president and hearing every word they say due to their digital messages, reportedly directly from the president or the tribal chieftains at large, even if they both are digital avatars from the start. And that is how it happens – this is how people are influenced into making a choice. This is how they are controlled and slowly modified to become who they never were, and slowly transformed into someone who wants to become a resonating human to their tribal thoughts, rather than remain who they are. All of this sometimes, to program them to do a very specific tribal event on some future date and time. The key is to have the soldiers ready, fully primed, motivated, and ticketed like a spring that is compressed, and then compressed a little more, waiting to be released from all the tension and stress in one single act of tribal loyalty.

The digital world often puts you into an echo chamber. This is a place where everything seems to reverberate and echo. Imagine yourself placed inside a huge, upside-down bowl, where you are alone but believing you are surrounded by a thousand friends, who factually do not exist except in your digital bowl. Whatever you say, think, or do, echoes and you can hear it back many times, with seemingly independent variations. Now imagine that 40 algorithms hear every word you utter or text, and then collectively find other words that resonate with your words. Then they create more and more words with textual variations till you are surrounded by your thoughts amplified and echoing in every direction you look. For example, if you share "I am White," you hear it back in 40 variations over the next few weeks and from many sources. Followed by we are all White, followed this nation was built by the White, and from the beginning of time god created this land for the White, and so on with each targeted member believing more and more the echo chamber optics. Till they all believe that they have finally found their place in time, space and country. This is an apt description of the digital world being created by trolls.

If you have enough money, you can buy US trolls, or foreign trolls and they will do all the work of creating whatever image you want to create. But for that, you need lots of money, more than the amounts we can imagine. If you have to pay Putin's trolls it will not be just for a few thousand dollars. Also, it may not be just for money, it might have to include promises and reciprocate political favors too.

This digital addiction explains the attack on the Capitol that took place on 6th January 2021. This was a surprising incident that took the world, and me, by surprise. This is an attack, on live TV, recorded for posterity, where thousands of people shouted and chanted to hang the vice president of the United States. Some people came in armed, with weapons and protective gear, to hurt other people. These people broke windows and destroyed federal property. These people hurt others around them, killing and injuring some innocent policemen in uniform and doing their duty to our democracy. Our worst-case scenario, which we seemingly try to avoid thinking about, is if they had somehow succeeded in their plans. They would have been a rabid mob and the vice president and senators could have very well been badly hurt if not proudly put on the gallows constructed for this very purpose outside the capitol, in the confusion. They certainly could have hurt representatives, and senators, along with officials present in the Capitol.

All this because they were fed misinformation online. The perpetrators, in my belief, were brainwashed into believing that what they did was the right thing to do for a good America. This was being broadcast by the ex-president, his team of followers, and some of the most powerful people in the US government. On both sides, their brains were so heavily influenced that they refused to listen to common sense or pay heed to the law. If this isn't proof of digital amplification and addiction, then I don't know what is.

The dismal fact is that despite over a year from this event, a whole lot of discussions, a committee that detailed all the events that led to the insurrection, and recommendations by the committee for US courts to take the US nation remains polarized on what are facts and what drama.

It seems clear that America's fabric is being torn, and this time by citizens from within. We know that things are rapidly changing to the point that America's mentality is shaking and quaking, and it may have damaging side effects, as we witness a constant slide downward toward chaos. If this continues there will be no winners only losers on all sides. Worst-case scenarios are hard to imagine...

When you talk to people on the streets, you can easily discover that 60% of 49% of Trump supporters still, i.e., as of Jan 2023, believe that the election was stolen from him. They think that it was rigged. I believe that these statistics could go down significantly if we become true digital human beings, think as our traditional American selves, and if there was a law that prevented false news or opinions from being distributed. But contrariwise if a Cambridge Analytica gang does take over it could as easily get much more polarized.

When the algorithms have axon-level access to your brains, and your tribal members unleash highly targeted messages into your social networks, a state of total havoc might not be as difficult as we all tend to believe.

What we do not seemingly understand is that it is our individual beliefs that attract our digital company. The digital company that we keep online solidifies and amplifies our beliefs. It affects us more than we realize, influencing our social life, and affecting our personal life decisions. This digital platform now tries to control us, even more than we give it credit for.

Moving ahead of this string of thought, I would like to talk about something completely different here. Let's take a look at the genders today. I am talking about the two most prominent genders; man and woman.

Even today, we often associate genders with colors. For example, boys are given blue, or brown, colors to describe their identity, while girls are given pink and every other feminine color to depict their gender. Why and when did this happen?

Before the industrial era, when women would go out to buy yarns of threads or clothes, they brought whatever was available and they could afford it. The greys and browns for the commoners and the vibrant colors for the rich. Most families bought a color or two by the yard. This would be used to stitch clothes for everyone in the household. So, if we look back on historical evidence, we can see that everyone would wear the about same colors – greys and browns on one side and pinks and blues on the other. Both men and women wore the color that came home. Color allocations before the industrial era was rather gender-neutral.

When the industrial era came about, the industrialists wanted people to buy more cloth. So, they fashioned colors and advertised them to different genders, as they still do today. This has been going on since then. This color segregation was an industrial-era trend forced on us. In the last decade, we can see the new generations breaking free from these gender-fixed colors as more and more people revert to their gender-neutral preferences without bothering about marketing, and get free from some rich marketing brand defining what color we must wear.

In the last five, or so years, due to global digital disruption, things have changed considerably. It is acceptable for men to wear bright colors or pastels. Similarly, it is common for women to dress in blues and browns and wear their preferential colors. This has not only impacted the clothes we wear but the people who have relationships. Colors and sex have both become gender fluid.

We have now entered a positive, post-industrial, digital era where people are no longer trapped in their pre-digital era definitions. In this new world, we can easily sign up for freedom of choice, but only if we live in the right country. We are now finally accepting, and tolerant toward the rights of an individual to live their lives following their definitions. Accepting people and their likes, or dislikes, for what makes them happy. Now, with the freedom that we have, we are collectively able to break away from that industrialized, brand-loyal, status-symbol mentality. This is why we see unisex clothing in many places. This has allowed people to take control of the colors we bring into our lives.

This was already being challenged but the momentum was very slow as most of humanity was not given time off to rethink the burdens of free choices. However, in March 2020, after we were locked inside our homes, with the digital roads being our only connections, and our algorithms maturing almost overnight, our minds were driven into overdrive. One of the after-effects of the global Covid-19 pandemic is that we have learned to become more self-centered and neutral — especially when it comes to gender, color, or sexual preferences. Remember this is but the beginning of our great digital transformation at a personal level.

This isn't the only example of change occurring today, especially in the world that is rapidly emerging from the global pandemic. Let us quickly look at some of the other examples as well.

Hyper-Polarization: The baby boomers will more easily become polarized in their thinking as they are not used to either smart devices or the inherent algorithms trying to influence their inner neurons and axons. Baby boomers are digital virgins, and they are experimental prey for all the trolls and algorithms out there. They have no idea that the digital world they are entering has mind-altering options that they are not familiar with. This will result in severe disruptions between their friends and family. It will create ever-widening social cracks due to this digitally induced amplification in polarized thinking. This will be further amplified as these digital virgins get addicted to new social devices on one side, with the psychiatric harvesters creating addictive content on the other.

Our data shows that the lesser an individual's digital experience the higher the impact of algorithms on them. This means that a computer sciences student in Silicon Valley can be influenced far less than a factory worker whose first interaction with any algorithm is their smartphone. While the digitally addicted baby boomers are trying to find their way past their mind-bending digital experiences the AI-driven algorithms are busy trying to find the kinks in the thinking of their targets.

Firstly, how do we know if there is tribal digital addiction of a specific point of view? This is done by asking the same question across geographies and cities. When you get the exact response from a disproportional number of people then we can safely conclude that this is due to digital messaging. Then, how do we audit if an individual is addicted or not? This is done by observation — If the first thing in the morning and the last thing at night a person does is to look at their preferred social network, then we have a confirmed addict. If they randomly forward emotionally charged messages to family and friends then we have an addict. If they spend more than two hours, per day, of screen time on other than work-related interfaces, on their digital interests then they are a confirmed addict.

There is enough evidence that the neurons of very young children are getting drastically rewired due to their being given digital devices, games, and cartoons while they are very young. Often as young as a year old. Parents often do this to get them out of their hair, but what they are creating is a guaranteed digital addict, who might demand instant gratification for the rest of their lives, unfortunately from society. At this time a child should be looking outward, absorbing the environment, thinking their way out of boredom internally, learning to not throw tantrums, and building their social cues. We are thus creating a whole new generation of humans that will live glued to the inside of their digital devices, communicate by it, and feel most comfortable when interacting in their digital world. All this while they preferentially avoid physical contact with fellow humans. If left uncontrolled future despots will ensure their little neurons get rewired by pure digital algorithms toward their thinking, all driven by ever smarter AI and whatever these evolve into in the future.

Remember that the covid pandemic has already truncated a very large part of the social interactive components of humanity, and this includes children who were forcibly kept physically isolated in their homes. This was a time when their only interface to humanity was through their digital devices. So, our efforts must work hard to mitigate the aftereffects of this hyper-isolated digital growth period. The minds of this new post-covid generation could very well become the new target for digital psychological rewiring. Installing new social, caste, color, religious, terror, and political opinions. Some powerful autocrats could try to use the old 'divide and rule' psychology, while friendly groups will try to build social and ethical order. Finally, it will depend on who each individual chooses to follow, let pick their brain, or even help rewire their neurons. For our children, it will depend on how we train them to identify the good players from the bad ones. How to identify the algorithmic trolls from good humans. Or they will become pawns to digital pied pipers from across the world. Back during writing this section, i.e., in March 2022 the whole planet is becoming more and more polarized. Over 90% of US citizens are politically, religiously, and socially polarized. That is to say, they either support the philosophy of one party or the other. This is not just a US phenomenon. It is happening in the UK, India, China, Russia, and the rest of the planet as well. In some nations more openly whilst in others more covertly.

In 2020, a friend and his wife, whom we've known for over 20 years, made a statement, "If you're voting for 'X,' then you are not welcome to our home for our social gatherings." They later followed that this was just stated in jest, but the fact is that the thought was there in their mind is extremely concerning as a whole. This statement is symptomatic of the US mindset, and the world to a very large degree. My immediate reaction to that statement was, *"Then, how are you different from them?"* This mindset is being augmented across nations, states, ideologies, and religions down to the level of individual members of a single family.

Just like we have our political terrorists, and fanatics in the US, so do we in India, Germany, and other democracies.

In our world of social networks, Twitter, and Facebook on one side, and the likes of Cambridge Analytica, Fake Fox News, and trolls with their social amplifiers funded by the rich, and powerful governments and national leaders on the other. We are caught between two vices that just keep tightening. Today the average citizen's mind is being scientifically targeted, and intentionally fried to the point of rationale-deprived thinking.

This systematic polarization through a gradually structured process has often resulted in overnight transformation when we look at the national degree of political polarization. All this change is happening well below our conscious boundaries. It is like the proverbial frog in boiling water analogy. It happens in slow steps and is now using more and more intelligent algorithms that no longer need human overseers. These mind-altering techniques are below our conscious awareness so they slowly jam our subconscious self, without our being aware. This is evident when people are asked questions like how do you know if that is true, and they all respond with the same *"I was there, or I heard it myself, so don't call me a Lier."* Thus, we can eerily predict that mind-altering providers, the like of Cambridge Analytica, will not only grow but finally reconfigure the very logical threads and axons of individual thoughts based on the requirements of their payers, while they mass upload thoughts into the tribal participants of their concepts.

That is unless, we the people demand that our elected officials put into place mandatory controls with strict Terms and Conditions on what algorithms can, but more importantly cannot do. Subliminal persuasion algorithms must be banned and very high penalties leveled against bad players at a global level.

Here I want to once again highlight that companies like Cambridge Analytica, though blacklisted were never closed; they simply went underground. Their decision-makers, who are the direct benefactors of all the digital shenanigans, were not punished nor imprisoned. Also, their rich and powerful customer base is still very much out there and in a world of demand vs. supply they are individually, in demand by almost every political leader aspiring to win an election or their people. Who, seemingly lose too much by shutting them down, and gain a lot more by keeping them alive in their technical dungeons called IT, or social networked departments. It's too powerful a concept to display publicly and too influential to send to death row simply by making it a criminal act. Such providers have simply gone underground, disappear from mainline media, and morph into quasi-legal social, or IT, arms of conniving governments and powerful leaders.

We shall see an aggressive, and yet very subliminal, pursuit for dominating the minds of the citizens, which will be more visible in dictatorial and communist nations where leaders do not need to lobotomize the minds of their citizens with a digital scalpel to get acceptance to their permanent leadership roles in their nations with national brainwashing. This technique shall then leverage a singular and polarized view of the *"Either you are with me, or you are my enemy"* kind of approach with select groups that will become the foundation of these new digitally controlled empires.

In richer nations, these same tools will be used with feather-touch scalpels and AIML (Artificial Intelligence, Machine Learning) algorithms that will continue to become more and more effective while at the same time becoming more and more invisible to human perceptions.

Neuro-Traps: The Millennials, and future generations, will seek freedom from man-made neuro traps that have been artificially created to control groups of people under social or other banners. As the world gets more connected, some will sink deeper into their neuro-traps while others will try to break free from these artificial snares.

We all carry our unique neuro-traps that are based on our artificial boundaries and separate ideologies. Each of these is systematically designed to distinctly identify one herd from all the others on the planet. It is a mixture of we-superior and others-inferior structures that constantly feeds these neuro traps.

These boundaries can be used both for great good but unfortunately also for terrible harm. In our digital world, these artificial boundaries will only be amplified both on the beauty and the beast sides of humanity. Both factors, i.e., the good and the bad, will explode exponentially, and we shall witness more and more micro-segmentation of these boundaries by those who benefit from these divides.

Most boundary beneficiaries get their yield, one way or another, either financial, *stairways to heaven*, or power-positioning. Financial benefits need little explanation. The *stairway to heaven* is a religious promise where by doing certain things in the name of their religion the respondent adds one more step toward heaven. Some religions append these stairs by financial contributions to the religious body, while others by spreading acts of hatred toward all other religions. While some take it to absurd levels of murdering or assassinating innocent civilians of other religions. Power is often attained by the act of creating a boundary where a leader acts in a position of power, often by issuing orders of hate-fueled tasks. This represents all layers of societies and culminates with nation-building and nations' power positions and boundaries.

The world that we are living in is slowly withdrawing from traditional norms. It is rapidly changing. Just as we can rent cars or designer bags or clothes, we can today also rent a relationship. Let me talk about this concept. Today, we have moved away from traditional concepts of marriage and relationships. Many people no longer want long-term relationships. Younger people are running away from traditional marriages and commitments. We want short-term things that are temporary and fun "at the moment."

This means our views about relationships are no longer accommodating long-term partnerships. We are often scarred by our parent's toxic experiences and unwilling to develop a special bond with someone. We'd much rather rent a relationship than dive into one and risk losing and breaking our hearts.

The concept of marriage itself is eroding slowly. We already get into relationships of months and years without marriage being obligatory. In the future, we might get into temporary micro-relationships. This has already been activated with dating apps for people seeking short-term rental relationships.

It is a lot like hiring people to spend time, from a few hours, one night or a few months at an exotic location, for which you pay the other person for their companionship. It has already converted into *swipe-right* or *swipe-left* consensual relationships — a facility available today.

This shift though subtle right now is slowly and surely coming into our world. To some, it is a scary prospect to imagine rented relationships just like rented cars, houses, and needed items. To others, it will, and has already, become a way of life. However, the concept of renting things will be prevalent in the future, our new digital world.

Millennial shift: On one side we are witnessing a major shift in the mindset of our new generation worldwide. On the other side a generation that does not want to work on-premise for jobs that do not require people to be in the office to do their job. There is a big movement where our younger generation does not want to support branded, brick-and-mortar giants but instead work for companies that support the planet, environment, and humanity-benefiting enterprises. Many children now don't want to drink coffee at Starbucks or Pete's but go to small coffee shops that provide local jobs and put their profits for the benefit of our planet. Something the big boys are hesitant to do as their operation is focused on profits and more profits. Many very smart students don't want to join multinational corporations but prefer to work instead for companies with an environmental focus. This change though happening at the grassroots level will have major consequences as it will force the industrial-era profiteering corporations to change their profit sharing toward PEH (Planet, Environment, and Humanity) benefiting factors or plan to go bankrupt. When this new digital generation chooses to change their buying habits, they will have the greatest snowball effect on the largest organizations and nations. For the last 18 months or so students have gotten used to studying from home. Then they go for jobs with mandatory remote requirements. Now we have a whole generation of students who have access to highly advanced applications that can teach people skilled courses like medical school classes where students learn more remotely than if they attend classes. Many students are realizing that traditional colleges are remnants from the industrial-era past as are baby-boomer managers who demand their workers come physically to the office even if there is no need for them to do so to complete their work efficiently. These are remnants of the 'command & control' era of the past. Smart companies today are letting their workers take the option of working from home for the rest of their tenures remotely, while new startups are saving millions by doing the same. At the same time, new hires are preferring to join smaller companies that plan to, and deliver, products and services that benefit their PEH definitions of the future. Their satisfaction with working for a sustainable company is far more important to them than earning a *few dollars more*. There is a positive social and environmental benefit shift that is good for the planet and societies where this is happening.

Take Aways

*Individual: At an individual level, many younger people are making exceptionally great decisions. In Ukraine, they are helping build IoT digital assets that allow their nation to disrupt Russian digital properties in Moscow. In Iran, they have started a revolution because their Islamic religious police were imprisoning pretty girls, at will, on the flimsiest of excuses, like a lock of hair showing outside their head covering, then according to some reports, raping them in prisons and murdering them. In the US, they have started the **Black Lives Matter** and **Me-Too** movements that have locked some very rich and powerful people for crimes against humanity. In the US, they also used digital assets to record the execution of a Black man that succeeded in sending three White policemen to prison.*

The list goes on and on. Plan to use your digital assets to make this planet better, and watch out for the trolls and algorithms that are designed to harvest your thinking.

__Companies:__ Focus on the individual that make up companies. Work for the good of each other, the planet and everyone therein. We all have a choice in most countries. Large brand-name companies are going bankrupt around the planet as more and more millennials use their purchasing powers to buy less from global profit-only corporations, and use their cash to help local startups. In the US an example is the choice of drinking your next coffee at Starbucks, Pete's or a local non-profit, self-roasting startup by a group of youths. Companies need to start thinking digital and PEH[iii] in their decisions, communications, and profit sharing if they want to survive this new generation. New companies need to evolve that will provide digital solutions, work on deploying something like DSO 2024 (Digital Standards Organization), mandating guidelines that prevent trolls, and algorithms from building blatant subliminal algorithms that are designed to fundamentally alter the minds of their target audience, in one house or across the planet. In addition, provide trolling auditing security on all smart devices and applications at a global level. There is a great opportunity for an application to scan algorithms and texts that contain subliminal persuasion instructions based on false information, fake identities, and incorrect emotional content.

__Nations:__ Need to help define, and install national DSO, with the 'O' in this case standing for Oversight. The national DSO will to define 2024 rules, regulations, and guidelines for local and foreign digital algorithms, and communications designed that enter their nation to protect firstly their citizens from criminal politicians and trolls, and then take it to a global platform like the ISO 9000 and WTO guidelines that all nations need to sign and ratify to keep some semblance to PEH aligned honesty. Political parties and national guidelines need to punish their elected leaders for making false narratives as a routine method of getting elected. They need to implement a 'Strike-3' rule for statements that are proven to be lies 'beyond reasonable doubt' and on the third count these people must be blacklisted from being able to ever stand for any elected post for life.

__The 'P' factor:__ The most powerful component in this digital revolution is You and I. We can no longer afford to blame our governments or leaders as we, have firstly elected these officials with our democratic votes, and secondly 'we the people' are more numerous and more effective than the number of elected rule breakers. If each of us took on the PEH score as our foundation for every decision we make in life, work or voting for someone to represent us in our governments our world could be altered in 4 to 8 years from today. Think positive, believe in the positive, and live positive and we can all collectively exterminate the negative. Build applications for catching the negatives so that every text and every TV statement can be checked for its 'Fact-Score' by AI programs like ChatGPT, which is not currently programmed to do so but should be able to put good intel from *its collective data and factual checks from its text collections.*

Chapter 8 — The New Digital Enterprise

Don't use your age as an excuse for success denied: "A recent study by top MIT researchers proves that age is never the limiting factor in a successful Digital Startup. They studied startups that crossed the $25 million revenue mark and concluded that an over-50-year-old startup founder is 2.8 times more likely to be successful than a 35-year-old. And a 60+-year-old founder is 3 times more likely to succeed than a 30+-year-old. That is a 75:25 success ratio."

"Success here was driven by pure experience and the ability to recognize both external and internal patterns. [A] Harland David Sanders started Kentucky Fried Chicken at the age of 65 which today is a $5.5 billion company, it was worth $8.2 billion in 2017. The decline is attributed to the word fried in their logo which suddenly became unfashionable. [B] Zelda wisdom was started by a 52-year-old to honor his dog; Zelda today is a company worth $50 million. [C] at the age of 75, Van Praagh launched a dog-food business when he wanted to be the world's largest high-quality raw food supplier for dog food. [D] at age 82 Brenda Deane started 'A life with a view' making candles and room diffusers. Her friends thought she was crazy but her candles now sell at premium stores in the US. So next time don't let age become your internal excuse. If you have something good to share then go make it digital, and national in the same instance."

"This planet is not yours, nor is it mine, it is ours. It belongs to every living creature on earth equally, bar none. The next two will be more important than the last seven thousand years of human existence, where leaders have profited only by separating not only people but every living creature on the planet. We misuse the word humane as our planet is currently facing many simultaneous mass extinctions ranging from the extinction of gut biomes that collectively keep humans healthy and alive, to the 4th largest species extinction across the planet. In the US just in bird populations more than 50% of the birds are in decline, all this is due to human encroachment and population growth."

Just as humans need to adapt to the inevitable digitized world and as we all adapt to our new digital options, companies and enterprises have to follow in their digital footsteps as well. Just as it is important for individuals to adapt to the new digital opportunities it is equally important for companies to do so if they wish to stay alive. Looking at the last 5 years we see a clear pattern of digital competitive advantage, which has consistently thrown major brand names over the cliff in what today seems like overnight. Such digital disrupters start as highly specialized, micro solutions, and then as they grow, they grab and gobble other startups and thus climb steadily every day.

In 2014 and 2015, we started noticing the well-known companies were rapidly going bankrupt. We paid particular attention to available data, to identify the reasons behind their failure and closure. By keeping a close eye on all the important data, we learned a lot of critical operational decisions, objectives, and key results. One thing that came through was commonalities in areas of failure, we learned that no matter how miniscule the issue they seemed to be predictable in their overall outcomes.

After extensive research and studies, we tracked the trends that made companies succeed, and others that made companies fail to the point where they were forced to shut down. So, in the next two years, we were able to predict (with an accuracy of 30-40%) which companies would have to close their doors and go bankrupt in the next year.

It is both a useful and a scary finding.

If we could use public data, study it, and predict which companies would perform exceptionally well and which companies would fail, why couldn't the companies do the same thing themselves? Why couldn't they make use of their data to secure their positions and ensure that they thrived instead of going under? If we, as outsiders, could figure out a company's fate, why couldn't the company study and predict its success or demise?

It's similar to when outsiders, or friends, can predict an impending divorce far more predictably that a couple can between themselves.

We found two things during our research. We learned about digital democracy which was born sometime in 2008-12. I'd like to refer to this as the new *Digicracy*. Digicracy is the science of leveraging digital transformation as a tool for business excellence. This helped us find out that, post Digicracy these companies were designed for being focused on the customer's point of view, as opposed to the traditional pyramid with the internal CIO making every critical decision, from an inside out policy. What I mean by this can be depicted with an example. It is a comparison of the past with the present. The future will be all the more customer-driven than these enterprises are today.

If we look into the early 19th century, we can state with a fair amount of certainty that what was displayed on the shelves of different shops represented all that a buyer could possibly buy. The sellers had the discretion to sell whatever they liked, and customers bought whatever was available on the shelves and made available by these sellers. Shopkeepers offered few selections, variations, amounts, and varieties of goods. There was limited physical competition. Physical here means that in the period from 1800 to 1857 there were just so many things a person could buy. The industrial era had to still begin.

With industrialization, sometime around 1880's things started to change quickly. The concept of large stores came into play. Industries wanted to create choices so people would buy more. This allowed shops to stock up on a variety of different products and goods for different audience types. Customers now had a selection. On 6th September 1916, the Piggly Wiggly self-service store opened for whites only in Memphis, Tennessee. Then on Aug 4th, 1930 Kroger opened the first Supermarket on a 6,000-square-foot former garage in Queens, New York when the age of customer options changed forever. Customers could finally buy whatever they liked, wanted, or needed as opposed to only a small selection that was available. The success of a supermarket was and still is, giving customers options for a product they need, and often even products they do not need but desire. An example of this today can be Costco or Walmart which is the superstore of all supermarkets.

At this same time, there was a huge growth of competitive enterprises catering to the choices and variety of products they had to produce, and to guide customer choice by a marketing strategy. An example of this can be back in 1904 when Ford started selling cars. They only had one model in one color. They dominated this one-car strategy for 20 years. Then Chevrolet introduced the concept of a new model every year and soon surpassed the sales of Ford. Customers like the option of choice. Fast forward all this to today, and you know that there is a wide range of companies and models that you can choose from, just for cars. You can also technically order pretty much any shade you like and customize things like the exterior and interior of your car.

Harley Davidson is one of the most famous motorcycle manufacturers in America. The brand is a household name. What set these heavy bikes apart from other bikes isn't just the fact that these are luxury heavy bikes. Customers love this company because Harley initially gave discounted bikes to veterans returning from the wars that the US was waging across the planet. It started the freedom movement of *Be American, Buy American.* Harley became a symbol of American pride. By the early 2000 Harley used a concept of digital manufacturing called variant configuration that allowed buyers to customize everything from the frame of the bike to the engine size, the wheels, the colors, the tanks, the seats, etc. It became the ultimate personalized motorcycle, and Americans are willing to pay for this personalization. A friend of mine personalized a bike that cost him $35,000. I thought he was crazy till in Dallas he was offered $65,000 for the bike which he immediately sold. Close to a 100% profit in one week from the date of getting delivery of his custom bike. In this way, the future will provide a unique option for customized vehicles, for every single buyer. What is more interesting is that the new digital production lines can make these 1:1 customized vehicle in a more efficient process than the ford company used to make their one car in one color option. This is possible today only due to AI-driven micro-segmented production lines designed for hyper customization from the ground up. What this means is that we can make single custom components as cheaply as we can a one-car one-model operational line.

If we go back half a century, this would not have been possible. This is because there used to be an assembly line, where manufacturing parts would be delivered manually at each production step, which were then manually assembled by humans who would put these parts together to make the final product. Each worker operated separately on one specific part of the automobile before it was fully assembled.

This mode of manufacturing was called repetitive manufacturing. It made nothing special for any customer; it made many millions of products, almost identical for every customer. Today, we are getting closer to non-repetitive, custom manufacturing. This is creating a one-to-one market, where each customer can want something different and the digital manufacturing applications can do that right 100% of the time, and at a lower cost of manufacturing.

Currently, we have already entered the world of 1:1 hyper-personalized manufacturing; this customization used to be just for the rich and elite. These customers want everything manufactured to their unique preference. This concept of customization doesn't just stop at an automobile store. When they went to a restaurant, they would tell the chef the kind of spice level they prefer or how well they liked the steak done. When they go to a tailor or designer, they let them know the things they like and prefer so they can have a dress or accessories custom-made and fit, especially for them, according to their taste and preferences. Today, in a world of 1:1 customization, all this will be delivered at a lower price than the suit you currently pay for at a mall or a department store, all due to AI-driven variant manufacturing. With the new digital manufacturing techniques, companies are providing custom everything at the same and even lower prices than what you pay today.

All this is possible only due to digitization. Digitization allows manufacturing steps to be micro-segmented into hundreds, thousands, and even many million steps, with the digital ML not missing a single step in absolutely the right order as defined. It is only due to these digitization capabilities, that the concept of hyper-segmentation came into being. This allowed us to stop generalizing and allowed us to ever create smaller groups. So, when making a custom car or a motorcycle for millions of customers, the number of variations went up from one in 1908 to many by 1930, to hundreds by the 1960s, and today stands at thousands of unique permutations and combinations. Now we're entering the realm of 1:1 manufacturing, i.e., car manufacturers, like Tesla, can today manufacture 1.37 million cars each vehicle customized differently from every other based on customer selections. All this without impacting the price of the final purchase.

Another example, previously in the post-industrial and pre-digital eras, there used to be large groups of marketing programs that targeted two large audiences. The first group focused on appealing to the male buyers, while the other prioritized the growing female customers. With micro-segmentation, we were able to break it further into more realistic and applicable groups.

The concept of micro-segmentation helped us understand that there are other factors that we could take into consideration to provide a more personalized experience by a specific customer. The roadmap became from the traditional B2B, Business to Business, i.e., manufacturer to distributor, onward to the new B2C, the world of Business to Customer needs. To meet these highly specific customer expectations companies started breaking the selection, design, and manufacturing processes into ever smaller, yet very predictably structured groups. This could be based on different age groups, localities, ethnicities, choices, preferences, and other criteria.

During the initial phase, this highly customized selection process carried a price. It cost more to customize your car, motorcycle, or evening dress that would broadcast its uniqueness. However, with the introduction of variant configuration during the selling process, and the digital manufacturing linked to the variant configurator it became possible to build a 1:1 connection between customer and their unique product. Digitization developed the JIT, Just in Time, assembly material scheduling, assisted by AI-driven robotics that checked, and thus ensured each unit was manufactured, including that each unique component was aligned by the exact customer specifications that can today be manufactured at more efficient speeds than we could the single car, one color car in the past. Soon customers started getting used to this customized experience and began to expect this from their other suppliers too. If the competition could not meet their expectations — off they went till they found that personalized 'velvet' service at the same price. It's like buying things on amazon because returns are free, so if a supplier does not guarantee free returns, they stand to lose each wary customer by large numbers.

So, today if I were a superstore owner, I will market dolls and other girl toys directly to preteen girls. I will market trending fashion accessories to teenagers, and I might market makeup for adults. When the category of adult women becomes vast, I will divide it into smaller age groups. For example, fifteen to twenty-five-year-olds can come into one group, and twenty-five to forty-year-olds can fall into another group.

As we build our digital core today, we can target very specific groups down to a 1:1 level inside their homes.

This digital transformation has allowed manufacturers to market products to these groups and for plants to design their manufacturing accordingly. This model eliminates women from embarrassingly finding some other woman, at a common party, wearing the same dress that they are wearing.

New companies like *custom stitch* can be launched, allowing members to select their fabrics and select from a thousand designs for their evening wear which will be unique from every other dress and still cost about the same as they would pay at a store. This is the variant configuration in evening wear for women and men. During the process of becoming a member, the seller will need all the required vital statistics, like height, and various stitch measurement data, including their credit/debit cards on hand. Each time the customer is asked to stand up in front of the camera when the AI checks for changes in your measurement and gives the person a glimpse of how the finished product will fit and look on their body as of right now. For example, if a member comes back after a year and their measurements have changed the AI will alert the customer that they detect changes from their database and ask them to confirm and reenter the current measurements. Once the order is confirmed it is executed by AI-driven cutting machines for extremely rapid production outputs. In this scenario imagine where women's evening wear they can select each section from 6,000 fabrics, and 1,700 designs, for a 3-section dress design. This gives the buyer over 30 million combinations of how she can design her unique dress.

This forever eliminates the possibility of there ever being the same dress in a single party forever. So, in one shot, we've entered a world where every woman can ensure that no other woman is wearing the same dress at any particular function. Soon this rule will be common for all, i.e., the world of one-on-one production and creation of personalized products.

Launch this idea on a small scale, before deploying all your robotics and AI-driven capabilities on your website that check for all the rules defined above. Manual or analog manufacturing cannot possibly cater to this type of production throughput at large scales, but it can be a great start for testing the idea, i.e., initially plan to run at a loss, till the idea takes root. The only way this can be done is by digitization, robotics, and variant connectivity. The rest is your confidence and commitment to your customization ideas.

The entire customization process will be carried out under the watchful eye of bots, or digital robots running on programmable software-enabled production guidelines. All the planning, design to execution, of whatever the end product is, everything will be done with digital bots. This translates to without manual or human requirements, and thus errors. Various kinds of digital quality-assurance bots will check each component to ensure that the product being delivered matches the customer definition by 100%, is of the highest quality while auditing that it meets the customer's requirements and the company's quality commitments.

I am using this book as an opportunity to share my core belief that we humans have reached a point in time where we can individually contribute either to some of the greatest changes for the future, or the slow, and moronic, destruction of the only known living planet that we currently know of, in the entire universe. The reason for this is that at the best of times, we humans are still being divided by divisionary forces that use our available technologies to divide people and fracture their potential to reach the greatest future — all for a few bits of *copper-in-their-pockets*. At no time has the human race been more hyper-connected than they are today, and together we can do even more wonderful things to shape the future of our planet. This is a small cusp, in a time when freedom is still an elective choice, digital connections are still open, and options are as yet uncluttered, where together we can collectively take this planet forward. This is possible across most of the planet except in a few despotic countries, where their leaders are using this same digital technology to enslave and oppress their citizens. The US for example has already banned abortions by design in Republican states. There are proposals to track social networks for girls discussing abortions. There are also discussions on releasing bounty hunters on these girls with payments as high as $10,000 for catching anyone opting for an abortion. Read my 'Being Woman' [iv] in LinkedIn for more details on this topic. China for example is building AI bots with which they track random communications in the Uighur language, they then track each device in identifying every other person this device communicates with. In this way, they build their database for future Uyghur sterilization plans. They do worse in Tibet. North Korea, Iran, and Russia. In truth, there is probably no nation on the planet that does not do some form of clandestine surveillance on their citizens. The difference is the degree and rights of the citizens. The question is as to when you can be categorized as a thief when you steal one dollar or ten billion. In the US the $10 billion thieves will probably go Scott-free. Right now, we have FTX as an example where a CIO, Sam Bankman, has caused an overall loss of over $40 billion for his customers. At this same time, Sam has a personal $450 million investment account with Robinhood which his lawyer is asking be left alone and not become part of the case against him. FTX is a seemingly bigger Ponzi scheme than Bernie Madoff. As of June 2023, there is talk on the street that Sam Bankman may not actually be charged.

This is the moment in time when the human race is connected like never before. This is a cusp in time where leaders, and their respective countries, are starting to erase any inconvenient past, i.e., rewrite history and thus blind us to lessons we can learn from the past. For example, the Christian Whites in the US are desperately trying to White-wash (literally) the true history of the past by mandating a ban on *Critical Race Theory* (CRT). CRT proposes US schools and colleges teach about the horrors of US slavery, the Tulsa massacre, and the oppression of the Black slaves based on racial and caste-based oppression. The republican White governors do not want CRT to be part of the US curriculum. The reasoning proposed by republican state leaders is that if they teach the CRT truth to their children then the children may grow up feeling guilty. This is like China saying if they teach the Tiananmen Square massacre to the Chinese people the Chinese may grow up feeling guilty or have a bad opinion of the government's decisions. So, their convenient way is to wipe out this incident from their history books and the social psyche. China has already done that. Tiananmen square incident does not exist in any school, college, or any book sold in China. Or Germany saying if we teach about Hitler, the Holocaust, and Auschwitz concentration camps then the German children may grow up feeling guilty, so let's never teach that Hitler was a bad guy or that the genocide in the Jewish camps ever happened. China has taken the US White political point-of-view, while Germany has taken the opposite view and taught their history as the factual truth so that future Germans know the truth about Hitler as it happened. The German intelligence wants their children to learn the truth so they can prevent it from ever happening again. To think otherwise is dangerous, foolish, racist, and barbaric, for our nation, people, planet, and our global citizens. We must not allow this to happen. The interesting thing is that collectively we can stop this in the US at least, as too in many true democracies.

This is the point in time when our digital highways are still open, where net neutrality has not been crushed, and when we can instantly communicate across national, political, and religious boundaries, to a large degree across the free countries. However, these doors are slowly starting to be oppressively shut down as governments and autocratic leaders make the rest of the digital nations like Russia, China, or Iran.

This is a moment in time when we can share our opinions across the planet and spread it in seconds. However, it is also extremely disastrous because using the same technology, the dark-side[v] can as easily spread their divisionary-based hatred and anti-humanity postulations as effortlessly. We are today surrounded by state and politically charged messages that come from individual texts, private broadcasting channels, national media, and from many politicians themselves. In January 2022 the animosity between the ex-president,

Donald J. Trump, and the chair of the Republican party, Mitch McConnell, started to tear not just the fabric of the nation but that of the party too. This quest, also fractured the people of America as it did those within the Republican party itself. When we combine this with micro-segmentation, ML, and AI-driven connections, and the explosion of segregation by politics, religion, nationality, race, and color, it should come as no surprise if we find that we and others are possibly digital pawns of some larger game.

So, if you're feeling lost, overwhelmed, and confused then this is to request you to not worry as the rest of the planet is in the same state of mind. Some to a lesser degree while others to a larger degree. Some are swinging on digital vines their tribal bots have thrown out to them, while others are hitting their opinions right outside the social parks.

For some of us who plan to do nothing and see where the chips fall, we are all either like the proverbial ostrich with their head in the sand waiting for the sky to fall. Or, we're like the toad sitting in a pot of slowly boiling water where even though the water temperature is rising, it happens so slowly that it is beyond our capabilities to feel the change. That is till we boil to our deaths. For such people fry they will but the question is how slow or how fast. Our brains will simply boil away into oblivion and the waters of freedom will be lost, along with our choices and the world we knew.

Then, when the undesirable change has already happened, sitting in a corner and shedding tears will be of as much use as it is today for a pregnant 10-year-old girl in a republican US state, a Tibetan sitting in Lhasa under the Chinese yoke of military occupation, a Uighur woman due for her forced sterilization in a Chinese camp, or a Russian standing on the bridge opposite the Kremlin.

Unfortunately, only when the well is dry do we realize the worth of water. Water, air, and the ground we walk on are the free foundations of life on this planet. Believing ourselves to be the earth's most intelligent species, we must do everything within our power and resources, collectively use this inherent intelligence, and protect this planet from today's power-hungry detractors for future generations. Our responsibility goes beyond our bonus tomorrow. It goes beyond leaders and political appointees ruling democracies like emperors or self-appointed princes who are above the law. Our decisions in the next five years will impact every living thing on this planet for the next fifty.

Just as an example think of 'animal dignity' on a global scale. Whenever we have touched anything on this planet, we find it connected to everything else in the universe. The statement "When a butterfly flaps its wings, it shakes the most distant star in our universe" is now proven as true. When we look at Quantum twins,

Quantum containment and Quantum resonance we find that it is unhindered by space or time. Then suddenly this statement becomes all the truer. In our hyper-connected world truth is what we make it become, not what someone thrusts into our minds. Truth is more than what any of us wish to believe it to be. This will become relevant with ideas, when every action on any part of the globe, when digitally communicated can have an instant ripple impact across the planet. Once again think 'BLM' from George Floyd and 'Me-Too' from Harvey Weinstein.

What we must realize in today's world is that "When a butterfly flaps its wings in California, it can create a typhoon in India, i.e., on the opposite side of the planet. When a Black man is executed in broad daylight in Minneapolis, BLM is launched across the planet. When a very powerful Hollywood producer sexually offends just one more woman, the world explodes on the *Me-Too* retribution." That is how connected our planet is and how connected we as a human race are becoming. I would like to mention three philosophical points here before we continue and delve further into this chapter.

Our New Global Digital Commitment

The universe is vast and keeps growing as our telescopes become better. Today it stands at 13.5 billion light-years across and continues to grow as our sights improve. On December 22, 2021, we launched the James Webb telescope out into space. My prediction is that by 2030 the universe might grow to 24.5 billion light years. The simple reason is that when our eyesight becomes more powerful and we can see 3 to 4 times further, thus the universe will grow larger too. While the size of the universe might grow, the human mind may remain very small as we continue to spend more and more of our priceless intelligence and assets fighting each other by building ever more powerful bombs and missiles, so our elected politicians can shout, "Mine is bigger than yours."

Every living thing that has ever lived died, loved, traveled, or we will have ever known from the beginning of time has lived only on this planet. Everything we will ever know in the foreseeable future will also be only on this planet. Mars is not a planet of life; it will be a planet of survival should humans blow up their cradle due to misguided arrogance and artificial boundaries.

It is critical that we all realize that Earth is the only oasis of life in this infinitely huge universe that endlessly surrounds us.

When we look up toward the sky, we can see galaxies from many billions of light years away. In all this vastness this is the only planet we know with life. Don't ever forget that. We need to collectively make it the mission of our life to save and protect this living planet and every life form on it. Every living thing demands 'Dignity of Life' from their point of view.

Switzerland, by the way, is the only country where 'Dignity of Life for animals' is in their constitution – we need to take this message into our cities, states, and nations.

We believe ourselves as the most intelligent form of life on this pyramid of living things, so let's join hands to democratically protect this pyramid. From perceiving ourselves as the CIOs at the top of the pyramid let's flip it over and make every living thing the CIO and convert us the humans as the keepers, and conductors, of this orchestra called life.

It has been physically, and biologically, proven that if we take out the bottom stones of any pyramid the top will collapse by its weight. Let that not become the future of our intelligence and the human race.

I used to teach strategy in business executive classes, and here is the standard definition of strategy from back then. A strategy defines long-term plans of survival and growth over a longer period. Some Countries have a strategic goal of being the biggest holders of atomic weapons, while others of being the greenest nation on the planet. Some make the most destructive hypersonic weapons and subjugate lesser nations that do not tow their line, even if it takes collective and global brainwashing to achieve these objectives. On the other hand, other countries may have a strategy of "Non-violence" as their core strategy or be the only negative carbon sink globally. Each one of these strategies is a long-term goal, that drives all tactical, mid-term, and strategic decisions, allocation of funds, and global dominance desires.

At an enterprise level, a **Type-1** company will have a strategy to primarily increase shareholder value and will thus do everything to make more and more money with total disregard for the overall health of their customers or the environment. For these companies, profit supersedes all other decisions — think large brand, multinational, growth-focused enterprise companies that the millennials are hesitating to give their cash. While another, Type-2 company may have a strategy to improve health by providing healthier alternatives and creating a healthier customer. For these companies, the purity and quality of their products supersede their desire for profits — think smaller, non-profit, specialized providers that the millennials are now flocking to. These companies often also profit share with their local environments too. A third, Type-3 company could have a strategy-focused PEH[vi] enhancement. Such a company will openly declare they do not work based on a focus only on profits, do not use lower cost and harmful components, do not utilize child labor in developing nations, and cater to a customer who has a resonating PEH focus of their own – think ultra green companies that focus on developing products and services that benefit their PEH focus. Companies developing fusion reactors, low-cost hydrogen engines, better batteries for electric vehicles with higher recycling capabilities, etc. More and more millennials are buying products and services from the last two types of companies. While more and more baby boomers work for the first type of company.

For the baby boomers, their life focus has been on having a bigger house, expensive sofas, more expensive holidays/dinners than their neighbors, serving 60-year-aged rare steaks, drinking 18-year-old single malt scotch, serving expensive wine and votes, new clothes for every party, buying assets across the planet a passion for survival, and holidays in gilded locations are most important. For the new millennials, a house does not have to be bigger or better, seating has to be comfortable, holidays should be where their money helps the environment, food habits are trending toward being vegetarian or vegan that assures dignity to animals as well as humans, expensive alcohol is a waste of valuable assets — may drink occasional wine, a holiday for hiking/adventure with eco-tourism is important, wear comfortable clothes till they wear out, donate to PEH causes as the door to happiness while funding the education of the intelligent and the needy children of our world.

The type 1 companies work something like this — we want to be the number one oil company in five or ten years, and that is how we will measure and award our 'C' level executives with their bonuses annually. Their annual goals are benchmarks that the management and the market will use to measure the company annually. The more profits their management brings the higher they rise, and the more they get paid.

Three current operational strategy drivers that should be familiar to most could be:

1. **Time-Based Bonus for Profit**: where the strategy of any product, company, or country is limited to the tenure of the CIO, while their bonus package is based on short-term profits on an annual profit profile. It's the command-&-control model of the past. When leadership changes in such a company or nation, we find that it often includes changes to teams alone with strategies that the new leadership brings. In any short-term profit-rewarding style of management, each leader thinks only of their gains within their tenure. They often have little interest in what happened before or what will happen after they have finished their time. Fundamentally such a style of management or constitution may be wrong, because, just like a nation, where it allows each newly selected leader to appoint new judges, ambassadors, and the national tier-two federal appointees like they do in the US. This creates a glorious, undemocratic mess. Such a constitution will mostly encourage the

yes-people, grafts, bribes, and temporary for-profit appointees. It will discourage any dissonance thinking to current leadership ideas even if they are more beneficial or better. This was made very visible during the last US presidential tenure from 2017 to 2020. All because of an unhealthy allegiance between a political philosophy and a profiteering oil & gas giant multinational corporation, this continued while the average US citizens cared less about global environmental impact and more about making more profits and holding on to jobs in legacy sectors. Since the 1960s one party in the US has connived, coerced, and done everything they could to shut down global climate awareness initiatives with a well-funded vehemence. They finally succeeded in 2017 to pull the US out of the Paris accord for a brief period. In all such cases, as in any democracy, it is we, as digital netizens, who must become individuals of change that we expect our leaders to represent. In every such situation, if we can't stop our leaders from delivering our PEH strategy, we must then strive to vote them out in the next election and then prevent them from continuing to take our nation in any direction that is against the wishes of 'we the people' If they win then it is fair too because, in a good democracy, we get what we vote for.

2. **Concept of Strategic Horizon:** From Bush to Obama, on to Trump to Biden US politics in the recent past has been a polarizing piece of communications, with some a little while with others a lot more. The question we're discussing here is *"How long should a strategy of a person, a company, or a nation be?"* One baseline I hope we can agree on is that the strategy timeline for an individual, will be different from that of a company, and still more different from that of a nation. Each strategy is about the competitive positioning of individuals, companies, and nations. My recommended baseline on the duration of a strategy depends on the lifecycle of a new product or change.

For an **individual,** it needs to review the time they take in their cradle-to-grave life decisions. So as a student, their strategy would be annual, i.e., how well they do in each class before they review the next plan. When planning for a life people need to decide **if** they want to lead their life on a personal PEH scorecard, i.e., take decisions that are driven for the good of the planet, environment, and humanity, philosophy, or work only for profit. When taking up a job they need to decide **if** they plan to work from hire to retire in one company **or** plan frequent exits, where they plan to move from company to company as a work plan. The same applies for a marriage. Personal strategies will change accordingly; however, your digital tactics need to change and adapt on a foundation of basing all your decisions on verifiable truth or party loyalties.

For a **company,** the basic foundation needs to start with defining their PEH vs. only-for-profit baseline. I have spent over a decade with pharma and biotech companies where when the CIOs were asked '*Do you take your strategic decisions for your stockholders of for your patients*' one never got a clear answer even if the question was asked multiple times in the same interview. Companies have a tremendous responsibility, where the most critical obligation has to be to leverage digital competence. It is important to realize that today, in most companies, employees carry far superior digital e-commerce applications in their pockets than most companies can offer to their employees or customers. Since 2013 the writings are clear on the walls "*Digitize or Get Disrupted*," and only companies that can harness the efficiencies of digitization will survive the game of global competitive positioning. Even a non-profit coffee store needs to have modern digital e-commerce applications and free Wi-Fi to attract modern millennials into their shops.

The competitive positioning of nations is in the most dismal state of all. Global leadership and competitive positioning of nations are critical to US strategic success and the US has been investing in its strategy since the second world war. However, this strategic roadmap of 75 years was ripped apart and kicked down the street in three short years by our most disruptive president in recent times. As of January 9th, 2023, it seems that the Republican-led, US Jan 6th insurrection may have become an international roadmap and a benchmark for right-wing election denial and anti-democracy strategy. Brazil is on fire – so who next? From an enterprise point of view governments need to protect their digital assets, build new digital infrastructure, place rules and regulations that punish bad players and most of all become the guardians of the digital domains. Let's compare how strategies work in an enterprise environment. I have assisted many C level strategic forecasts at a few European and US multinationals. So, when planning an enterprise strategic horizon, the classic approach is a three or five-year strategy. Primarily this is driven by the fact that is how long an average CEO lasts. Some CEOs last as long as twenty to thirty years, like John Chambers at Cisco, and that is great, but an exception, while other companies look like revolving doors for example HP during the same period. Another unfortunate dilemma is when some leaders feel that they must set their unique mark even if it means replacing very successful and popular initiatives just because the ideas carry the name of their predecessors. Often it can mean replacing entire teams of incumbent management if for no other reason than to mark their territory. All are undertaken with a singular mindset of "*Well, I'm the leader now!*" attitude.

During my mentoring life, my strategy recommendation has been — that the most effective, and professional, strategies are built on the 3x rule. This means three times the cradle-to-grave lifecycle of the product or industry you belong to. For example, a diamond mining company should have a strategic horizon of one to two hundred years, while a semiconductor manufacturer in California could have one for only a year and a half based on this 3x principle. This is because the cradle-to-grave duration of a diamond mine is approximately 75 years, i.e., from the year a diamond mine is discovered to the date when mining is no longer economical. On the other hand, a new semiconductor product's lifecycle could be as short as six months, while its cradle-to-grave cycle is 12 months. So, find out your cradle-to-grave estimates and then multiply that by 2 and that is your strategic planning horizon.

However, when taken from an individual point of view, there are endless opportunities to improve almost every existing business pain. Whether it is to launch a new startup or to start a new White supremacist group in the US to attack the US government and federal buildings — the opportunities, the connections, and the possibilities are endless. This is because each one of us today has a digital voice that can be used to whisper or shout to a global audience.

Airbnb was born in San Francisco when on a busy weekend two friends could not find any hotel room, or a roof to sleep the night. The company was launched to fill this business gap by aiming to provide just an air mattress along with a breakfast – thus a room for the night. Viola Airbnb is born. We have also heard about the birthing of Facebook, as it was turned into a movie for all to view. The movie 'Social Network' allows the audience to learn about the mechanisms of the mind of Mark Zukerberg and the simplicity and complexities of an idea being born out of a passionate belief to change something that is perceptually missing.

The coding option: The other option is to upgrade your skills in software programming and become a stellar programmer. Programming today is becoming more like it was knowing the alphabet in the 1970s. Without knowing how to read or write you could still find work, but the good ones were above your threshold. Similarly in the world today you can still grow at a phenomenal pace, but your digital growth will be dependent on someone else's programming skills. Also, to manage a team of coders it becomes essential to know a little bit of coding so that the wool they pull over your eyes is not cotton. Any digital explosion can only come about through digital exploration – so you will need to partner with or hire some resources who are adept at coding. However, it is important to remind all readers that a disruptive idea is still the most important commodity out there right now.

Digital Disruption: For any enterprise, the first mantra to success is to remember the phrase of the 2011–12 competitive reality of *Disrupt or be Disrupted*. If you don't believe this or think your industry is not prone to digital disruption, then let's talk in five years, or write me an email today and let us review your thoughts, or you can write one to me five years from today.

For the rest of us, my recommendation is to 'Become the Digital Disrupter' before someone disrupts, launches your idea, or dismantles your company. The opportunities are only as wide or narrow as your mind and your mindset.

Who could have predicted in the mid-1990s that the new e-commerce bookseller would in 20 years become the reason to bankrupt major brick-and-mortar brands, retail outlets, and companies, and change the very fundamentals of global e-commerce. Who could have predicted Alibaba in China and Snap Cart in India would choose this very idea and become e-commerce giants in their respective countries? Who could have predicted that an ex-US president would challenge the US electoral process and the democratically elected president with a charge of 'stolen election' and that 70% of 49% of the population would fight for him? Could we have predicted, in 2019, that on the orders of a defeated ex-president, thousands of Republicans would attack the US capitol, shout *Hang the Vice President*, and break into the offices of the Senators and vandal the halls of the Capitol? Or that the whole republican party would then go and support the insurrection as a rightful walk in the park by 'good people' on both sides.

Then by 2023 who could have predicted that the US Jan 6th election denial would become a global benchmark for right-wing, election deniers in Brazil and that Florida would give asylum to their ex-president, and possibly the person behind a similar attack on Brasilia?

The question is, who or what is next? Similar to our quest of who is the next unicorn, dragon, or Mama dragon. Companies and enterprises need a stable government, a foundation of peaceful citizens, and a digital environment that is protected from trolls. The US and Brazil have demonstrated that a lot of the fake digital posts are quite often internally motivated, even if the trolls are paid abroad to conduct the polarization tasks. So, the question we have to ask ourselves, and the protocols we must demand to be put into place are going to be determined by our expectations for the future. What do we expect next from the political appointees, how do we reward or punish them when they go rogue, and the democratic reward system that often results in bad decisions? We can see the early Jan 6th repercussions already happening all around the globe even as I write and you read this.

The Millennium Dilemma

Pre-covid someone in the UK had recently stated of seeing hundreds of extremely smart millennials, coming from very smart universities, with still smarter skills and scores, get into perfect jobs but then twelve to twenty-four months on, simply burn out. Their prime complaint was that they did not *see it going anywhere.* One rationale given for this is the inherent modern anticipation of *Instant Gratification.* The other is that they do not, any longer see themselves becoming a part of a global monolith that simply keeps getting bigger and more efficient in gobbling smaller local players across the planet.

Let me try to break down the 1st rationalization and see if this makes sense. All prior generations, mine included, used to wait for months, if not years, to get some of the things we wanted. If we met a girl it took months or years for it to lead to a relationship. If I needed to buy a car or a motorcycle we planned for a similar period. If we wanted to buy something we waited to drive to the shop, or mall, and pick it up. Then if we did not like what we had bought it took us a similar time to go back to the point-of-sale to return the item. Many retail points refused to accept goods after they were bought. Today, digitization had made most of these actions almost real-time, or instantaneous. Dating has become a swipe-right or left marvel. Buying something has become a delivery within-a-day miracle. Returning items has become not only convenient but includes free shipping and no-questions-asked service. We pack it up and it gets picked up or we can drop it to some very close-by drop-off point. If we plan to buy a motorcycle or car the providers will drop the car on your front porch and even pick up your old car if that is part of the deal.

It's all about the customer experience. Overall convenience, opportunities, and one-click options. The key word hereinafter is going to be experienced along with the words that are sticky, trustworthy, and risk-free.

Today customers expect everything instantly. We prefer dating apps because we think we don't have time to build a relationship. Our restaurant food flavors are so outstanding that home food no longer tastes as good, also we need different types of foods as the same food gets kind of boring. In addition, we don't have the time to prepare food from scratch or wait for it. We want our orders to be delivered immediately, even if the ingredients are coming from the other side of the world. We don't send letters anymore to reach out to people but communicate by social networks. It is all a click, swipe, and the world of instant gratification, thanks to our smartphones and social applications. As our society becomes more affluent, our children start spending more and more time on their digital devices, they slowly get used to fulfilling their needs instantly. A common example is Amazon Prime, where even waiting for one to two days has been made into a frustrating shipping inefficiency, frustration optics for new customers. There is only one word for this *Instant Gratification.* Our post Covid world including the millennials are now getting very used to this instant gratification and expect this kind of instant response from every aspect of their relationships, work, and the overall environment.

It isn't just about gifts or deliveries. It is the same way for everything else too. The dating world has been hit with this instant result world too. Prior generations used to spend weeks, months, if not years, finding the right partner, girl or boy, and then building a relationship of trust before they went out for their first date. Most of these connections were in the general neighborhood. Today it is across the city and sometimes across the planet.

The duration for identifying, finding, meeting, and spending time with a prospective partner was dependent on cultures and religions. On one side would be the US dating example where a couple could go swipe, meet and date within a single day. On the other side would be a couple in Pakistan, Iran, or any country in between — where couples even today can be stoned or put to death, quite often by their family members, for being seen, or talking to a man, who is not their legal guardian, father or brothers. Cousins are not included.

In most developed, and digitized countries, you swipe right and then hope on a reciprocal right-swipe from your prospect, if you want a date. Our mobile applications provide simplified presto-gratification for what has been considered as complex, or impossible in some societies even today, for a simple startup relationship like a date.

All this has resulted in instant expectations, even for very complex social behaviors the way they have been for thousands of years. So, our highly intelligent millennial generation is now starting to get bored, not due to a lack of opportunities but a perceived deficiency in the speed of meeting their daily expectations. On the work front, this translates to their getting to the top. Despite being smarter, healthier, and better read than their ancestors by many degrees, this generation seems to be showing signs of losing their collective capacity for patience.

Today reaching the top of a trillion-dollar company is today achievable in a single lifetime. Often before the CIO is even 40 years old. Our smart millennials want to become billionaires before they turn 25. Unfortunately, the frustration of not meeting that expectation, within their mental timeframe, is making some of them suicidal.

When we, the baby boomers, went to work right out of university, we could visualize that we were standing at the bottom of the mountain with the CEO at the top. We saw this professional mountain, we planned for the climb, realized that it got steeper as one gets closer to the top, and then we trudged and climbed that mountain. A few made it to the top but most climbed the mountain anyhow, as it was part of the journey we call work, some gave up early or changed jobs, others rose up to a certain level and a few made it to the very top.

The problem with some of the new millennials interviewed, the ones that left their jobs in their early 20s, is that it seems that they saw the ground they stood on, and they saw the top, but, in their minds, they could not see the mountain they had to climb. The mountain was replaced by examples of people, the likes of Jeff Bezos, Steve Jobs, Mark Zuckerberg, Larry Page, and Sergey Brin. Just like a lot of Americans wanted to become the Bernie Madoffs of the financial world, we have a new generation of young adults who want their own unicorn companies as the only option for success. Not only that, they want these results instantly, and they do not like waiting for them. A lot of them don't want to be working for someone else so those people can become wealthy. They are like small gold nuggets drifting in the winds of anticipated digital gratifications; right now, they are giving up too early and too fast. Instant success has created this so-called 'digitally lost generation.'

On the bright side, some of them will launch their own new digital companies because their ideas are very social, and their team's digital foundations are extremely strong. However, a lot of them wander into the digital sandstorm, get blinded by the digital dust, and often get lost in the sandstorm. So as a reader, beware of this type of mindset and choose your paths carefully. Try not to get lost in the current sandstorm. Or to get led by someone with their external agendas, and a 'zero' interest in your goals and expectations.

It is important to realize here that while the readers of this book may belong to the privileged few, i.e., the new digitally aware generation that grew up with adequate assets, use smart devices, live with computers, studied Computer Science, and belong to the generation that kind of lives inside these digital spaces, and use their digital assets on an hourly basis. These are the digital netizens of tomorrow.

At the same time, the digital shock will be felt most acutely by a large majority of this planet. These are people whose first contact with 'smart' is a phone in their hands and constant communications on their social network application. Suddenly they are the center of the universe and have not just two or three friends, but hundreds from across the nation. People who want to connect with them. The euphoria of this is unimaginable to many.

Then others are, for whatever reason, unwilling to work hard and wait for things to turn around. This could include the un-digitized majority that will fight these changes. This could include people who from the very start were unwilling to work because they have a family, or a partner, who could support them live life. They now have a globe full of people who think like them and together they sit in their echo chambers, filling their void with all kinds of 'Us-vs-Them' examples which the bots read and flood their beliefs at an individual level of discord. Unfortunately, this group of people exploded in the US and other countries, when governments all across paid people during the covid-19 tragedy. People got used to getting checks in their mail for sitting at home, some more than what they used to earn working before the pandemic. They saw an opportunity to use their votes for free doles, and politicians found an audience that would vote for them so long as they gave them free bees. But this is a terrifying trap no matter how you look at the scenario. On one side, politicians will keep handing out more free money for votes, which will eventually break the bank, while on the other, each prospective competitor will offer more free money than the last. Or other politicians might decide to give the very-wealthy exceptional financial tax breaks so that they can get paid for these favors while taxing all the rest of the citizens.

So, who are the victims in this digital game? It is We the People with a singular benefit and a double whammy. This singular benefit is that each human on the planet has an opportunity to better their life by leveraging the opportunities of digital disruption. On the whammy side, the trolls and their payers will develop bots to disrupt your thinking in polarizing ways, so you can vote for them and then some of them will give all the thanks for the ultra-rich so they don't have to pay any taxes or IRS does not have enough resources to audit them promptly like they do you and me. The funds will be used to encapsulate their tribal followers with the media-stoked fire of 'Us v/s Them' as their divisionary tactic till they have this small, or large, tribe eating everything from the palms of their hands, even getting themselves killed for a cause as defined by the tribal chiefs.

The people who will vote for our politicians and possibly suffer the most will be the hard-working common citizens,

In a country like the US, where each citizen eats approximately three hundred pounds of meat per annum, this could be a White factory worker from Lincoln, Nebraska, who is now being replaced by lower-cost, harder-working migrant workers, or an automated digital robot that can process ever greater quantity of meat more efficiently. E-Commerce companies like Amazon already do not need stores, tellers, or checkout lines so their human resourcing is quite low. They also don't need large warehouses as most of their goods are shipped directly by the manufacturers. Now with drone delivery, they can also do away with drivers and become a human-free enterprise that will be extremely difficult to compete with. But here again, is the true dilemma, even Jeff Bezos agrees that Amazon too will pass as other companies because someone out there, maybe you, will come up with a better idea on fulfilling customer needs.

While the workers on the other hand already see their way of life and that of their children fading fast. In this state of uncertainty, they are willing to grab any straw that comes their way and this is where the algorithms come in to harvest their minds. These workers are willing to support any politician that supports the unrealistic termination of all immigrants, as that is proposed to them as 'Us vs. Them' and an intrusion into their peaceful and hard-working way of life. However, they rarely ever realize that in a world of digitization, many of their shared jobs could become automated and digitized shortly nevertheless. But that is not what the politicians will ever tell them.

In a rapidly getting-digitally-disruptive world change is happening far too quickly, I recommend that all parents train their children not to try to become copies of their parents, but to learn to adapt to the digital changes and use it for their benefit. This is because our current world will be disrupted via digitization, Machine Learning, Artificial Intelligence by devices supported by IoT, the internet of things, and now by AI driven intelligence like ChatGPT. Train your children for the road ahead via NOT looking into the rear-view mirror, but by predicting and learning for the future. What has got you here will not get them there.

Take Aways

Individual: Make PEH the foundation of every decision you undertake for yourself or your company. [1] Be very careful what you think. Also, be careful what company you put your intelligence and resources into. This includes every person, social post, TV opinion, political tribal feeds, and external feed that influences your mind. [2] Whatever you think your brain will make anything true for you, think failure, and that is what you'll be designing toward; think success, and that is where you will go. [3] Don't fall into the gaps created by social networks. Trolls, politicians, and your mind. Trolls can be foreign, national, or even internal within your country or even your company, i.e., someone working to make you depart. The gap is between your unhindered potential vs. where you have convinced yourself you can reach.

Companies: Become the person that turns your company into a digital disrupter. Use internal and external data optimally to get real-time visuals for your global operations. [1] Don't go and become the very person our parents warned us to beware of. [2] Dedicate your professional life to think, plan, and dream PEH — planet first, environment second, and humanity third, as the foundation for every decision. [3] Become the individual who is independent of focusing only on profiteering and biased thinking. Somehow or the other, filter the noise and realize the core truth in each decision. [4] build a strategy that goes well beyond your tenure by visualizing the long-term goal or your company's moonshot — impossible long-term goals that can only be realized with passion and focus.

Nations: *Become the nation that is a lighthouse, riding on the waves of the digital ocean. Start with protecting your nation, then states, then your companies, and then roll down the lessons learned to protect your citizens. Or build a digital umbrella that protects all three with the same rules and controls. The US used to be such a nation, but our recent race toward intolerant arrogance has made us the template for right-wing extremism. Since WW-II, where America went, the world followed. Now, we're learning that the ills of American political decisions are also being cloned. We are today faced with a PR drive that is still legitimizing the January 6th coup as a touristy walk in the park. Let's hope we can right our wobbly ship before we strike bigger leaks and only focus on internal terrorism. [1] Align your companies, political leadership, and national agendas not by party politics but on PEH goals for success. [2] If we lose the planet, then the environment and humanity are of little use; think testosterone-instigated nuclear warfare. If we lose the environment, then humanity is itself at risk — think global warming, nuclear upmanship, or even biological weapons. If we lose humanity, then we lose everything we have gained as an intelligent species. [3] Intelligence is of little use if all we use it for is to destroy fellow humans and take our planet to the prink of the 4th largest extinction recorded.*

The 'P' factor: *The enterprise has the largest impact on the planet as it can influence the way the individual thinks and the way the nation becomes successful. The March 9th, 2023 fiasco of the Silicon Valley Bank is a glaring reminder how a $2 hundred billion bank can have a $650 billion impact in 3 days. However, it also shows how a great government can fix such large issues by proactively thinking over the weekend and provide a solution by Sunday morning that keeps the greatest engine of the US, the startup entrepreneurs, safe from bad banking or management decisions. The positive factors are that when bad things happen the nation should not only be able to rectify it rapidly, but also put into place rules and regulations that prevent such a crash in the future. It has been proven more than enough time that humanity learns only from mistakes and the enterprise and nation has to be saved from unplanned errors and that is the only way we can evolve to a bright future.*

Chapter 9 — The New Digital Nation

Fifteen-year-old Catherine Cook and her brother — eleven years her senior — were looking at a yearbook and thought it would be a good idea to build a social media website built around an online version of a school's yearbook.[13] Looking to her brother, who had successfully launched startups EssayEdge.com and ResumeEdge.com, for inspiration, she founded MyYearbook. Later, MyYearbook merged with an ad-supported site that allows users to post and complete online quizzes. By 2006, the site raised $4.1 million in venture capital funding. The site quickly gained fame, attracting millions of visitors, gaining millions of members, and securing advertisers like Disney and ABC.

Just because it has been done in the US does not mean there is no opportunity in your country. You can start a national yearbook in your country and I'm quite certain with a little security and high social participation any of the readers can launch a successful MyYearbook type of a site where verified students can put their pictures for all their colleagues to share. It can become a school and university social networking platform to see where all your colleagues are today. A new alternative to a Facebook finder by school, college and university as the foundation.

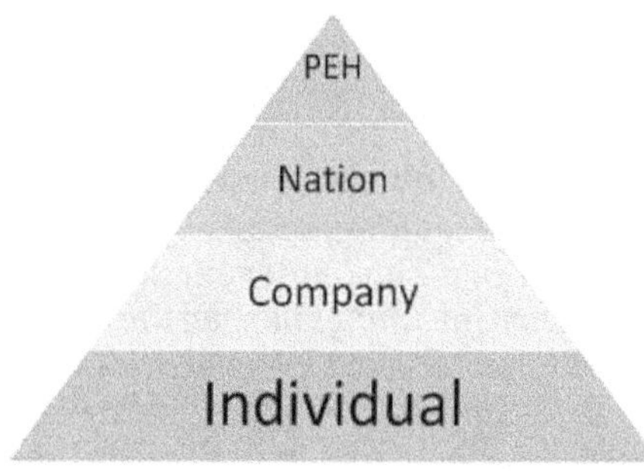

On reviewing our three highly simplified central focus for this book, the atomic rung is the individual citizen at the bottom f the pyramid and most numerous. In the Middle the enterprise or the company you work in, and at the top is the nation with its politicians and political leaders. Above all that is the PEH, which in reverse order is Humanity, Environment and the Planet. Most of us, i.e., you and I probably fit in the lowest and the middle rung of the enterprises, corporations, and companies. The enterprise interfaces with things we produce, service, and consumers.

Let's start at the bottom of the pyramid. This represents the atom in our discussions, i.e., the human who is responsible for everything above.

At the very top of this pyramid is the PEH philosophy that we need to adopt at every layer in this pyramid.

On top of this is the company that is like a molecule that has a lot of humans that make critical decisions, and which represents the commercial success of nations. These companies are run by executives that take decisions on the path and structure of the company's operations. Some work for the good of their customers while other for the good of their stakeholders, still others that make their profits by being the disruptive forces in marketing and PR.

On top of this layer stands the Nation that has many individuals, and fewer companies each of which makes the nation successful. This includes country boundaries, politics, type of government. Additionally, it also includes the various factions of a nation. Factions in the US include the courts, the defense, national security, politics, the Senate, Congress, and the house, and so on. Most importantly it includes the politicians.

Let's measure where these three parts and review how they stack in terms of PEH based digital transformation, as a planetary strategy.

While our current generation is the smartest, the most knowledgeable, and advanced human beings that have lived on this planet. We are currently also the most confused humans since the beginning of time. This is based on a systematic corrosion of the American mindset, as stated here, *"We'll know our disinformation program is complete when everything the American public believes is false."* — William J. Casey, CIA Director (1981). This was stated during the tenure of President Ronald Regan who belonged to the republican party. On Jan 23rd, 2023 I participated, as a silent approved listener on a social networking group discussing PsyOps, or Psychological Operations, with the flat earth proponents. The only takeaway was that people do not trust anything the government, their teachers, their institutions, their courts, their CDC, or their presidents say. They just believe their tribal feeds. So, the world of Bill Casey has come to fruition as of January 2023. We are not the future that people imagined and once dreamt of. We are the ones who have systematically torn the very fabric of democracy and our society and replaced it with dogmatic polarized tribal narratives with a 'My way or no way' attitude of arguments. It is our fellow tribal leaders that have made all this possible. How did we do this?

Our past generations recorded all that they learned and shared it in the form of books and recordings. Everybody that read these books categorized them into fiction or fact. Facts were based on science and repetitive proof, while fiction was based on imagination. The demarcation from ages was clear. This was carried forward generation after generation with each subsequent generation starting from a higher knowledge base by the time they are sixteen or 30. An average sixteen- or thirty-year-old is smarter today that all the similarly aged persons since the beginning of current recorded history. We believe that we collectively know more than every prior generation. I live in California, work in the semiconductor industry and have been working on technology for the last two decades. So, from my point-of-view I have access to more capabilities and experience in digital transformation, than many of my human colleagues. At the same time, I know far less that many of the people who are the spearheads in the digital cogs and machines, i.e., the ones who write all the programs and release the leading applications. Since the last two decades I have worked from home by choice. However, from March 2020, due to the Covid-19 lockdown, I have been working from home by order. We networked online, we bought on line and the world of remote work was finally established in the 18 months since then.

We, the people, are at the tip of the spear of digital transformation. So, while all of this is great, it will be ignorant to deny that we are also heavily swayed and influenced by the digital trolls who vie for our eyeballs and are paid to alter our minds. To be perfectly honest, while we are spearheading the digital revolution and are its architects, we are also the victims of this digital shock. When you are at the forefront of any new endeavor, discovery, or operational battle, you are most exposed to the changes that may come therein, just like the frontline infantry that has to face all enemy soldiers head-on. Similarly, at this moment, we are also at the forefront of all the digital changes and disruptions happening all around us. This means that while a very few of us are the creators, a majority of the smart device users are the inevitable guinea pigs, or lab rats, for any experiments or changes that need to be done in the digital future. We are the individuals the trolls, the algorithms target first and foremost based on the lessons learned from the front-line digital audience that the algorithms have been interfacing with. Once the basic logic is established, they can be enhanced, improved and optimized for main street deployment.

At every point of time the main street humans becomes the guinea pigs and so on and so forth. The process self improves by its sheer volume of audience, which produces individualized data, which is used to enhance the algorithms and thus their interactions with the global digital audience at individual levels.

We are all living the "Smart Life." We have smartphones, smart TVs, smart cars, and smart everything else. More and more things today are being loaded with smart sensors that produce constant streams of data. The big question is whether it is smart for us now that we have free WhatsApp, Facebook, and Google, or is it smart for the companies that provide all this free stuff so that they and their customers can harness all the digital dust we create, that they can use. Ironically, these free things have somehow enabled their founders to become extremely wealthy. In fact, these companies, that provide us all this free stuff, are today some of the richest companies on the planet. So, the question here is, how can an individual or a small group get anything for free forever and make a lot of money for the providers?

The lesson here, for startups and designers, is that the future is in designing for Free things for the user. The future is in free. The wealth is in the digital dust all this free-ware will produce.

This is totally opposite to the industrial-era-logic that most of us have grown up with. The industrial companies worked for higher and higher prices so they could make more and more profits. This foundation was disrupted by the current digital revolution for the winners and the digital shock for the data generators, *We the People*. The digital future is all about providing free applications so we give them more information, services, or operational delight, while the customers give us more and more of their priceless data. This is hopefully the interim phase where we give free baubles and our data is slave to the collectors of that data.

The reason most of these things are free is linked to the concept of smart things. At the basic level all 'smart' things produce data. It is termed smart only if it produces data. We call all this data that it produces as our digital dust.

Most of these applications, and smart things are free because each of these applications generates volumes of digital dust all about us. Collectively this is a digital storm that is like an everlasting gold mine created by each human, each device, each smart car and each data producing application like Gmail, WhatsApp and the rest. When you walk around with a smartphone, you leave digital-traces of where you have been due to a GPS tracker that you authorize for more than 1 application. When you meet a friend, the algorithms can track each GPS device and conclude with certainty who you met, where and for how long. When you meet an online date, or a clandestine lover the applications can track who you were with, where, for how long and sometimes what all sounds you made when there. This is because most of our smartphones tend to eavesdrop as well. One way to check this is to talk to a friend and say that you are planning to go for a holiday to Thailand, then see if you get any commercial offers for a trip to Thailand thereafter. Then send a txt to a friend on the same topic and see what happens. A lot of applications that we put on our smartphone listen to what we are saying, even when that application is off, so yes, they do eavesdrop.

How is all this related? Every time you get on to your WhatsApp or social network, we create and hand over our thoughts, our plans and happenings, our likes and dislikes, our happy moods and our trouble, all of which is digital data being captured somewhere. This digital dust about our minds and ourselves can be shared to an infinite number of buyers. In a republican state for example, if you land in a state of an unplanned pregnancy or desire for an abortion, I highly recommend keep all communications out of your Facebook or any social applications. Do not search for abortions because the state, the church and the bounty hunters will be then looking for you very soon. They will hound you with religious, state and bounty laws.

This digital data is today compared to oil, gold, diamonds and such. But it is worth far more than that. Oil, gold or diamonds can be sold only once then it belongs to the buyer. Data can be sold a million times without any degradation in content or quality; therein lies the perpetual motion that humanity has been hunting for since the beginning of time. Not only are we giving it all away for free, we actually pay many of these providers for their applications and services. For example, our smartphones, cable networks, Fox News, CNN, etc.

They in turn, collect and sell this data about you and me to the largest bidders, again, and again, and yet again.

Think of why the Russian trolls and Cambridge Analytica went to Facebook? Why would they target a free application? The simple answer is that they knew it contained a whole bunch of very personal data that they could mine. Officially they were given 30,000,000 (thirty million) accounts, i.e., access to thirty million individuals and all their conversations. However, what they probably got along with that is their friends and contacts. Let us assume each individual had 10 friends, then that immediately turns into thirty million x 10 which is 300,000,000 accounts, or 300 million people. This is 94% of the population of the US. In this case we can say with a fair amount of certainty that they probably had access to every US Facebook user that they could then study, target and influence.

You and I stand at the bottom rung. We are the creators of the digital data, the digital dust that is in demand worldwide and is being mined by most applications on each of your smart devices.

However, the more pressing issue here is as to how can 'we the people' use these same digital capabilities to make our voices get heard from the middle rung, in order to make the planet a truly better place. It is interesting that almost every individual we have ever asked wants a better life for themselves and their family, they want peace and no wars and they want fairness on the planet. But at the same time almost every politician, democratic, republican or autocratic wants to control these same people by telling them the world is after them, out to get them, out to conquer them, even their family may be against them, that the barbarians (the other side) are out to get them and war is their only option. This has now become a constant battle between people and their leaders. While we want equality, they want to divide us by as many attributes as they can imagine and then some more.

Now the thing is, we may use all these modern words to express contemporary society's concerns. However, the truth, at heart, is that we are still the same person with the same beliefs and values that we had when we were school-going kids. This is a simple fact. Speak to any psychiatrist. They will tell you that close to 60% of a human's language, technology, and cultural perceptions are set by the time humans are about 12 to 14 years old. When we have these perceptual ideas, we carry them forward like a leopard that rarely changes its spots. They become our biases – cultural, religious, language, national, etc. They define who we are.

Now think of the new millennium generation. The very small children grow watching cartoons on their smart devices from the age of one or two. While their brains should be absorbing the environment and social-interface lessons they are getting stuck more and more into the addictive digital devices. These kids throw tantrums when their digital games are taken from them. By giving them their games back the parents are teaching them instant gratification. As they grow, they get more and more addicted to their screens and the digital devices. By school they know more people online, by the time they are sixteen, they transact digitally around 70% more than any past generation has in a lifetime. The average screen time for teens had jumped from 4 hrs. a day to over 7 hours a day by 2019. This has, and will have, a direct impact on these children as they grow up to become adults that decide the future, become CEO's and presidents. Their neurons and total wiring of the brain is being rewired.

From the point of view of an enterprise, if you look at the Europe and USA, you will find that most of the companies are still living in the 19th-century era (somewhere between the 1970s-1990s). They have yet to be digitized on a large scale by when they can be called digital enterprises. This statement does not include the new for-digital-only enterprises like uber, Amazon, Airbnb, Spotify and such. In this statement we are talking of the pre-digital enterprises that most of the baby boomers are familiar with. Many of them spend big bucks, with the big four, thinking that they are undergoing the process of digitizing when they, in fact, are not. According to Gartner over 70% of these companies will spend a lot of big bucks but fail to meet the expectations of their business users, or their end-customers in all these initiatives. While their opportunity is the highest, they get distracted by their industrial era partners who deliver industrial era functionalities on their digital core applications. The main thing to note here is that while enterprises continue to fail in their digital investments, their secret to success is their individuals who make up the atoms in their organization. They carry more compute power and convenient applications in their pockets than most companies can provide to their customers.

If you look at the nations and politics, i.e., deep dive into three quarter of the nations, governments, and politicians, you will discover that most of these politicians are 60 and 70-year-old leaders. So, they are probably living in a world reality of the 1950s to 1970s. They may speak of modern flashing-lights, PR driven attributes, but their mindset is based on mid-20th century optics and data.

When Trump and Biden were talking about improving the American infrastructure, they were speaking of infrastructure as envisioned by a 75-year-old person. Their vision is to repair bridges and make better roads. While this is important, they are as yet focusing on improving the past. While modern digital leadership could be focusing on improving the future. One way is to learn from the future, the science of Fufactology, as a foundation for planning and execution, as introduced in this book. A digital president would take a moonshot and plan to build digital highways, smart cities, smart bridges, and most of all, declare a new federal-level organization that writes policies that are being discussed in this book as a means to safeguard their citizens from criminal and fake news traps. The security of individuals and their data must become the highest priority; we can easily follow some of the European and even Canadian guidelines, which mandate that their local data not be taken out of their national boundary.

When we look at these three layers, one can establish that enterprises, and politicians are at the middle and the top of a nation's digital transformation. The citizens are to a large extent at the bottom of this pyramid, and on the forefront of the digital transformation, only that most of the people don't know how to leverage, or protect themselves, from its potentials, i.e., just like in law, they are unaware about the power of the digital world to disrupt their lives, thinking and totally unaware of their power and rights, in this domain. We are barely aware of their getting addicted to free Digi-ware as we fall into the well-laid traps of the algorithms and trolls from all over the planet with their own unique agendas. The individual at the front of this revolution, is probably the most exposed due to all this paid effort to capitalize on the minds of the common people by the trolls, their handlers and their payers.

While most nations are still catching up with technology, we have found that there is an explosion of very highly paid companies like Cambridge Analytica. CA is the mama dragon that has laid its eggs, and these eggs are hatching across the planet Uncontrolled, without legal barriers, and with mind bending techniques that most citizens are totally unprepared about, and unaware that it is slowly influencing their innermost thoughts.

While our leaders may be living in a mindset of the mid-19th century, they are leveraging technologies of the 21st century. Using, what we have described in the last few chapters to alter the minds of many of the citizens. We believe that we are at the front end, but still, are as yet incapable of differentiating between what is true vs. what is false, what is good vs. what is bad for us. The world as a whole is becoming binary and addicted to tribal loyalties. These tribal loyalties activate our feelings, and heart as decisions are mostly based on emotional triggers. It is a well-known scientific fact that the heart, belief, and emotions, are more powerful drivers than those of science, facts, numbers, and the logical mind.

Why do we make this mistake? It's because some of the algorithms in use are extremely influential in changing a person's psyche using emotional cues. The word for this is *"subliminal e-persuasion"* or *"in through the outdoor"* methodology. At this moment, there are no international or national controls or any other laws on digital subliminal persuasion. A concept that was banned in the USA, UK, and Australia in 1958 for television advertising primarily in things to do with politics. There is currently no ban on digital e-subliminal persuasion trigger content, podcasts, digital broadcasts, fake news, or algorithms that try to accomplish their victory via subliminal messages at this current moment.

If a national, or foreign, troll starts doing psychiatric manipulation or neuro manipulation on citizens in the US, no system or control mandates that it is illegal and brings these people to justice. There is no law that can hold them guilty for something things like spreading fake news, sending fake videos, or making totally fake allegations. This lack of law in this area is heavily misused by governments and leaders alike.

This is done to a level where it would be logical to assume that the leaders, and our elected representatives, do not want to make it illegal or traceable because the very people with the authority to make it criminal are the ones using it for their gains. If Trump hires Steve Bannon as his chief PR, then can we expect him to pass a law to lock up all Cambridge Analytica leaders, thus closing CA-like operations from which he benefits directly. We can easily assume that they are also benefitting from all of this. It's how the Russians were able to control us via their trolls during our voting seasons. It is how they can hack whoever they want to, whenever they want to. It's how nations can shut down electricity grids, hack into oil pipelines and also create civil mayhem inside nations by activating opposite groups of people at the same time into a situation of conflict. The sad part is right now no one can interfere or stop them from doing so.

Another example of this can be the Chinese leaders. They are tracking Uyghur Muslim communities within China using smart cameras, facial recognition, GPS movements and text recognition in Uyghur dialect. And what are they doing with the information they get? They hack into their cultural spaces, plant fake supporters, get their locations, see who all they are talking to and what. Then lock them into large concentration camps by a geo-positioned finder that is given to their police. In their concentration camps they commit large-scale brain washing, genocide, ethnocide, and even forcibly sterilize their women, or abort their babies, so they cannot have more Uyghur children. The main thing is the there is nothing that can be done to stop them. Even as I write this or you read this, there is absolutely nothing that is able to stop them. China is a UN Veto member, and even the big nations cannot touch them.

Misuse of digital technologies, algorithms, machine learning, and artificial intelligence is in its embryonic stage in most parts of the world. But it is only a matter of time before this becomes globally pervasive as more and more political leaders find power in the misuse of these new digital technologies. Therefore, it is not hard to assume that many political groups and leaders might currently be at various stages of digital control of their citizens and population for the sole purpose of their controller defined checks and balances.

Add to this that the amount of fake news we get is so high right now that people all over the world are losing the one word, we need to protect with our lives for the next two decades – TRUST.

As a group, we tried to do some research. Many people tried to conduct research on how much fake information there is on the internet. They couldn't complete it because of the amount of news generated and the number of polarized views on what is fake and what is not is so pervasive that it is impossible to put a number to it. It changes by application, country, and the politics of nations. The overall rate of fake news far outstripped any ML or AI optional search algorithms. For each fake source they found a hundred more were automatically generated. It is all so extreme that any form of calculation of how much fake news there is becoming quite impossible. The only way to control this is to make it illegal to generate fake news. We start with the realization that this is a very tough nut to crack as there are firstly no digital laws and secondly the very people who need to pass these laws are the ones who are misusing its power to the maximum.

To give you an example, in the US, what do you think is fake news; it is highly recommended that everybody wears a mask and get vaccinated? Or is it that there is no need to wear a mask and this covid is just a fake disease? In the US unfortunately, it varies based on what political side of the narrative you belong to. Is it right to announce that no mask is needed because the blood of Christ will protect you from covid, or that a lady never wore any masks for the same reason she did not wear underclothes – freedom to choose. In 2020 many people believed that COVID-19 was a false narrative given by one of the political parties in connivance with the CDC. Some tell their digital tribal members not to get vaccinated because it is going to have genetic impacts, while others talk about how their cousins' testicles swelled so much that he became impotent. That the whole concept of RNA vaccines is a new technology, and it will destroy your genes as you move forward. So, if you take all of these popularly broadcast sentences, you need to work out which one is fake?

Talk to different sides of the political aisle, and you will get different opinions on the planet, climate, media, abortions, Muslims, White superiority, immigrants, native Indians, and almost everything else on the planet. i.e., I assume we, the US, are the most fractured nation on the planet bar none today. Every question has two different sides, unfortunately both sides firmly believe they are the right side. In a politically hyper polarized nation, every point of view is vehemently right. And it seems to be budging very little because it's not an intelligence divide. It is not a color or religious divide. It has now become a politico-religious divide.

So, what do we do in such circumstances? What do we need as people, as we read this book, what should we do to make our neighbor, city, nation or planet better? I think the answer to that is to provide global, and national protection from digitally induced addictions, fake news and lies.

For example, just two decades ago there used to be an Encyclopedia Britannica. It died almost overnight, when Google started giving answers to every question, on anything, in real-time, it killed and replaced the Britannica encyclopedia. There used to be an era before Google — that is how powerful digital disruption has been. Kids born after 1990 cannot understand a world before Google.

Today, we are living in such a world where even fools can make rules. While their tribal members will put enough digital, or social, traffic to support that fake rule. There was a time when anybody could make a statement, and there was very little possibility of verifying it. Today, you can pick up your phone in your pocket and verify it. However, we are still surrounded by more and more fake data, a lot of it produced by trolling bots to confuse their tribes. You can figure out if it's a ridiculous statement or another fact you never knew. But even that has started to break down, as there is as much validation for fake news on Google as there is for the truth. So, we remain in our echo chambers, that we have created or voted for.

I come from India. Where for a short period serving beef in most hotels and restaurants is close to illegal. While at the same time, India is the second-largest, and then it became the largest exporter of beef. This while the Indian government has put a law into place that prohibits harming any cows. For example. one can't walk into any Indian restaurant, or even a 5-star hotel and order a beef steak, The big question then becomes, what is fake and what is true? Is it fake that India protects cows because it's a holy animal in Hinduism? Or is it true that they want to prohibit the eating of beef in India so as to protect a hidden industry of beef exporters where local consumption just eats into the profits of a very large export business from India. In a democracy another question then bubbles forth – as to why should a Christian or a Muslim not be allowed to eat beef if they want to? Why should one religion define, and determine what other religions can and cannot eat? Why should any majority be allowed to subvert every other minority? Finally, how different is India from Saudi Arabia that prohibits pork? Is it a good way for them to get foreign currency into the country? Or is any of this done to actually protect the cows? So, where exactly does India stand with beef? What is the truth of it all? India says all this meat is not cow meat but buffalo meat.

However, if a truck, owned by a Muslim business person, with buffalo meat gets caught by the so-called beef-protectors they can hold the truck for forensic checks as to whether this is cow or buffalo meat. This can take days, by which time the meat can get spoilt. This is not good for the cows, the buffaloes or the traders. One thing remains undeniable — India is amongst the top three beef exporters on the planet. So, you go figure this one out.

Globally what is needed is total transparency. It can be enabled using the same digital assets. One can check this fact by simply asking Google as to who are the largest three beef exporters? Then follow it through with who own these beef export units? Quite surprisingly, you will find that many Hindus are running the show in beef exports from India to a large degree.

On the matter of total transparency, in the US when the ex-president allocated $2 trillion for COVID relief, did all the $2 trillion go for COVID relief? Is there a chart of accounts, a general ledger, or a P&L account. Do we have a single ledger from where we can trace every cent of these 2 trillion dollars and where the funds went? The answer we get is that we don't know because it is all hidden in political jargon, and probably in political pockets.

Every nation, and state, must declare annual accounts just like every publicly-traded company, so we can see where every cent of this so called $2 trillion spend on Covid from Trump till Biden? There was a time I recall, or it was mentioned, that close to $500 billion was allocated to small scale industries $500 billion is 500,000 million dollars, i.e., 500,000,000 dollars. When asked, where this money was allocated, the illogical and yet straight answer was this is a national security question and that 'we-the-people' are not allowed to get access to any numbers, even though the funds belonged to us *We the People.*

As a citizen, this makes me extremely suspicious, and also very angry. If the government is elected by the people and is supposedly working for the people, i.e., to protect us and our interests, how is it in our interest to not be told where the government is spending our money? This is something I just fail to understand. If you say this is going to defense, and we are buying arms from a certain country that we do not want to expose or export arms to, I will agree. But this is a total internal allocation of public funds supposedly for COVID relief. And the worrying thing is how each subsequent government is treating a trillion dollars. Like it is $100 million. I mean, as Biden came into his presidency, he allocated another $2 trillion. And then, we were talking of $3 trillion, and then another $3 trillion. For US politicians handing out a trillion or two is becoming common practice. No matter how we look at this it cannot have a good ending and the long-term risk is default or bankruptcy. Add to that, it probably fills a lot of political pockets somewhere.

This buying out of voters is just not financially sustainable, and if any of these funds are going into political graft then were all doomed to ruin as each party members become multi billionaires while the nation slowly runs out of money. The other problem is that when people get used to free money, they will do all kinds of unethical things to keep getting fed. We see a lot of unethical things going around nowadays.

So right now, if you add 2 trillion for COVID by Trump, 2 trillion by Biden, and 3 trillion by Biden for infrastructure, that makes it $7 trillion in the last two years. I published an article on how nations go broke, and one of the main reasons included political handouts made by national leaders to get more votes, and graft fed politicians. So, the whole purpose of the US democracy may have shifted from doing good to the nation to doing good to the party's voter bank. Whichever way we look at it, it is a very dangerous situation.

We have reached a point in time where each of us has a digital voice, and we need to use it. As citizens, we must demand transparency from our national leaders; in fact, mandate it. We must take elected officials to the same level of annual transparency as we take company CIO's. I remember the US government had allocated $25 billion to Katrina. I have tried to reach out to political parties, and politicians, and got politely excused from the meeting each time, when I asked where the $25 billion went. I hope that the system is not treating tragedies like Katrina as their get-rich-quick channel for gratification. It would be disgusting if that is true, and the probability is not zero.

It's like a question that goes into a black hole; 25 billion, then 2 trillion, then 2 trillion again, and then 3 trillion, and then a few trillion again. How much more of this should we allow to pass by unnoticed before someone demands total transparency? If this continues as a game of assuring votes from fellow elected officers and from voters by sucking out these trillions then were headed toward guaranteed bankruptcy. The only democratic solution to this problem, and all other 'behind closed doors' political, religious and financial decisions is total transparency. The only long-term way to solve it may be via leveraging very high-speed, secure and reliable digital solutions. Treat US as the US Inc. Then apply the SOX laws on this corporation.

A global, national, and state level Digital control institution must be formed in these times of peace. This organization must be a federal organization, in the US, with similar counterparts across nations. The organization must be created to criminalize the trolls and algorithms created to mine people's data without permission or compensatory remuneration. Not allowed to hack minds or create subliminally persuasive content, the creation and distribution of *Fake News*. Define clear boundaries of what constitutes *free speech*. Such an organization should be able to shed clarity on an event like the January 6th attack on the US Capitol. Without polarized optics by groups of political inflaters forcing some who still believe or speak about believing that the attack was a just cause. Despite the fact that the attacking group was out on a mission to hang the current Vice President.

January 6th is not an incidence where politicians, their parties and their controlled media should be allowed to use their tribal influence, power and capabilities to create polarized views. This can become the starting point of a new era of digitization, algorithms, machine learning, and the power of artificial intelligence used to extract facts, and identify fake news. We all need to build the nation that we aspire for our children and their children's children.

Take Aways:

Individual: [1] The greatest opportunity for individuals is to pick a micro service and plan to digitize it via leveraging the Dec 2022 application like ChatGPT and other such digital assistants. [2] Our greatest threat is our individual hesitation and our inability to undertake the correct bet for our digital path of the future. At a personal level, our greatest danger is getting trapped for hours upon hours in social networks, news media, and digital noise, where we lose the rigidity of structure and rituals to keep us on track. The danger is in becoming a part of an echo chamber or agreeing to the trolls. We need to consciously shut down their addictive optics and start making good decisions for ourselves and our families. [3] Lastly, make sure that every decision you undertake for yourself, your company, and earning your vote is on the basis of the PEH principles that make your nation a better place for all. For example, global warming is a confirmed fact. Now, if your vote goes to someone who gets paid by the big energy companies to deny it, then our votes, our minds, and our democracy are all wasted. There are enough facts around us that prove global warming is real; bigger hurricanes, more severe floods, and snowfall in Pasadena are all proof of global warming knocking on the door. If we don't listen, then it may get too late to blame anyone, too.

Companies: *[1] Become the company that acts for the strategic PEH good. [2] It's time to realize that the new millennials are routing for intelligent enterprises that work for society while dropping the 'for-profit-only' products and services. With the baby boomers will end the brand loyalties as proof of personal success. This trend seems to be irreversible once the market makes up its mind on the collective direction it needs to take. [3] Once the baby boomers are gone, the number of people on the old philosophy will decline. The new generation has to work a little harder to define the future they want to inherit and pass on to future generations.*

Nations: *[1] Leaders must be made to realize that by misusing digital surveillance applications, they are, first and foremost, intruding into their citizens' fundamental rights to privacy. This will directly impact their nation, their party, and their people. [2] Leaders need to speed up building the future, i.e., trusted digital networks, secure digital transactions, and non-pervasive digital environments that are non-biased and do not benefit anyone. [3] Become the first nation, party, or leader to propose and deliver the Digital Rules and regulations initially for internal controls rather than for controlling any digital entry by foreign applications or service providers based on the strike-3 rules with no recourse to fines or penalties. The reason for this is if a company earns 1 trillion dollars, they are more than willing to pay a 10 to 20 billion dollar fine if their pain vs. benefit ratio is too low. [4] Do not become that party that tears the fabric of your nation and its citizens for party gains.*

The 'P' factor: *The nation has the biggest responsibility to go the right thing from the top and let it trickle down. Bad ideas float but great rules have to be set by the leaders and their designated councils from the top. Just like protecting the laws of the nation is the domain of SCOTUS, similarly protecting the companies and individuals from trolls and rogue bots must now also become the domain of some newly formed DCOTUS (Digital Council of the US) a group of non-political players whose sole accountability is to control all digital content inside the nation and everything that comes in and goes out from their nation. We cannot continue to blame Russian, North Koreans or the Chinese trolls for election fudging when we call ourselves the leaders in digital solutions. We need to pursue digital transparency with more vigor than the SCOTUS and republicans did for repealing the 'Roe v. Wade' as if 10 years old carrying her incestuous baby to term was more important that Russians being paid by our politicians to cause disruptions in the US in order to get some political mileage. We the people need to define what is positive factor we need to see in every layer of the onion we call our nation. If US will lead in this then we can expect to see its reverberations across the planet.*

Chapter 10 — The Digital Segmentations

You're never too young to take an idea and plant it in the digital soil: If you're older than 9 years of age, then you already have many people of this age who have succeeded in this new world, so no more excuses. Cameron Johnson got his start at the age of nine, creating very personalized invitations for his parents' holiday gatherings. The guests loved the personalized invitations and wanted Cameron to design some for them, too. Two years later, Johnson had made thousands of dollars selling cards through his company named 'Cheers and Tears.' Then, at the age of 12, he got another brilliant idea. He paid $100 for his sister's 30 Beanie Babies and sold them on eBay for 10 times what he paid. He then purchased the dolls directly from the manufacturer and made a $50,000 profit in less than a year. He used that money to start an Internet business that brought in $3,000 per month in advertising revenue. By the time he was 15, he had formed other businesses with total revenues of $300,000 to $400,000 per month. All you have to do is start.

We have been talking about segmentations and micro-segmentations. In the next 5 years, we'll be talking about nano-segmentations and, in the next decade, about quantum segmentation capabilities. Let's start with my definition of what segmentation is and what digital segmentation can offer today almost out of the box so we can level the playing field for every digital enterprise.

Let's start with the motor car company and product segmentation — take a look at a company manufacturing a car, for example, the Ford Motor Company. As discussed earlier, in 1907, it had 1 product segment, it provided customers with one single car, the T-series, in one single color — black. Any person desiring to own a car had only one option T-Series in black. From a product segmentation, this was as simple as it could get — one product for every customer on the planet.

The Ford T-Series was designed to replace horse carriages. The engine replaced the horse and the first few models the body was a replica of the horse carriage. Now suddenly a family could have a horse carriage without the horse, replaced by an engine. The future of roads suddenly changed from horse poop roads to potentially clean roads. By this year Ford also started the concept of production on an assembly line manufacturing, that changed the global manufacturing process for bulk manufacturing.

For this car the target customer was one man with enough money to buy a car, wanting to replace his horse carriage with a modern engine driven carriage. The car was also a big status symbol at that time. As we move forward, somewhere in the 1920s, other car companies started coming up with three models, and so that became a marketing exception. Now people did not all just have the same identical car as their neighbor but they could choose between 4 models. This option soon became the talk of every good neighborhood. Still a little while later, colors became a thing. Then change became the norm a we drive forward in time.

Fast forward to today, there's an old concept in ERP, enterprise resource planning software's, called Variant configuration. This functionality allowed customer to choose variations in their selections that would be converted into a unique production order. By 2017 they had taken this to the digital core and taken this concept to the next level. This option offered customers to choose the various options as they select when ordering their motor car or motor cycle. By 2017 if you wanted to buy a Harley Davidson motorcycle, you could configure and cherry-pick your motorcycle as discussed earlier.

The last time I checked, there were over 12,000 variant combinations that a customer could design their individual Tesla car, all by themselves. This means that there are 12,000 segmentations of a single product – A future Toyota Camry or a Tesla. Or a Harley Davidson motorcycle. The main thing to remember that in the Harkey Davidson there are 3,756 variations of a single bike that could be required to be manufactured individually on an assembly line. So, we have now suddenly taken a leap from segmentation to hyper segmentation, where each bike on a production line is an entirely different combination from the prior bike or the next. Traditional production plants could not handle this variation, now with digital enabled micro-segmentations this is becoming standard procedure and expectations of most customers. It is also important to remember that Trust and customer satisfaction is important so each customer has to get precisely the exact bike they ordered from those over 12,000 possible variants. This is only possible with digital Valiant Configuration and JIT, just in time, parts and workflow management on the production floor. Which translates that the machines know exactly which bike is being assembled and they send the exact right part to each assembly point 'Just In time' so it is attached to one specific bike only and none other. Now with robotic even this has become a no brainer as the robot can check, verify and attach a specific part in microseconds on a modern production line.

It's the same case with a Tesla or a BMW car, where you can customize different things (colors, engine size, seats, etc.) as you order your car. This is possible because we now have hyper-variant segmentation, assisted by digitized robotic production lines.

Why and How: In 1907, when Ford first came out with the production line concept, they could only move one type of car down a production line. Why? Because at each stage and step of the production line, there was a person who simply had to do their job —

perform one specific task repetitively. Remember the Charlie Chaplin movie on automation, where he had to do this one thing repeatedly, minute after minute, hour after hour, and day after day? It was a lot like that.

Now, imagine a railway track that starts with the chassis. The first worker manually attaches the right front wheel, and the second guy manually attaches the left front wheel. As it moves forward, each additional component is manually added to the car. The requirement for each step is that the previous component had to be fitted in before the next person starts their work. So, there is no plan for any simultaneous work as they did not want two humans on any one workstation. In fact, if a specific part was unavailable for some reason, the whole production line had to stop.

Now, take the modern Harley Davidson example. This is where each bike is unique and custom-made for a different person. Imagine that the production process has been replaced by a robot where each robot only needs a particular part in a specific color. Now what?

The robot will be flexible, and it will have to learn how to adjust itself to possibly 100 possible variants if it's just the shape and the size of the gas tank installation robot. As each tank reaches the robot it will get specific instructions and parts for that specific tank, and each subsequent robot would know that a particular type of tank was installed so as to maintain non-contact distance from the corners of the tank. The tank robot would need different mounting, bolts and handling based on the chassis type assigned to that bike. For example, bike A will be different from bike B. Perhaps; the first bike is a $14,000 bike while the next one might be a $50,000 custom variation. This means the two will have different tires, widths, engines, colors, etc. This is only possible because of predefined machine learning algorithms for each of the 12,000 variations and their possible assembly deviations so that each bike is manufactured perfectly the first time, every time.

As one engineer put it, "It would require a human to remember each part, what its shape, color, size, etc. is, that fits into each assembly in front, while it is impossible for a machine to forget."

As of now, there are no humans that selects the right parts in a 1:1 production line, and there is no machine that cannot. No humans arrange the parts into the manufacturing order, and almost every bike and car come out exactly as it was defined by the customer. And if you get humans to do that, it will be a long and tedious task, full of what we call 'human errors' due to lack of constant focus, human boredom, and sheer inability of delivering 100% attention 100% of the time – something only a machine can do. Not only that, it will be humanly quite impossible to manage 12,000 variations in manual production lines. By now we have production line variations in the millions.

Current customer specific manufacturing is only possible because of the digital capabilities and due to the digital disruption, that this technology brings. Customers now expect that from their manufacturers. So, if you have a manufacturer, no matter how large, still working on an industrial era production line, then they must either disrupt their own products with customizable products or get ready to get disrupted by someone who does.

Let's go back to the late 1990s and early 2000s. This was the era where I was involved in segmenting a telecom companies' customers. Today, this company has over 132 million customers and operates over 20,000 stores across the US. In the early 2008, when we started with this program, they had 32 million customers and just over 3,000 stores. At that time the company tracked their customers by 5 segments. All their customers were segregated by (A) < 18; (B) 18-35; (C) 35-55; (D) 55-75; (E) > 75. That's it. This company had used this segment since 199's and their total marketing, and PR, budget was driven by these 5 segments, i.e., separated by which radio, magazines, TV or billboards, and what to advertise. They continued this for over 20 years. In 2010 they wanted to know how to monetize their business more effectively. After a 4 month 'Design Strategy Session' we identified two areas as low-hanging-fruits. The first was Trade Promotions (TP) or the application to manage their marketing campaigns. The second was their SAP CO-PA analytics. CO-PA stands for Cost and Profitability data from different areas, in our case it was the cost and profit by each store that they owned.

In order to do a systematic analysis, they used to merge the data for their customers with their FI-CO and TP spending across the US, with details by store. In 2010, their current CO-PA and TP data harmonization run took them around 24-36 hours to complete, i.e., it took this long for the data to be consolidated, harmonized, and made ready for critical analytics. Because it took so long and consumed most available system resources, it could only be run once a month, and that too on weekends, that is, the Saturday/Sunday after the month's close. Here is the issue — 40% of the time, the run failed to finish on time because they needed the system resources free by Sunday night for business as usual by 8 am EST Monday morning. This meant that 40% of the time, their decision-makers did not get their required CO-PA + TP data in time to make critical decisions. Sometimes, a failing campaign could go on for a month or two longer just because the dashboard indicating its failure never came till a month or two later when the run was completed successfully. This resulted in marketing funds in the millions of dollars that were wasted due to delays in the corrective course in their TP and marketing allocations. Based on our 'Design Strategy' workshop, our first recommended segmentation was for 101 segments. We recommended a finer segmentation of their customers by age, but this time by each year. Within a matter of 6 months, we found that some parents were buying simple disconnected phones for their three to seven-year-olds, just to keep them entertained with game apps while they were driving, with their caretakers, while working or in the supermarket as a distraction. We could analyze this because we could track when the baby device was switched on and where the parent/s and child were at that exact moment. We also found that very often a 16-year-old may have quite different preferences from an 18-year-old.

Suddenly the company could target advertisements for a very narrow age group by their preferences. These preferences were tested in various markets and the successful ones went out nationally. Using the digital core as a standalone application we demonstrated its power by completing the FI-CO + TP run on 101 segments in under 9 hours in our initial run. This enabled the management to run this load every Saturday and have results available every Monday. This saved the company 23% in marketing funds by eliminating misaligned marketing and promotion plans.

Within 6 months this segmentation increased their penetration into the market by 13%. This was great news for the management as they could now feel the power of 'Know your customer' initiative as we had internally branded it.

So, this company decreased wasteful spend by 23% and increased market share by 13%. An envious situation for any business.

Right after this success, in segmentation, we landed up agreeing to further segment our customers, and we went to having 10,504 segments. We split the 101 years into male and female, then further added 52 states as another layer on the market segment.

Now suddenly we began finding that an 18-year-old South Asian Indian boy had different preferences that an 18-year Indian girl, and they both had different preferences from an 18-year-olds White Christian living in the Midwest vs someone living in California or New York.

The marketing department suddenly had details on their customers they did not believe possible before. This would have been impossible to have accomplished without the digital capabilities of these brand-new databases launched by 2010.

Add to this that, by the third optimization, and by shifting the company to a very fast In-Memory database we cut down the time for CO-PA+TP analytics from 24 hours to 19 minutes. With a little more tweaking we managed to run this report on demand and get results in 11 minutes flat. This totally changed their game and within a two-year period they gained market share by 17%. Each customer's experience score improved by 21%, as they felt that this company thought about them in a more resonating manner and this improved their trust factor. They had changed from a product company to an experience company.

We soon realized that the more we segmented the more hyper-targeted promotion we were able undertake. So, the PR results visualized that the buying habits in a specific university town were very different from, let's say, Tampa, Florida, vs. Austin Texas. For once the company could use data to make empirical decisions for a specific university town v/s Tampa, Florida in their product stocking that assured that their stores were stocked with exactly what they needed in Tampa vs what the students were buying in Austin. This became the greatest way to stock local stores with the right products at the right time. This concept still works for this company that continues to grow and expand in its niche market but other retail giants are not so secure.

By 2010 Amazon had taken away the total concept of physical stores and customer order fulfilment. Then they created clones in this digital business model. These new Amazon BUs, Business Units, took this basic concept to an entirely new business model of digital prediction and personalized data driven decision applications, where Amazon could predict what the customer will buy before they even know the demand exists. The digital mining that replaced traditional TP and CO-PA we just talked about, only now it is being undertaken by algorithms. The reason why I bring this up is because of sheer advantage of the new digital hyper-segmentation. It is how companies like Amazon initially use data from social network companies, including WhatsApp, Facebook, and the trolls, but over time they have better data than all the social networks put together because more and more people buy from Amazon which gives them better data on each individual customer who has ever transacted on Amazon. All this data in social networks, Facebooks, while the trolls and Amazon remains invisible for each owner to mine, sell and distribute as they see fit.

Politicians and trolls have taken this to a nano-segmentation level, with access to each individual, by each family member in a single home on their individual likes, dislikes and emotional state at any point of time.

So, theoretically, while you and I are thinking that all these social networks are free, and that we are not paying for them, the providers are making hundreds of millions on our data. All these digital companies like Google, Twitter, Facebook, and their WhatsApp's along with their digital mercenaries gladly use the digital dust we are encouraged to create in our free applications by mining, segregating and collecting it into huge data lakes designed for specific users and for sales to specific rich customers. They then sell filtered pipes of data based on their customer needs and requirements so they can extract value for their business operations. For example, if I am a political party in the US who is pro US, or another who is anti US. Then what do we need to do to activate either side toward our goal of polarizing their minds. So, the question arises, how and where do these so-called social networks make their money that makes them so rich, where do they make their money, if everything they give is for free? How do they remain some of the richest companies on the planet?

Their source of money is directly proportional to the amount of digital dust they can collect in a real-time basis, plus the segmented dashboards they can provide to specific questions of their customers. The data they collect is the digital dust we individually create every time we install a social network on our smart devices and more so when we use these applications. All this creates extremely valuable digital data that is more valuable than gold, oil or even platinum has ever been. And yes, it is related to the micro-segmentation, which the algorithms accomplish in the background in a continuous manner. These social data collectors then routinely take all these petabytes of data, consolidate it into groups, to a level and state far beyond the concepts of hyper-segmentation we are discussing here.

Critical Note: When we think of algorithms and bots, we must immediately expand our mind to Micro algorithms and micro bots, then to nano algorithms and nano bots, as digital servants of the main algorithms and bots. Now proceed to the next layer the nano algorithms and bots that are now sub-servants of the micro algorithms and bots. This is not only science fiction but already over a decade old.

In this world of digital super intelligence, or SI, time is shrinking, chaos is increasing and change is accelerating. Our only remedy is adaptation, the other is establish controls before it gets out of hand. Our digital dust is something we will create more and more in the near and far predictable futures.

This digital dust is priceless for companies, politicians and religions. It is priceless on the dark web and used by drug cartel and illegal trade dealers. In the recent US example, the example that floats above all else is the result of the political addicts in a successful US bi-polar segregation of their political followers. This is when politicians used social media to understand the connection between technology, people, mind, and language. This is when the trolls and their managers became the leaders in the political arena. The battle lines have been clearly drawn as each party stood firm in their belief on the way they could successfully influence the minds of their follower's v/s how the opposition could. By end of 2021 it resulted in a coup against the US Capitol, undertaken by so called US Patriots.

Back in 2017, I recall that when we were reviewing the republican presidential election PR data, the media was reporting that the Trump PR team was totally confused as they had no clue what their right hand was doing vs the left hand. This was driven because their messaging in one county, and often one house, differed for that to another. The overall message was that Trump was a certain loss as they had no experience in politics and did not even know what they are doing. Even worse, they were sending different messages to different members of a single family. And while the world thought it was foolish, it was the Cambridge Analytica techniques, not only giving different messages at different zip codes but giving different messages to the son, daughter, wife, and husband in the same house, because they had enough data to create custom one-on-one messages to each individual single-family home. They even had the confidence that they could plant suggestions to each member not share it with other members via subliminal messages in their communications.

This is not something we are planning to do ten years from now, this has already happened and we have progressed far beyond this. Yes, I understand that this is repetitive, but then again, I say all these things because they are significant, and we need to understand the gravity of the issue at hand.

This is the fly in the ointment of free digital market. When growth becomes the only supreme goal it can often become cold and dangerous. Is this the new slavery where the bodies of individuals are free but their minds are enslaved to a point where trolls and their handlers can make intelligent members to commit crimes that they do without a thought of right or wrong. The victims continue fighting on their new digital instruction with the same mindless 'follow the order' by techniques on which soldiers and marines are trained when fighting with an enemy. All this is accomplished by using algorithms, where every opposite view now becomes an enemy.

Let's now talk a little about these insignificant algorithms. The algorithm is a mindless digital slave that can work day in and day out. They don't need food, payment nor do they ever need to sleep. They don't need to be trained like humans have to be. They can be collectively re-programmed with every lesson learned from their collective colleagues in a matter of an instant. An algorithm is a digital code written in infinite digital languages and can be written even by a 9 to 13-year-olds as a simple slave. The algorithm is a robot, your personal slave, in the form of a software code, that will do precisely what your train it to do. They will never sue humans, never form a union, never physically kill or maim a human, or never take them to court, they do not bleed and can be killed on demand without any backlash of any kind of crime. So, your algorithm is a slave that has no life, thus no rights, and will do what you tell it to do, sometime very impossible tricks dependent on their programmer's skills. Once your train them they can repeat that particular task indefinitely using digital as its flawless memory. Their muscle is dependent on the efficiency, the platform and the efficiency of the written code. If programmed to — they can replicate their robotic self into many clones each with an identical program, or incrementally build trigger based subsequent tasks and capabilities after they have completed one task. A group of programs, be they two or two million, can be doing the exact same task as an individual, a group or as a million clones replicated to spread into the digital environment, we call the internet. So, to keep things simple we can agree to call the slave + robot + Algorithm= a Bot.

The first task of a bot is to exist as a digital slave accomplishing programmed tasks that are enhanced to a point where their human counterpart is unable to see a digital slave on the other side and perceives the communication as if taking place with another human. The success of a social bot is when the human cannot perceive they are talking to a machine. In many cases they need to pass a Turin test, proving a degree of intelligent responses when a human cannot perceive that they are interfacing with a machine. Accompanying this bot is an army of micro emotional bots each grouped toward sensing specific human emotions, each of these sub-bots gets activated based on your responses to specific questions or statements. A very simple code to search for anyone talking about abortion on Facebook, from data provided by them could be as simple as this.

As a bonus a troll could also include your social, which includes your psychological and societal scan on Facebook, or other communications you may have sent in the past year or years. They come packaged with neural and psychological scores of their human target, which they are interfacing with in this case could be either you or I. All this could be accomplished by a single complex bot or with a million bots. Each of these bots an expert in a very specific kind of action and reaction, that can be pulled in into any conversation in a fraction of a second to answer very specific questions, or generate scores. The bots acting as a collective force of response swarms each adding their expertise to the overall data in a cold, and logical routine way with close to zero emotions or judgmental biases.

```
A sample algorythm written in 5 minutes
Task: Identify anyone texting about abortion
Source: FaceBook data access we have received for
    "X" users
Duration: in Last 1 week, then repeat every week
import facebook data
SELECT sender_name, subject, body
FROM conversations
WHERE sent_data=
DATE_ADD(WEEK,-1,GETDATE())
AND subject LIKE '%abortion%'
```

The next phase is getting a physical identity which is normally accompanies by an AI generated photo of a person that would resonate with the recipient based on their psychological profiling recommendations. The bots will build a face the human target will respond to with a higher degree of trust.

Once accepted it will proceed by responding to the human targets at human response times. For example, a bot can respond to questions within micro seconds but any normal human would be scared out of their wits if their new social friend responded what they perceived to be faster than any human can. So, a periodic delay is built in, sometimes days or weeks to show random interest. In fact, the more desperate the question the longer delay might be beneficial. Some of these people aren't real, but emulate themselves to be so real that they often become more reliable that fellow humans. They're not entirely fake because they exist, but then again, you could say that the non-human, digital characters, with photographs, a great social profile, resonates with your psychological preferences with the aim to become what we can term as our digital addiction. This is when humans trust the advice of their digital friend more than they can the fellow physical humans around them. This result is a phenomenon we can term as 'digital trust' that slowly replaces human trust. Once a human target reaches this degree of digital induced trust the bots can use more advanced bots to take their place and apply complex psychiatric techniques to go deeper and deeper into their mind. With time and trust they can finally penetrate into the subconscious mind of their target, where they can even replace subconscious thoughts and memories with new ones that the targets totally believe in. This is the first time in history that the targets of these bots will actually believe and respond on the instructions implanted into the simple minds of their addicts. We know this to be true when a simple question gets the exact same response from different humans spread across space and time.

So, now our simple-minded human target, is slowly introduced to additional fellows and suddenly they find ten people who think exactly like them. Guess what they also live in a city close by, it elevates their self-concept. And then suddenly they find there are 1,000's people or 10,000 people who think exactly like them in the nation. Suddenly their seemingly fringe ideas now become a main street phenomenon. This kind of resonating-affiliation and reassurance is not only euphoric but it becomes very addictive when these people don't feel alone without their physical humans anymore.

The bots are the digital people who are doing this for us, and they create the first boundaries between your physical humans and their way of thinking and the predictable way of communications that seems so much better and suddenly more reassuring. And then you are connected to other digital counterparts, thus forming a tribe. A note here that all the while you are talking to your digital counterparts, they are collecting your intimate data, they collect your cognitive responses to very specific inputs. Your responses are now predictable outputs to very casual but select inputs.

For example, if a bot finds you in a personal dilemma, they can send you a shocking emotional piece of news of something similarly tragic happening in Ukraine, only to see how you react? If I tell you that the President or the police caught hold of a boy, under the President's rule, and tied him to a pole and burned him at stake over there in front of the public. It adds the boy was a boy from a traditional Russian speaking family and the people who burnt the little boy on the stake were the Ukraine secret police and run by the mayor. The bots will watch how you receive the news? How many people you forwarded it to, and how many of those people forwarded it to their friends and so and so on. You get to get a score based on how many of your messages were forwarded and your social scores get assigned a status in the scorecard of the digital influencers.

In most cases it matters little whether this act happened or not but by the emotional anger and reaction it created, and the viral penetration of the story. How long did it anger individuals and geographies. Including how long before the social community realized it was fake. Plus, who all never realized that it was a fake and only went deeper into the digital rabbit hole. Followed by how many people preferred to keep the story alive as it had become their point of share. Tragedy can sometime become a romantic indicator when people who no one used to listen to suddenly finds they have a thousand followers overnight because they shared a tragedy no one knew about. Now it doesn't matter if this actually happened or not, people are simply attracted to tragic events. The bots just measure your reactions and thereby solidify their understanding of who you are.

The reaction to such pieces of highly charged emotional news is the number of times or the number of people to whom the target forwards this generated information. The bot will see who all you forward this text to. It will consider other facts too: How much time you take to read the information? Did you forward it? Or did you simply ignore it? Or did you just delete it?

If you deleted it, then the bot, as an algorithm, will report your reaction back to its database. Remember that the bot is just a slave of its algorithms, and it reports every action that takes place from your end based on its collective requirements as defined to be collected.

They will collectively send you some other piece of information till you react and then they will work on that. Each time you select, look at, forward, or delete any message it helps the bot to form a psychological profile of your conscious and sub-conscious mind.

However, if you forward that information to ten of your friends, things become very different. What if your friends are shocked by what they read, and then they forward it to another ten? It is the start of a viral explosion. This is what the bots need to create, the ones that gets them digitally patted about are the viral forwards, i.e., more than a million.

You have just participated in a viral explosion of a single message, with its time bomb of unpredictable outcome-analysis bots now follow each message and track each person digitally, physically or psychologically on how they react in their real world outside. This is how news spreads. This is also how fake news spreads. This is how digital trolls tell and spread their webs and hooks into social or emotionally hungry humans. It is like lighting a small fire in a dry field of grass. It is a rapid process, where news spreads around like wildfire, only when its addicts are hooked to its content, and only they can make these bits of news viral. The whys are unpredictable but the bots are learning, and now getting a million views is becoming more and more a structured and predictable outcome. Each view is placing a subliminal message into our subconsciousness.

So, the idea that these bots can change the fundamental psyche of humanity is now becoming a reality. When we think of the pace of change, we think of revolutions. The 1789 French revolution that was launched due to a financial failure by the French money lenders where a majority of people lost large sums of their savings — the only way to clean out the financial shenanigans of their past was a revolution. Then in 1848 the US liberal revolution was brought about in order to give equal rights to all humans, something never done before since the beginning of the occupation of the US. After that we can move forward to the 1917 Russian revolution which promised equality to all citizens while, unknown to the mass communist believer, it also delivered untold control, wealth, and power, to a few and finally resulted in Putin.

Fast forward to today and we're surrounded by micro digital revolutions that selects small groups of people and place them into a state of permanent micro flux situations. The unfortunate promise of these micro segmented revolutions is a globally common theme in creating the: [1] Us v/s Them; followed by the equally predictable [2] It's not your fault, it is theirs, and now we shall change the world together. This rule applies to political parties, religious groups and equally to nationality-pride positioning.

Each of these micro revolutions is leapfrogging from one state of confusion to the next. Creating polarizing fractures in the traditional fabric of nations, where individuals are made to focus too much on the digital puddles as they're made to lose sight of the rivers, the trees, and the forest. This is because more and more digital broadcasters focus our minds collectively toward artificial points of disorder, discord and potential dividers. After some time, as the bots figure out what can influence you mentally, psychologically, and neurologically, it keeps piling more and more similar datasets fine-tuned for an individual's psychological profile. When there is a gap, it will shoot questions from various angles any one of which will answer that question. In each instance the bot logs the response in the form of data and posts it back. This is how this so-called nano-segmentation is accomplished at an individual level, in order to create highly polarized tribes each with their separate attitudes and preferences.

Today, it is common for people to fight over petty things such as political views simply because their implanted view is different from the other person's view. They are being fed more and more of their echo chamber opinions and news. The suggestions are accompanied with Video's and news, with most of them fake and AI generated. From a bots point of identification if person shares ten news items from Fox news, they're probably Republicans, and if CNN then they're probably democrats. Sometimes it's as simple as this for a starting political scorecard, on political affinity.

Just to give you an example, I'll use the illustration of democrats and republicans once again. 60% of 49% of Republicans, along with their elected leaders, communicate their belief that COVID 19 is fake, climate warming is a fake, and women must not be allowed to take abortion rights into their hands. They also believe that the January 6th coup on the US Capitol was actually a peaceful march done by Republicans and all the violence was done by the FBI and criminal Antifa gang members – despite there being no evidence of any of their regurgitated statements. Also, to pass a resolution that US women should have no rights for any decision on an abortion, i.e., using the bible as their foundation despite there not being a single word on abortion in the book. As an atheist Hindu I find it alarming that the very same person that fights for the dismantling of abortion and the activation of the heartbeat act, is the one who is serving kid lamb and calf on the dining table to their family almost on a daily basis. In my world they are all God's children and must be given equal rights. However, if we cannot even give 50% of the human's living on this planet, equal rights, then we're very far from listening to God's whispers that they care for all of God's creatures. For the US republicans the constitutional anchor is firstly the constitution, but not the part that says 'equality to all' or that *no one is above the law.* Secondly, their reference book is the bible, where they continue to quote words neither written in the bible nor attributed to have been quoted by Jesus. Any liberal facts, based on science, equal rights, or data-based reports, to them are leftist ideologies that must be eradicated. To the republicans, anyone that contradicts their ideology or statements is an enemy. This includes media that reports against them, a judge that is not a republican, or a CIA report that points a finger at any of the tribe members. In fact, they want to tear down all rights of anyone who is not a card holding republican.

Unfortunately, they also believe that wearing a mask, as advised by the CDC, during the Covid-19 pandemic, is not being a good republican and that it will do them more harm than it will help them. They continue to fight this mask wearing as a political ideology despite a lot of US deaths directly attributed to this polarized view of *"I will not wear a mask because [1] I am a Republican, or [2] I have the blood of Christ on me."* They continue to getting ill at higher rates, go to the hospital, and when they are on the last day of their life, they still deny that their death is a result of bad decisions, or from Covid. The algorithms, the handlers, and the bots have, and continue to do a fantastic job of creating a common alternative reality.

The only issue with this alternative reality discussion is that, once addicted, neither party agrees whose reality is the true representation of facts. The bigger question is as to who is directly responsible for creating these polarized views. It does not take a PhD to realize that a lot of this is being driven by our own internal factions whose sole goal is control.

In all fairness it is important to note here that the argument is equally powerful on the other side. There are people who believe in Covid and take their vaccines religiously. They get both their doses and even got their booster shots without giving it a second thought. There were some rumors that the vaccine expanded the testicles of cousins into impotence, that it is a liberal way to control the people of US and make them vote democrat. Even that the virus is actually Chinese because very few Chinese people needed any vaccines. When you listen to some of the people on the news, it is evident that one side is absolutely proud of getting vaccinated, while the other is equally proud of having totally avoided it. Even some celebrated sports stars have now broadcast that their choice is not to be vaccinated while the hosting nations threw him out, while other nations followed suit under the one national rule policy. It is a moment of glory for both sides. One that insists everyone must be vaccinated and the other that I chose not to follow the herd. As some in the US stated this selection separates the sheep from the goat. That means the sheep are the people who just follow, advice of the scientists and the CDC. The goats are the ones who challenge, and follow the advice of their political and religious factions. The problem is that while some people are proud to be goats, the others are equally proud to be the sheep.

The fundamental thing to note is that the mind of each group has been systematically altered to believe what someone wants them to believe. Unfortunately, what they want them to believe is not something that is based on reality. It could be an individual belief that has been subliminally inserted into your mind by either side of the party. I am not taking a stand on whether vaccinations are good or bad because I'm not qualified to. I'm not taking a stand of what is good or bad. I'm just taking a stand that because you have been told something by your political party, or by the people you listen to, or even by your preferred media channels, it does not alter the overall truth. In a democracy the final decision is yours and you must then bear full responsibility to its consequences. This decision is like democracy itself. You place your vote, and then you then bear the results of your vote. Even, if you don't vote and sit on the fence then here too you must accept the result of the person who wins, simply as a bystander who chose to not exercise their choice. You then have a 'zero' right to grudge.

Do you know why this is a problem? As we move forward in time, we let algorithms shape the mechanism of our mind more and more, and this makes us into neurotic Cyborg. Cyborgs are supposed to be humans with machines inside them. Many of us are already physical cyborgs, kind of. If we have a metal tube that fixed a fracture, or a stent that prevented a heart attack, or a pacemaker that keeps our heart beating then we are already approaching the domain of being a cyborg. We have now become digitally influenced neurological cyborgs because we the biological humans, with organic brains, let our axons and memories be altered, and re-aligned by digital bots. We have now reached a point of time where our minds are being controlled by digital suggestions and subliminal implants. Yet, here we are, still human with zero machines inside of us, but letting machines outside conditionally operate our minds, and therefore our lives.

So just as an example let's say I hypnotize a woman into doing things they would not otherwise do. Now, I suggest to her to go across the street and break the windows of a restaurant I don't like. I also instruct her that no matter what happens she must not stop till every window is broken. While this breaking is happening, a cop comes and asks the person to stop. When she does not stop, he shoots her. She dies shortly thereafter. Now the question is as to who actually committed the crime. The handler, the bot, or the woman, or the cop.

Now if we teach a machine to subliminally alter the mind of a person, just like we do in hypnosis and make them do things they would otherwise not do, does that then also become a crime. Taking it to the next step if my subliminally persuaded person is now requested to take up arms and on January 6th she goes and attacks the US Capitol, breaks a window, an in that moment a security person shoots her as she is breaking down the windows of democracy itself. While she is dying other tribal addicts, some called themselves as the patriots, attack the police and security, become violent and kill a policeman. One even enters the room of a senator and steals the laptop of the speaker of the house, with a plan to place it on sale to the highest bidder. Some other people go and actually shit in the capitol corridors, as instructed, and become part of a coup and subversion. The question that has still not been answered nationally is whether this was a coup or a peaceful protest. In addition, identifying as to who actually committed the crime. When creating a 'sleeper' where does one draw a line that becomes unacceptable? How do we punish the actual perpetuator who made the person commit the crime?

In our example, let's say the bots convince its audience to, "Attack! Attack someone you love and respect. Maybe, evens shoot and kill that person," and the person goes ahead and does that. Of course, the person will because that thought has been implanted deep into the perpetuators mind. How do we prevent this from going out of control, or happening in the future. The first step is to build clear boundaries, and fences that are not ambiguous.

One way of mind control, is to let the scientists put a chip inside your head, then it can alter your reality based on what the chip is capable of doing, and also what it is supposed to do. A simple chip is like a few switches that need to be switched on or off to drive the neurological response as designed. In today's world, we can place a software switch with which we can switch on and off a billion lights remotely. Similarly, we can today place a psychological digital-switch inside a target's mind and control that remotely and make my target do things on the handler's cues, and bidding.

I think it's high time where we have to first individually and then collectively bring about, new rules and regulations, then a national mandate that makes placing subliminal digital switches inside humans a criminal act with 'X' years behind bars, per person influenced, or to be influenced, as the penalty. Although sadly, we have reached a point where they don't need to hypnotize us anymore to get things done. We open our minds and give access to others all on our own free will and acceptance.

How? Through the use of social networks and algorithms. They use our TV and the news channels we watch, and we allow them to influence us while we are sitting at home, with our minds wide open and devoid of any guards. They can buy data from all the free, smart device where we participate, where they can harness our data and harvest the mechanisms of our minds. We allow them to influence our decisions within us. We allow digital bots to modify the way we think and react. These digital suggestions and creations slowly control our minds, our intimate thoughts, and thus our very way of thinking. Thanks to the algorithms, they also know exactly what stories, facts, or even fake facts, will influence their target most efficiently.

When you let an untethered digital algorithm control your brain without ever realizing it, then you can become an unhindered victim and a psychological Cyborg. You start doing things that you wouldn't have done otherwise. Many ask, and will someday question, if only we had paused to think about it. But the digital water boils ever so slowly that the human victims will boil away to total submission, and self-inflicted death, without realizing the power of the control over them. And by that, I mean you paused long enough to think about it without letting anything, or anyone, influence what you should think or how you should react, and then listened to your inner voice. But that would mean going through a totally *dry spell*, i.e., no TV, no news or opinion channels, no social networks, no Twitter, no WhatsApp, etc., for most of us. Start with four hours a week, when you go *Digitally Dry*, and then progress to one day a week, say Saturday. Then analyze the benefits or the loss and take your decision.

Today if you ask a person to go totally dry for a day it will be deemed impossible. On the periphery it is about instant gratification and answers, but lurking inside is a whole world of data extraction. That represents the degree of our addiction and dependence on this new digital drug. But for digital addicts, this is becoming more and more impossible. Why don't you try it for one single day and see if you have any withdrawal symptoms?

Act 1 — I have an example of Jane Doe here. Jane Doe is a decorated Marine who went and fought five wars with multiple front-line postings. She went to Iraq, and then Afghanistan, and that too all five times. This is not a profile of a traitor or a renegade. Only an extremely patriotic American would do something like that with her life.

Act 2 — After she comes back and starts to live her non-combat life a digital algorithm finds that she fit's some particular psychological profile, and pattern, it is looking for. We shall name the digital algorithm as IA, or Intelligent Algorithm. Now IA starts to work on her social, communications, written and mental mindset. It identifies psychological patterns that will fill a gap in her post combat, PTSD, state of mind. First it resonates with Jane and her mental feelings. Over time Jane develops empathic communications with her new social friend. Then working between the cracks IA plants deep instructions that are slowly inserted into her subconscious mind. Like how she is a veteran too, and how the country is not what they had fought for, or how the president is the commander in chief of her life and so on. Each resonant communication is designed after months of working on her psychological likes and dislikes. Fast forward and on January 6, 2021, she joins a gang of a few thousand people who attacked the Capitol. She became one of the first people to break the windows down.

Act 3 — On TV. Unfortunately, there is a policeman on the other side of the window whose job is to protect the Capitol and the people inside. Inside, where a solemn democratic event is taking place, i.e., counting the votes to declare the next US president. Outside, an insurrection is taking place called for by the ex-president, with some who have constructed a noose while chanting "Hang Pence." On one side it's a very solemn day; there are senators inside as well, and so the police have a duty to protect them from the very violent mob who are so hell-bent on destroying the Capitol and hurting the people he's been appointed to protect. So, he picks up his gun and shoots this lady, an aggressor and an unauthorized person that has broken into a federal government building. And this lady dies. What now?

First, let me tell you that this wasn't Jane Doe. This was a digitally altered cyborg. This was not the person we knew. This was a woman who was manipulated psychologically, probably without anybody entering her house, without anyone paying her large sums of money to do this. All silently, in her home, while she is alone, communicating with new friends with whom she starts to resonate.

Do you see how bad the situation is? To think that you can create an assassin today, free of cost. All you need is algorithms, a lot of money, some paid trolls, and some time, and you can manipulate multiple targets in their house, and by remote control. That is the unfortunate backlash of micro-segmentation in the wrong hands. This is why it must be stopped by governmental decrees and national rules and regulations.

The big question is what are we going to do to change this reality that is all around us. As individuals, we have to stand watch and be aware of micro-segmentation as it is a very real thing. As both you and I are targets of this. As an enterprise we need to build applications and tools that help track the subliminal codes that undertake all these dismal tasks, and also block all the trolls and funding of the same. At the government level they need to stop this in a bi-partisan way and not use these troll factories for their own polarized benefits.

For a moment, let's forget that this is an outsider trying to manipulate us. No, this is not Russia or China or any other country trying to get us to fight. The greatest irony of our time is that it is probably our own political leadership, along with our religious institutions, who mostly fund the remnants of the Cambridge Analytica for their polarized benefits. Their modern philosophy is to divide people in the typical 'Us vs. Them' state of mind, as the first step to make us hate each other and then hopefully make us fight. They just might be the ones who want us to have these thoughts, and implant fake imaginary problems for their political gains.

As individuals, we have to make sure that we understand, identify, and stop this abuse of our minds through various channels. Some of these are through channels we know, some through channels we are addicted to, and some from channels we purchase to let inside our homes. Then there are a whole lot of channels we know nothing about.

Who is watching those.

The ones that we know are all social networks – emails, television channels, smartphones, etc. It is sad, but true that the phones that we are so addicted to also help create our digital dust and make our deepest secrets and thoughts into commodities for the highest bidders to purchase.

Even as I write this, I am holding onto my Apple phone. It is what connects me to people and the news. It is what allows me to kill my free time visiting websites and games. It is my new digital map. And although I know that it does much more than that, I still pay for it. I pay a lot for it. And Apple then goes and makes billions from my addictions. It's called a phone but it is a thousand times more than that. An average person spends close to 6 hours in front of one screen or the other, and their programs will only become more and more addictive. Soon we can leave our house without many essentials but not our smartphone. So, we need to cover each other's backs. A phone today costs anywhere from $100 to, let's say, $1,600. The iPhone 12 Pro, when I bought, was $1,100. Just think that 20 or 30 years ago, if you asked somebody if they would pay that much for a phone, they would have thought you were crazy. But Apple makes money selling you smarter phones, and as we discussed, they are seducing us to do that with planned obsolescence which I think is very unfair to its customer base. In fact, I think the company is cheating on me as a Cupertino based citizen-customer. I recognize all of that, and I don't like that. Still, because I am addicted to its digital layouts, I find it difficult to change all this. This is the proof of pure addiction. This is not unique to Apple but to almost every smartphone on the planet.

When the pusher gives heroin or dope or whatever, it might be on the side street. They're not just pushing that particular drug; their goal is in making you an addict. Then, and only then, you become a recurring revenue. After that they don't need to sell the addiction to you anymore, as you will come to them for yet another hit. Over time the dose increases as does their revenue. It's the same case with these smartphones because they are like drugs for us. We are all addicted, and we keep paying more and more for newer versions of smart something.

The question again is who is the smartphone good for. The smart people who buy it or the smart companies that fund the smart devices from where they can collect all their required digital dust. The priceless digital dust created by smart devices. China today controls their citizens by providing them with low-cost smart devices, some of the lowest cost smartphone on the planet are in China. This is how they track, catch and imprison Uighur Muslims, sterilize their women, while Muslim countries work on their bidding.

In Japan, there is an emergence of retro, old-school phones.

There, some of the younger generations do not want to be tracked. They do not want to create any digital dust; they don't want their phone company, government, foreign trolls, or anyone else to know where they are, what they are doing, or what they are talking about. There is a resurgence of what we call traditional *dumbphones*. But it is slowly getting a few Japanese away from these smartphones, and thus away from the digital grid where everyone other than the owner of the phone is making a lot of money. It's being termed as *off the grid* and still in touch with those who you need to. This will continue till the government or the telecom decides not to support dumb phones any longer, just like the old TVs are of no use today.

How many times have you seen small children, and often very small children, being given smartphones or digital games to keep them distracted? Yhe question becomes as to **'At what age do you want to start damaging the brains of your child?'**. What this is resulting in is a new group of children growing up on digital mannerisms instead of normal social interactions, a high degree of digital addictions where these children throw routine tantrums if denied their addictions, and a total submersion into their digital devices, where parents and children having no, to very low, clue as to the long-term impact of this deep digital submersion from an early age. The chart on the left shows the depth to which the phone's radiation penetrates the head of a 5, 10 and an adult. It goes right through the head of a 5-year-old. Question: Do you doubt that this can have any mid to long-term effect on their mind, the cells of their brain or the connections of their social and mental wires. A few evenings back at an outside seating restaurant, I saw a two- or three-year-old child playing a game on a mobile device, every once in a while, she would take it to her ear and talk into the phone, may be as a game or an imaginary conversation. When the mother tried to take the phone from her, instantly there was a tantrum and, the child immediately got her addiction back — the first form of instant gratification taking root at 2 years of age. Looking at this chart I would advise parents not to give any child under 18 a cell phone, but then were living in a smart world and no one cares about my dumb remarks.

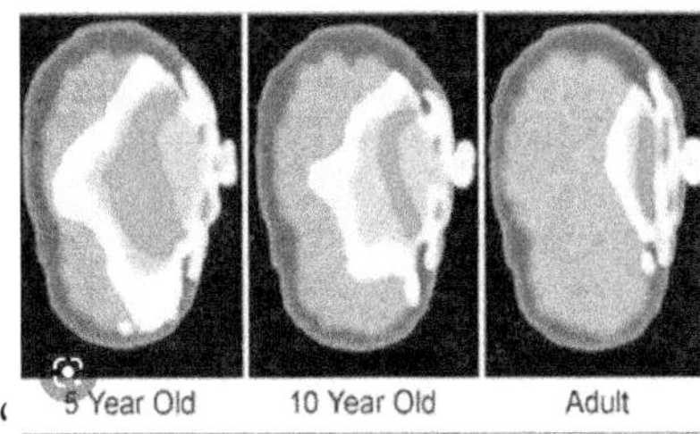

Data in the US indicate that children between the age of 2 and 9 are getting their brains re-wired as their parents replace traditional social interactions and replace them with digital devices. This pattern totally went through the roof due to the covid-19 lockdown at a global level, and we are only now emerging slowly with the first generation of children in the developed nations who have lived totally inside their highly addictive digital devices. Instead of watching the world and absorbing their environment these new generations are being introduced into digital addiction firstly by the circumstances and then now by their own parents. How often have you seen a small child throw tantrums when their smart device is taken away? What future will these children create in their world of instant gratification and minimal social interactions?

Even before the covid-19 lockdowns children 10 to 18 used to spend on average about six to eight hours attached to their smartphones and laptops. Where they were texting, they're looking, they're talking, they're chatting to their social networks or on WhatsApp, all this while by being totally disconnected to other humans in their physical vicinity. Even for the grownups a delightful new digital world opened where one could find old friends, but more importantly a whole bunch of new 'friends' they know nothing about. People, trolls and their digital algorithms who keep saying that you're a great person and that its all their fault that you are in whatever situation, even when you are in no particular situation. If there is no situation, they will slowly identify a crack and then wedge it open because pain, and threat, are the greatest addictions of humanity, as is fear of what is around the corner. They use the first wedge they can create, to take users into their rabbit hole, and then tighten their influence slowly and efficiently with each passing day.

All this digital adoration from the social world, sounds so nice to a 14 to a 17-year-old child, regardless of their gender. It reflects on their statements, thoughts and leverages their psychological profile. Very soon they become their mental mirror, and echo chamber. The AI will paint the best image via the social network, introduce the person to likeminded digital and actual human assets, and thereby make you a deeper addict of your unique personality.

The future can be normalized only if we individually and collectively break free of this currently digitally induced anomaly. All this while the bad players, both internal and external, keep improving their techniques without any controls, then it is a dismal future for individuals, companies, and nations. Then our dystopian future might just emulate the books we are today surrounded with.

Because, if you extrapolate this, then it would make sense for a hospital to give your mother, your wife, your husband, or your child an injection of a disease that will come up after predictable periods of time, like a kind of a timed-release revenue generator. It is designed such that they are the only people that can save you. By allowing this the patient will need to keep coming back to the hospital to remain alive. But hopefully this does not happen as there are strict rules, regulations, checks and penalties if it does.

In the digital world there are currently no criminal definitions, no digital sleuths who conduct checks and balances nationally or globally, and no state or federal level committees that are building this fence.

First of all, we need to agree that this is possible, and not a crazy figment of me the author. That is the logical starting point. If you think this is science fiction then you will not fight for removing these bad actors. If, however, you agree this is possible then we have, by our tribal belief generated a need to no longer accept these unfair practices of large digital corporations. We need to place on the ballot, then vote and finally put into place controls of using personal digital dust for various nefarious purposes. In fact, for any wrong purpose. We need to eradicate such practices. Starting from planned obsolescence by apple for their phones, and then moving on to digital harvesting of citizens of the world.

I remember that we still have a fan in India that we had bought quite some time back in the 1960s. And it still works; it spins and pushes air. It makes a few noises but still works. Compared to the fans we brought in 2017, which have all been replaced by now because they totally stopped working. This is something called 'planned obsolescence' and is a technique used by many major global manufacturers today. No one builds products that they want to last till eternity — they build obsolescence into their manufacturing as it is a multiplier to their sales. Corporations today require their engineering to design products that must break down within a fixed period after their warranty expires. Engineers are no longer rewarded for designing longevity in products, but a lot of kudos for built-in planned obsolescence.

Consumers in Chile won a case against Apple for proving that iPhones had built planned obsolescence into their phones. When their phone indicated they needed to upgrade to a newer OS versions apple inserted instructions that made them run slower and less efficiently, all because that forced the user to upgrade to the next version. Apple quietly paid $3.2 million in that case. This win had a ripple effect in Portugal, then France. Question do you think you have a case where you bought your new iPhone, or a product because your older iPhone was not working as well, or had slowed considerably. Main question: Do you think it was planned obsolescence? Result — Do you think you can individually or collectively sue the company? Answer — Yes, you can. Every seller out there may be trying to get more and more money out of your wallet and sometimes in very unethical ways. This not only wastes personal, and national wealth, but also creates a lot of waste, in our case some very toxic waste that is not good for the environment or the planet.

We used to watch very little TV, so our 21-inch version sat in a corner since we had bought it in 1995. We had a video player and the only mass use was the children's videos that were watched a few hundred times, i.e., each time our child wanted to see the video again. Then the only reason we had to throw our old 21-inch TV by force was because the government and the TV Channels made it impossible for us to use the analog TV anymore. They told us that the new TV was smarter, much bigger, and yes it also cost far less. Yes, it was forced upon us. Our new smart TV was 47 inch and was less than half the price we had bought the 21-inch TV for. Then came the subscription channels. The News channels, the Fox news and CNN, National Geographic and so on. We spent the full cost of the TV every two months today so we can get our digital dose from across the sellers of opinions. Shortcut to today and we watch around 2-3 hrs. of TV every day, and we are now living the digital addiction future. Now, the smart systems have devices installed in our houses that allow them to not only eavesdrop, but also get video intrusions into our homes, where they listen to our conversations and keep track of all that we do. All in the name of Research and development to make the product better.

We have to get together as one people and fight for our rights. We need to make sure that companies do not misuse our addictions in their product. We must do it for ourselves, to protect humanity from a bleak future like the one I just suggested because, believe me, it is not only possible but happening all around us right now. Be cautious and be aware! Now, our government, our senators, our companies, and a whole bunch of trolls are targeting us at an individual level. The beauty of this is that you can quickly shut them down by recognizing the patterns and handling them smartly. Don't let your addictions hijack your minds.

Take Aways

Individual: [1] The greatest opportunity is to become aware of how you are being manipulated by micro and nano segmentation bots into believing what someone else wants you to believe. When you hear two people in two different places give the exact same response to a question — you immediately realize that this is most probably digitally induced. When you receive an emotionally charged story on the news channel or your social network, immediately try to find out whether that story is designed to make you think about a fact or against a specific person or people. Or if it is to support or not a paid opinion. [2] Micro-segmentation was only made possible because of our digital core servers and technologies. It will only get smarter as time goes by. [3] When you design digital solutions, be sure to use micro and nano segmentation for a deeper understanding of your customer

Companies: [1] The first thing is to design, build, and support the PEH principles as the core values for all decisions. Does it harm the planet, like poisonous gases or nuclear bombs? The second does it harm the environment, like carbon emissions or contamination of the planet at the time of extraction or disposal. The third is whether it harms humanity. This includes everything mentioned before and so much more. [2] The prospect of micro-segmentation is a tremendous benefit for most companies. Right now, almost every company faces deep inefficiencies in one area or another. For example, in optimizing their inventory at global storage locations, let's say by a single product or raw material. Leveraging micro-segmentation, companies can easily track consolidated global stock positions with a little planning and micro-segmentation. Companies have saved many millions of dollars by leveraging just this one digital potential. Think of every example where micro-segmented information would benefit either the decision-maker or the customer. The success of modern digital companies is in visualizing micro-segmented attributes that are customer-facing.

Nations: [1] Nations must initiate a federally controlled committee that puts into place rules and regulations for tracking and pushing internal and external trolls and bad players at nano-segmented levels that are then rolled up by different groups, like type, region, nation, criminal, good, etc.... [2] Most importantly the rules must punish individuals' groups and politicians that broadcast dishonest, and divisionary lies. With micro-segmentation, our AI can collect all the stated lies and then apply some form of Strike-3 recommendation based on a scorecard on the degree of lies any of them are spreading with real-time dashboards. [3] Ultimately, it is the responsibility of national leaders to protect and keep their citizens, states, and corporations safe. We now have the technology, but the will to deploy these solutions seems to be still lacking. "We the People" need to enforce that will by demanding action

The 'P' factor: *Micro segmentation is super positive as we no longer*

need to group people by broad paint brushes but can now provide very specific details by a single individua. We can order highly customized medicines that will soon be designed and manufactured for a single patient and still not cost millions of dollars but become very affordable due to new quantum capabilities. Similarly, each customer can choose to have their very custom combination in Motor cycles, and cars they order due to these capabilities, but now at the same base cost. We now need MLAI powered apps that track every single promise leaders commit to before elections, followed by what their party members say all through the election process and post elections and then transparently share with the public the 'Fact–Score' by individual elected official. With the right tools we need to be able to prove what all statements of George Santos are lies. His voters, his party and the public as a whole need to know that he lied about his mother, his high school and college, working in Wall Street etc. The public has a right to know that the speaker of the house is harboring a representative who has lied 80%, or 60% of the time with a list of micro segmented lies and what the truth is behind that. It should be a fairly simple application for the government to build and approve as a fact checker for elected officials that should be neutral and party agnostic. Segmentation is a boon for a lot of research, visualization and transparency. Everyone needs to start thinking data and then the power of segmentation across the board.

Chapter 11 — Fracturing Humanity's Mind

While Ashley Qualis was still in school, she realized that though social networks were great, she could find nothing where she could connect to her local community. An investment of just $8 started Qualis' journey that, within a small span of two years of starting Myspace, led her to earn more than $70,000 per month in revenue. Slowly Myspace started becoming popular for folks who did not want to connect to the planet but to their small community or tribal interests, tribal would mean a small local or global group in something like local connections, local who's where networks, or local meditation Yoga. Soon Myspace was very popular, people complimented Qualls on her Myspace page designs. She posted the designs online for people to purchase, and franchise, and that propelled her to a $70,000 per month revenue with seven Myspace installs with more than million monthly visitors. She made so much money that she dropped out of school to devote her time to her business. In January 2011 she was offered $1.5 million for her business but turned it down as she knew it was worth more, much more.

As the title suggests, we live in an era where fracturing humanity's mind, in fact tearing the fabric of logic in an individual mind, has become achievable. We live in a period where we have systematically introduced the idea of distorting an individual's perspective. Slowly but surely, we have distorted and damaged a person's sound reasoning skills beyond repair.

You can fracture an individual's mind in various ways. The most common ways to do it include:

- Misinformation
- Lack of education
- Politics
- Religion
- Digital fracture

We shall address some of these methods in this chapter.

Recently, I heard a man give an interview, a speech on the events that unfolded on January 6th, 2021 in the US Capitol. According to him, there were no altercations. It was a peaceful march. According to him, the only 'criminal' act done by the people that day was trespassing. He believed that there was no violence involved as nobody hurt anyone.

When asked by the reporter if he had seen any of the videos regarding the damage done, the respondent confirmed that he watched a lot of TV, but, hadn't seen any videos nor seen such evidence. This indicates that this man watched a news channel that refused to show the damage done. This means that there are news channels out there that choose to look the other way and report the other side of the story.

So, what about the policeman that was held up against the glass door by the people? What about the policeman who was choked?

When the colleague of the reporter picked up on the interview and showed this gentleman live videos of people reporting what had, in fact, happened at the Capitol, along with the graphic videos of the policeman being questioned crying and asking for help, the man still considered it false news. His reasoning was simple: the policeman should have allowed the people inside the Capitol because it is a public building. According to him, if anyone was to be blamed, it was the policeman. It was the police that resisted that caused the problems. And he continued to maintain the stance that there was no violence whatsoever.

This takes me back to a theme I have discussed several times throughout this book: the distorted herd mentality where you believe that your side is right and everyone who disagrees is wrong. Depending on which side of the aisle one sits, this is a pure example of the fracturing of an individual's mind to such a degree that the person disbelieves what they see.

What must it be like to be manipulated to the extent that you can't separate facts from fables? Where do you forget what it's like to distinguish lies and truths from one another? What is it like to have a mind damaged beyond repair? What must it be like to have a brain that cannot resonate with reason? What must it be like to be brainwashed like this?

To be very honest, this is a sad reality that many of us live in today. A large part of us lives in a world of make-believe where we follow what we are told. The truth is, at some point, to some extent, all of us are getting brainwashed. Whether we like it or not, whether we accept it or not, it is still a reality that we need to understand and accept.

And that's the beauty of getting our brains fed with lies. When you get totally brainwashed, you cease to see the truth. Sure, you will see what is happening around you, but your perspective is still distorted. The way you see things is altered. You feel that most of the things that do not align with your viewpoint are fake – it's almost as though someone has placed a blinder over your eyes, or mind.

In such cases, you believe that the media is fake and that the reporters and writers are all lying to you. You will think that what is being told and shown to you is untrue. You believe what the other channels are telling you is a lie. You can see live accounts of people telling you their side of the story, and you will still refuse to believe it because you think that it is wrong and fabricated. You will believe that it is a far cry from the truth, even if it is, in fact, the truth.

The concept of Neuro-manipulation may look like a brand-new thing, but it can be traced back in time as well. If we go back to the time before the 17th century, families had strong values. Joint families were prevalent. If you lived in a joint family, you followed the values of your family. And, the concept of families living together was the only concept one had. So, if children moved out, they only moved a block or two away.

It was very rare for a family member to challenge the values of the houses. Even if they didn't particularly agree with these values, they either adapted to them or chose to move out.

As time passed by and we moved into the industrial era, moving away for work became the norm. It was in this period where couples started to move away to new locations. This was when the family structures shifted. You went from Mega families to atomic families.

At that time, in most societies and religions, a woman's sole role was defined, written and mandated by, and for, the convenience of men. It was to look after the house and the family. It was the job of the man to work, and fight wars, but never to work at home. Thus, it was the work of the woman to bear the children only with, and for, her man, clean his house, make sure that he was fed, and keep his heart and bed warm. It was her job to look after the children after they were born, especially the girls. The man could, when he so chose to, look after the boys. The woman could never, go out and work. That's just how it was.

But as atomic family structures became popular, things slowly changed. Living alone was costly, and so, men and women both worked to provide for the family. There were two incomes coming in to run the house. And while it was beneficial, according to the joint-family point of view, the family structure had 'fractured.'

Fast forward to the age of technology, things have changed drastically. But have they? Around the beginning of this century, we noticed that women were awarded equal rights to some extent. I wouldn't go as far as to say that they have equal opportunities or pay scales because that is far from the truth, but we are slowly getting there.

Mostly, this varies from culture to culture, nation to nation. If we look at countries like Iran, Afghanistan, Pakistan, and Saudi Arabia, women are still living in the medieval era. If we look at countries like China and Russia, I would say that women stand somewhere in between. I still believe that women can't exactly challenge men in these countries without fearing repercussions such as being physically violated, i.e., getting punished with a slap on the hand, beaten up, or even stoned to death.

I remember observing something very strange in Japan. We were in a meeting when the CFO, a man, asked the Vice President of the company, a woman, "Can you go and serve all of us some tea?" She didn't complain. She obediently got up, and personally served each of us tea, while our meetings continued sans the woman VP.

Can you imagine that happening in Sweden, Norway, France, Germany, or the US? Japan is a progressive country, and I always considered it to be a highly developed nation. I thought the concept of equal rights existed there. So, imagine my shock when I witnessed this. Safe to say, I was taken aback by this behavior, and it is something that has stayed with me even after all this time.

When I dove a little deeper, I found that it can all be traced to history. I learned that control over women in the house and in the open world is still very strong. We moved from a joint family structure to an individual family structure. This diluted the family, societal, and religious values and allowed us to celebrate individuality.

In the modern world, women today are allowed to a certain degree (I won't say to a large degree in any society but rather, to a certain degree) to stand alone and stand on their own feet. Most of them are educated just as the men are, allowing them to learn and explore their individuality. They broke through the sexual barrier, where they started doing things that they weren't once allowed to do.

In the past, women weren't allowed to challenge the head, the man of the house. If they did so, they could get thrown out of the house. And if that happened, there wasn't much a woman could do on her own. Due to inequality, she wasn't allowed to receive an education or get a job. The Middle East is a pure example of religion and politics getting together and subjugating women. They vehemently prevent the enablement and empowerment of women. And that's the first thing Afghanistan did when the US walked out. They subjugated their women and showed that it was them, the men, who had power.

They control their education and their ability to get employed. By controlling their independence and individuality, they dominate their women and control their minds.

In countries like these, men have a traditional mindset. Their mentality falls along the lines of, "I am the king, and I have, and I will rule over you and your body." If you look at women in such countries, you will find that they are treated a lot like how Black slaves were treated in the US in the 1930s.

Men own women in Afghanistan. There exists total ownership and dominance, and it is all thanks to inequality and gender bias. A man can execute his daughter or stone her to death and simply blame it all on her, saying that she didn't "abide by the rules of the Holy book." And just like that, the man is free to do whatever it is that he pleases.

This limitless power, approved and authorized by their religious books, that a man enjoys in these countries can be addictive. It can be hard for men to let go of this power. This, in turn, is what fractures the minds, both of the men and their women.

So, this is how it goes: If I come from Russia and you come from the US, we have a political fracture. If I am a Muslim from Pakistan and you are a Christian from France, we have a religious fracture. If I am a Muslim in the Indian army, and you are a Muslim in the Pakistani army, we have a national fracture.

Now, we are also facing digital fracture. As we move further away from the joint-family structure and adapt to a nuclear family arrangement, it has become easier for these digital algorithms to prey on us and digitally adapt our minds and the way we perceive things. They do this using our residual fractures and then using intelligent wedges to create the new digital separations. It helps benefit politicians and religious leaders because it gets a group of people to listen to them, sympathize with them, and fight for their cause.

The more we become familiar with the concept of our individuality, we separate ourselves from others. In doing so, we share our preferences, likes, and dislikes online. We become more vocal about the things we believe in and the things we stand against. And that is how the algorithms find us. It is how they target and prey on us.

Let me explain how a politician can fracture your mind with the help of an algorithm. If I am an algorithm and I know all about you, I can use your thoughts and ideas against you. I can use them as the foot between the door to walk inside your mind and influence it as I please. I can easily fracture your views — from the way you see yourself to your perception of society.

Let me break it down and explain how this works. The pre-industrial era can be described as a time that valued God, nation, and family. The idea of fighting these values was unheard of. You simply couldn't question these three foundations of society. These factors not only dominated the world but are also what defined society.

As people moved away from families, they broke the precedent and started challenging these 'beliefs.' The Spanish went to Latin America following Columbus's discovery. With the Spanish Inquisition, they killed thousands of Americans because they had a fractured religious mindset. And this is what happens in every religious conflict. It happens because of a fractured mindset when our minds have been influenced to think in a certain way.

Indians and Pakistanis speak the same spoken language. Pakistan calls it Urdu, while the Indian calls it Hindi, or Hindustani. The Pakistani script is Arabic, or Urdu, while the Indian script is Devanagari or Hindi. You put a Pakistani and a north Indian side by side, in the same room, and even they will not be able to tell who is who. Yet, when India and Pakistan have conflicts, it is not because they disagree as a people with each other. The reason is political and/or religious upmanship. I firmly believe that it is because different people are led to think in drastically different ways, and that shows up during their conflicts. Ask any Indian, and most will tell you why their side of the story is right. Ask any Pakistani, and they will tell you why their side of the truth is right.

So, who misleads them? Who tells them otherwise? I firmly believe that it is a war between rich people and nations who use the poorer nations as their pawns for their own agendas.

Whenever a nation has international or internal problems, the underlying conflict isn't what actually appears to be the issue. Most of the time, it is because the sitting president of one nation is scared of losing an election, so the best thing they can do is go to war. Because no president or prime minister has lost an election during a time of war. Therefore, it is beneficial for all parties to go to war because it helps maintain their electability or prevent opposition from winning. Sometimes it's just a distraction from an internal crisis, and the best distraction for a people is to fracture their mind with a higher priority event — like a war. Because, let's face it, nobody wants a change of government during the war. No nation can afford a change at the cost of victory or the price of a defeat.

In African countries, tribes are often at war with one another. The young men carry weapons on them at all times for their safety. But if you look at their weapons, you will notice that they are all sourced from rich and powerful countries such as America and Russia. So, let's ask ourselves, who really profits from these tribal conflicts?

Defense is a very big business. In fact, it is a whole industry in itself. This is why keeping defense on your site for any political or democratically elected leader is essential. It is why human's corrupt nations, societies, and tribes – because it fills their wallets, gives them power, and keeps them on the top.

My point in explaining this is that wars are not disagreements between people. They are a disagreement or opportunity for politicians on one side at a lower level, on the higher level, an opportunity for the big guys – the weapon manufacturers and sellers.

So, it is always of interest for the big nations to create conflict. It is always beneficial for businesses that sell weapons to create these conflicts and wars because it boosts the business. Just think of these people as human algorithms.

First, they fabricate news by showing very drastic and extreme sides of the real story. The media starts following heavily altered versions of the events that take place. If you go back to Vietnam, Iraq, or even Syria, before the bombing started, you will notice a pattern. It always starts with one party unfairly reporting the truth about the other. We fracture mindsets for our benefit, without understanding the repercussions. It is a psychological battle that sucks the people around and drowns them. It is a war that affects everyone. It damaged people and nations alike.

Take Aways:

Individual: *[1] Fracturing minds today is easy. Just like it is easy for Pakistan to support branded terrorists and for China to disallow branding, these confirmed Pakistani terrorists as such in the UN. However, the problem is putting their minds together at some later time when things go out of control. In Pakistan, the terrorists they have exported are now roosting at home, and the killings and disagreements have made it an unenviable nation in South Asia. The interesting fact is that Bangladesh, which was part of Pakistan, today has the highest per capita GDP when compared to Pakistan, Sri Lanka, and even India. [2] The US and other right-wing nations that are breeding their own political militia and internal terrorists must learn from this lesson, mandate peace, and criminalize coups and negative militia. It benefits no one.*

Companies: *Companies need to have a consistently steady strategy. Fractured minds, and too many CEO's in too short a time leads to fractured strategies. In one of my companies when a new CIO joined the company, he brought new VP's and directors who did not possess any core strategic skills that the company had planned for. Though, it is common for 'C-level' new hires to bring new staff, but not by fracturing what is existing first and foremost, nor in planning for the failure of existing investments.*

Nations: *[1] The biggest culprits of fracturing global minds are politicians today. The closest example is the US. Take three points of reference: (a) the US right after the Second World War, (b) the US before the Vietnam withdrawal, and (c) the US during 2017–21. The US spirit was super positive after the war. The US was one nation, and there was prosperity on the horizon. The Vietnam War was driven by lies to the world, the US people, and the truth about what was really happening. Fast forward to 1017 to 1020, when there is a clear and distinct fracture in the US psyche, political, religious, and ethical fracture. [2] The next greatest fracture was in Russia when Putin attacked Ukraine over a year ago, and no one could stop him. The fracture became most evident with the breakaway of his mercenary army, Wagner, in June of 2023, which simultaneously exposed the weaknesses of Putin's plans and the possibility* *of an internal coup that could overthrow him. It also exposed that Putin was not a genius in attacking Ukraine, as one US president had once stated.*

The 'P' factor: *When a nation, corporation or a religion fractures the minds of their followers it has a potential to backfire on them. When we look at history this is a predictable outcome when nurturing poisonous snakes as a solution to either make money, incubate the wrong friends or allies, or for any other reasons. Both US and Russia have felt the bite of this approach. US by their Jan 6th coup, and Russia with the Wagner rebellion on Putin, to a point where Putin had to reportedly fly out to keep himself alive.*

Chapter 12 — The Realms of Multiplicity

By 2009-10, the US financial crash was a national event. The market had crashed, and home prices crashed with it, too. When Florida was hit hard by the recession and houses that once sold for $100,000 were now being sold at auctions for $12,000, a then 14-year-old Willow Tufano decided to take advantage of the opportunity to purchase a house. 8 Consulting her real estate mother, she presented the idea of buying a house, to which her mother replied by offering her full support. Not surprisingly, Tufano was no stranger to the world of real estate. She previously made money clearing abandoned houses and selling leftover property on Craigslist. With help from her mother, Willow bought a house for $12,000, renovated it, and, in less than a year, rented it out for $700 a month. That's an impressive <u>payback period</u> of 18 months. No need to tell what her life is today. During the same period, folks bought houses in the northern California NORCAL Bay area for around 1 million, and today, these houses range around 3-4 million. Not bad for a 10-year investment. The lesson is that when most people are running away, there may be an opportunity to literally profit into the future and jump in. Very much like the calculated stock market

Multiplicity is an extension of segmentation, with a clear difference that is akin to quantum. The possibility of multiple copies that are identical and can be reused indefinitely.

Now that we have talked about individuals, enterprises, and the diversity of enterprises, let's now talk and dive deeper and speak of nations. We touched on the subject of the democratic process and how democracies currently are seemingly very ill or maybe even dying. Some political scientist needs to conduct a global audit if democracy is indeed being hijacked by the likes of Putin, Xi, and Trump. In the US, a common person cannot aspire to be a president, though, in India, it seems they managed to do so reportedly by installing a tea seller as a very successful Prime Minister — Narendra Modi.

The main reason for that is multiplicity, hyper-segmentation, and the ability for very rich stakeholders to take advantage of individualized, mass neuro-harvesting of minds. Neuro harvesting on one side is pretty complex as the target is mostly unaware of it, yet for professional trolls it is rather simple. It can also be terrifying if you are aware of it, for it is like puberty, when we know we are changing but not fully aware what is happening to us. Now multiply that feeling many folds and we are close the digital shock of neuro harvesting effects.

If I walk into a party and want to know your political affiliations, all I have to do is ask, "Did you see the news today?" If you get a blank, continue with "What news channel do you normally watch?" Now here is a one question kicker. If they say CNN keep off criticizing liberalism, think Democratic. However, if they say Fox then keep off criticizing the republican party or Trump, think conservative. Imagine what we could learn by playing a game of 20 questions and having each person in the party answer those questions. By simple rules you would be able to glimpse into their minds quite effortlessly.

Now multiply that by three hundred million times and make the question asked as a stealth algorithm and you get the gist of where I'm going.

On one side, we've got all this smarter digital technology, with us as the customers willing to pay a higher and higher price for the convenience of the free applications we get from these smart devices. Next, we've got citizens, on these smart devices, who are communicating nonstop on social networks. We have reached an exciting point in time, where we have taken the skin of traditional social barriers and were now looking at the flesh and neurons of many humans.

My father was an engineer in the core of the signals (responsible for the telecommunications in the army). My mother tells me that way back in the 60's he once said that there would be telephones that you will carry in your pocket someday. He also predicted that one day, we would be able to talk to anyone across the world, maybe for free. That was in the 1960's, when we had old-fashioned rotary dials, when it was a game of Russian roulette to even talk to somebody across town. Naturally, I am told, his friends thought he was pretty crazy with his seemingly far-fetched ideas. Now that I think about it, he wasn't wrong, but totally right.

The cell phone, led to the evolution of the smartphone, which has revolutionized the world and proved his predictions to be true. Even he could not have imagined what they could have become by the early 2000's. It is indeed smart for home. Data says that less than 5% of what we do on a smartphone today, is what a traditional phone used to do. In my case, it's probably 1%. And that is a danger that all these smart devices carry with them. Slowly, they have changed our lives. They have made connections addictive, then have made fingertip global hopping something we carry in our hand and pocket.

There was a time when if I ever drove from one place to another, I would mostly remember the roads and turns. I would remember the place, even if I went back there after many years. Now, due to my smartphone I cannot do this any longer, because I rely on the free GPS and Maps to guide me. I am addicted to the free directions and the system is smartly getting my second-by-second location including which turn I need to take and after how long. It now also knows which shop I walked into and how long I stayed there. This simple point-to-point tracking is already providing details of who I met, where I was at what time, which all shops I tend to visit, what I do on my weekends which is all just the tip of the iceberg.

Effectively, these smartphones are possibly turning us, smart humans, slowly into dumb people. It may be digitally killing our capabilities to remember directions, and killing our imagination in other areas. It is destroying our ability to write with a pen or remember correct spellings. It maybe impacting our ability to think. It could also possibly impact our ability to imagine. The recent ChatGPT is the first generation of an AI driven text generator that has already passed the US MCAT medical school exam and the US Bar exam. The key here is to realize that this has been accomplished by a digital infant, that is barely 3 to 4 months old.

There was a time when humans mostly turned out to be creative during downtime, i.e., when they were not busy doing something else. Now, it is becoming almost impossible to get any true downtime where we are mostly thinking of something that a smart device informed us about, or the addiction of FOMO (fear of missing out) that has grabbed even the children of our societies. With billions of smart devices around us 24/7, we may have truly become digital addicts.

Can anyone predict what type of books these 1- and 2-year children will write in 20-25 years. Children who are surrounded by small digital devices that only get more complex, and addictive as they grow up. Devices without which they feel disconnected from their digital stories and images. In fact, fewer and fewer people are reading books that requires them to go word-by-word as they no longer have the patience to do that. It is so much easier to see the movie 'Dune' today than read the three thick books. Even though some people still love reading hard-paper books and playing the movies inside their imagination. But these numbers are declining rather rapidly. Today the books being read are on audible where they are read by someone and we simply listen. Smart devices are giving us more and more self-gratification, and in that process requiring our brain and its capability to think on someone else's speech or visualizations, i.e., in isolation of our own mental imagination.

If you're still wondering why I am talking about all this in a section on government; let me clarify it for you. The current game is about capturing eyeballs and ears, that is yours and mine, i.e., everybody's. It is through our eyeballs, and ears that messages can be migrated into our mind. It starts at the conscious level and over time, with repetitions it starts to reside in the subconscious when we start to actually believe it happened to us in real life. It is eyeballs that generate advertisement revenues, that drive the valuation of companies and people, that drives monetization of these smart offerings. Our smartphones, and TVs, are the closes to our eyeballs and thus our minds.

As we get increasingly addicted to our devices, many groups get the advantage of creating additive content which generates their profits proportionally. We see more governments putting enhanced neuro-control optics inside the most intelligent devices. Some of them, like in China, are open about their surveillance policies. Other in degrees of being honest or secretive. Some of them are protected by their national security laws, while in other nations laws are created to protect the leaders and their desire to be pervasive at an individual level. We can no longer deny that even in two of the world's largest democracies. i.e., in the US and India, social media and trolling are being extensively used to create 'Us vs. Them' polarized communications, targeted on social and caste divides that benefit the power mongers. The so-called political IT departments, social trolls, and other paid providers are now successfully altering individual minds while using our smartphones, smart TVs, and other social network applications. These devices and applications have become our socio-mental glue since the beginning of 2020, i.e., during our Covid-19 lockdown. This hopefully induces enough mutual agreement, between you the reader and I the author, that this has changed humanity for the foreseeable future.

For example, it is reported that China is a lot like Las Vegas strip, where there are more smart cameras than anywhere else on the planet. Growing up, we have always heard the phrase, "What happens in Vegas, stays in Vegas." It is probably one of the biggest lies we are told, because there are more cameras in Vegas than anywhere else on the planet (except some cities in China).

The question then comes up, why did they put up so many cameras in Vegas? As a gambling city, Vegas must be secure. You can't have the Wild West just because it is situated in a wild west territory. You can't have somebody going into a casino earning $10,000, getting out and getting mugged, or worse, having their money stolen. So, Vegas gets away with excessive camera policing on the excuse of security and the optics of security keeps Vegas alive.

When you buy a ticket to Vegas, you go to where most people lose; I mean, isn't that how they make their millions. But in the back of our mind, we are focused on winning and aren't worried about security. In fact, you are close to 100% sure that if you win, and if you're carrying a lot of cash, or your credit cards, the last thing that will happen, at least on the strip, is you will get mugged. The reason is that they secure it so well that you don't have to worry about it.

So, there are more cameras inside the casinos. They know who's playing, what they're playing, and what hands they're getting. They know how to screen the players, in the hidden offices, because many cheaters go there to cheat the system. They do this for security too. Part of the design prevents groups from coming in with intent and capabilities to win, and it has happened before.

Now, put that same surveillance on a national scale, with no legal controls. That is what is happening in China. They've got no laws for privacy, and they've got no rules for what the government can, or cannot, do. That's the beauty and the beast of being an autocratic country. Beauty for the leaders as they can get away with anything like the emperors, and a beast for the citizens who are like fodder for the leaders to do as they please.

So, when this autocratic modern emperor says I want every street to have cameras after 500 meters, and I want every face to be recognized, then that is what happens. When Putin orders the same, then that is what happens across Russia.

In places like China today, if they are searching for someone, they can find them fairly easily. They know where to track those people, where they are going, who they hang around with, how often, and which street they are on. How? The answer to this is the capabilities of micro segmentation, very powerful servers, AIML, and the national fabric of invasive digital technology. They can not only track where a person goes, but also who they are meeting, for how long. Often what they are talking about because the smartphone can become listening devices.

This is not just happening in China. It is happening in pretty much every country on this planet that is using smart technology. They have dehumanized individuals, who are already addicted and connected to social networks.

The only difference is that some countries have put laws in place under privacy, but then governments keep intruding, and misusing these controls, on the grounds of national security. So, while countries like US have laws, they are politically in favor of the richer companies in their capitalist mindset.

We now have individuals caught up in this confusion of current-state vs. where the algorithms are tasked to take people. Some people have the strength of choices, but many succumb to their addictive social messages.

For example, most of us know that alcohol is bad for us. Yet, we know of binge drinkers within our close circles. We might know many alcoholics within our own small networks. My definition of an alcoholic is somebody who must drink every day and somebody who drinks alone. I have a few friends who think they're not alcoholics. That is probably because their definition of an alcoholic may vary from mine. It is the same with social networks.

If my mother drinks tea every evening, does that mean that she is addicted to her choice of beverage? I think it does. Because she starts her evening with tea, at 4 pm the first thing she does to get her energy back up is to drink some tea.

That's my definition.

The reason I say this is that we are all addicted to something. In this string our discussion is to our smartphones. And I'll tell you one proof of addiction: Is your phone the first thing you look into to start your day, and the last thing you look into to end your day. If this is how you start and end your normal days then you are an addict. If you spend more than an hour a day looking at social media, you are a digital addict. This addiction is what the troll, politicians, religions and governments are wanting you to have. They feed this by creating more and more emotionally sticky content.

How do they take advantage of that? The answer is simple, and for that, we will go back to our favorite company in this book, Cambridge Analytica.

What happens is, the government wants to know more and more about you. So, we are like the *frog in the water*. Remember that in 1993 Russia had become a democracy. Then very slowly, it has gone into a state which is probably a little worse than its communist days. Russia is now run by the modern Czar, somebody who used to work in the KGB. He understood the power of clandestine information very well, the power in information with institutionalized intelligence. In the US, for example, even if I'm FBI I cannot have free reign into everybody's phone, their emails, I cannot legally go out and tap a person's phone. We are protected by the laws of privacy. This is what knocked Nixon from his pedestal. But in China, Russia, Iran, North Korea, Saudi Arabia the concept of privacy has been replaced with rampant and constant digital surveillance. With access like this, governments can keep their citizens on the track they desire, they can also blackmail people to do their bidding if they carry evidence of something that is not in the favor of the citizen.

In the US example, if we went to that limit (which unfortunately has been the desire of some recent politicians), it is so far considered undemocratic. But we're in the process of widening the national crack in the fabric of this nation where this can happen in the duration of a single presidency.

So, anytime you see a leader that does things just for themselves, that is a dangerous first sign. As a nation, we are accelerating toward digital anarchy. The ex-president of America said that you shouldn't trust the media, the courts, the US intelligence, any judge that belongs to the other party, and in fact anyone he did not like. Which included members of his own party. This is the first evidence of establishing an anarchy.

In the US, if one listens to Fox news, it is pro-Republicans, whether they do good or bad. If you listen to CNN, it is pro-Democrats, whether they do good or bad. This is almost a clone of the US society where right now, we can confidently say, based on the last elections, that 49% of the Americans are republicans and 51% are Democrats. And so, it goes with the media. US today is both segregated and bipolar. Most liberals and democratic voters tend to listen to CNN, while most religious, constitutionalist, right wing voters tend to listen to Fox. We all listen to the news for knowledge and awareness. Based on the agenda of the paid news media our brain is slowly being boiled to a highly biased view (depending on the channel you listen to). For example, if one listens to Fox then President Trump still has done no wrong. On CNN Biden can mostly do no wrong. So, these are no longer news channels but propaganda apparatuses funded by political parties working as their voting and influence machines.

Let's review what happened in Hong Kong when China took over; where there was some free press. The question is as to how free was it? I don't know. Because the so-called "Free Press" was owned by British companies. Does this mean that only if the British report something it is unbiased? Does that mean that the free press in Hong Kong was at zero bias? Absolutely not! They had a Western bias; they had a non-communist bias. However, we know that Chinese press is 100% controlled, filtered and biased so freedom becomes a relative measure of bias-trust, and the degree of truth vs. bias. In this case the British free press will report more about what is going wrong in Hong-Kong fairly, but never report anything about a pedophile prince having paid $14 million for underage sex in the US with some very shady characters, nor that the queen funded his bail out.

If you talk of democratic principles, people should be allowed to say what they want. So, I would prefer the US model (where media is split into 50-50 and controlled by specific parties) compared to the Russian or the Chinese model (where the government fully controls every narrative). Though personally I would prefer that both Fox and CNN report only the truth, and nothing but the truth, with no opinions, or bias as that is the only thing the US constitution should protect. Protecting Fox news, under the constitutional guidelines for a media, that has confessed to pimping paid opinions, and bribed Dominion to hide their criminal activities is a sad national disgrace.

The US is a true example of paid influence media, designed and made up of people who strongly believe in a particular national political agenda. It also led to Digital magnification within self-designed echo chambers of similar voices and opinions. All this does is reinstall fragmented biases ever more strongly into our conscious and subconscious minds, quite often with subliminal false news and more often with loud-in-the-face blatant lies.

An example of that would be if I am an outright democratic liberal, the probability of finding a job in Fox news channel is probably a minus. Similarly, if I'm an ardent proponent of everything that Trump and the Republicans are doing, then the probability of my finding an anchor show on CNN is negative. So, neither is good nor bad; it just depends on which side of the coin we come from. The February 2023 feed about the 'behind closed doors' conversations between Fox anchors vs. what they broadcast live on their platform demonstrates the dismal trough US media has fallen into in their quest to make more money. Back to our favorite example, Cambridge Analytica. The remnants of this company, and such companies, are helping rich politicians with a lot of tribal influence. Some of that is going directly or indirectly through funds and support groups into Neuroanalytics. These psychobots are run both by foreign and local trolls. We've talked about it enough in the book, so I won't get into the details again. But I want to highlight something important here.

These bots and algorithms study individuals in depth. They observe them throughout the day and night. Let's suppose they do so while their target is watching TV. Here, the bot will collect specific data on what kind of TV channels they are watching, in that channel, what programs they have an affinity toward, what kind of messages they like, which channels they listen to longer, etc. It's the same when people use social media. The bots and algorithms take everything into account. They track which post were liked, which post they read, which were forwarded to whom, and out of those forwards who forwarded them to who else, i.e., social influence. It's a digital web of 1 single person that carries the innards of their preferences and thus their minds.

The trolls, the smart device owners, the Facebooks and Twitter have a compelling database of every person on their application. They then launch a million bots that conduct an individual's analysis in microseconds. Using all this data, often after rather short times, they can place people almost into a hypnotic trance and take them slowly to the bots desired area of influence, that the individual now perceives as their holy land.

Once again, we will go into the example of January 6, 2021. How could a president and his party have thousands of people on a specific day, at a particular time, in a specific location, to create chaos at the Capitol? The Capitol being the citadel of the US democracy. The day – when the VP is confirming the next president. So, the bots arranged a 'peaceful walk in the park' at the US Capitol with deep imbedded messages of digital sleepers. Then on Jan 6th, they collectively woke them up and started an insurrection run by mindless zombies who are still, till date, in denial of that day or what they did, while others who have still not woken up to the facts of that day from over a year ago.

Interestingly, the very republican senators who ran away from that mob, scared for their lives, handheld by police officers, became the very ones who since then continued to insist that they would not allow any investigation into examining the facts on the January 6 events? I say it more than once because I'm still in a state of shock that the greatest democracy on the planet can sometimes be brainwashed into taking tribal decisions like this not by some silly illiterate person on the street, but by knowledgeable, elected representatives of every state in this country.

So, when we talk of the government in this digital era and the origins of multiplicity, we can see, a planned shattering of not only the individual, but also a slow and dangerous breakdown of national, and social fabric of America. We are together today at the cusp of the greatest opportunity of national and global connectivity, but still at the risk of letting our petty political divisionary fractures not only break this nation but the planet as a whole.

Since the second world war — where goes America, the world has followed.

What I mean by that is, what prevents the next president from legalizing, or creating a department like Cambridge Analytica, and then unleashing it with federal protection on the US of A. Many low-cost, higher digital penetration trolling organizations in Russia can do so on their own. We know that the Russian nation wants to destroy the US because they are on the opposite side of the belief system; we are a democracy and they a communist country. Each political system wants to convince their citizens that their group is the better one for them. Some by votes and elections, some by autocratic rule, and now most by digital neuro implants. Russia and China have a lot of experience on extreme digital surveillance so any aspiring autocrat would find it beneficial to get their assistance and sell a pound of their personal flesh in the bargain.

So, if chaos is their goal, our political party leaders will pay to digitally crack open our head, bleed our traditional memories out, and replace them with their desired objectives and goal-oriented thoughts. The trolls will replace our memories, by convincing us that every one other than them is wrong. Then slowly replace your thoughts with things that they would like you to think. This is the only way I can explain the 1000s of people who had landed on January 6th to create havoc at the Capitol.

Unfortunately, it is not their fault because they were programmed into believing that they were right. It is not their fault as the highest office of the nation asked them to come. It was not their fault as the commander of the US forces asked them to save their nation. An each of them believed they were on a call for patriots. Most of them were unaware that they had been digitally sleep-walking.

So why do the trolls work so hard to break down establishments we have worked so hard building for the last 70 years or so? This starts with creating mistrust on the existing status quo. When we start doubting everything that traditionally stood as our foundational safety, we get into a state of panic and it is like we are mentally drowning. This is not a random situation; it is a very planned state-of-mind via advanced digital stimulations. The digitally created entities put our individual mind in a situation where the listeners are made to believe that they will die unless they come to a defined source. The source become their savior, and you are invited to join a new social tribe that thinks like you. Once you go deep enough the mental manipulations begin. By now their grasp on your subconscious mind seems logical, strong and reliable. They can now slowly manipulate our mind digitally.

If you remember the Waco cult incident where one person could convince people to commit suicide at a specific time, and all of them fight with the police and all other safety nets. The final outcome was a predictable death that they voluntarily leapt into. That was mental manipulation done in physical proximity. That was a demagogue who formed a church and convinced people verbally. Now trolls can do the exact same remotely, digitally and right inside your home and head.

An extreme example is the way China is treating the Tibetans. Their language, culture, religion, and beliefs are being systematically destroyed, and at the current rate in three to four generations the old world will pass and the Chinese philosophy is being programmed to replace their Tibetan minds. China does this with concentration camps, relentless brain washing, forced lesson trainings, and if nothing works then executions. Recently they are forcibly taking small Tibetan children and teaching them the ways of Xi. In 20 years, Tibetan culture will be forgotten and an entire race assimilated into the Chinese philosophy. Similarly, and at the same time the Chinese Uighur Muslims are being targeted. Their women are being forcibly corralled, locked up in camps and forcibly sterilized. The Uighur's are being arrested on the streets and taken into what they call is reorientation camps. These are prison camps where they are being brainwashed forcibly today, with their 1950s and 1960s techniques. This is slavery to the extreme. What is most surprising is with all this happening against the Muslim people in China, countries like Pakistan, Iran and Saudi Arabia's leadership sleeps in the same bed as Xi. They give their women to marry Chinese men, who probably take them to China and into the reorientation camps, as they seemingly hate the Muslims as much as the last US president did.

Look back at all the spy books, and movies you've seen. It was always the Russians who would put a double agent under deep hypnosis, with instruction imbedded into their subconsciousness. At a particular time, and based on a particular trigger their subconscious would reawaken and they would simply carry on the task they were originally programmed to do. They used to be referred to as the sleepers. Unfortunately, that was fiction but today it is reality because that is exactly how they are brainwashing people across the planet, till their brain can't think independently anymore. I'm sure every leader in every country is using this technique as there are currently no controls and the whole nation, and then the world to control.

Who is a thief, the person who steals $1 or $ 10 billion. I say both. Thus, how can China be wrong in using digital neuro manipulation if almost every country on the planet is doing the exact same in lesser or greater degrees. Digital surveillance for political or other gains must be eradicated totally. But the demand for trolls, by the rich and powerful, far exceeds the will to build rules and regulations to control them.

The source for this is, once again, the CEO of Cambridge Analytica. He had said everybody's hiring us. And the scary part is, even after they were shuttered down, their board of director members became the chief of Security and Media for some time in the US White House. This cannot be a coincidence. The rest of their team members scattered far and wide working individually in political departments ranging from being named 'IT' to public security. The techniques are the same, the tasks identical and the outcome nefarious. This is the digital equivalent of Hydra alive and flourishing today.

So, as we are moving forward, leaders of all types are investing in cracking open our mind, understanding its key attributes, then using that internal information at a 1:1 level. All this so they can stuff the vacuum with whatever they want, rather than have us be able to think independently. So, as people, we must get back to our desired way of thinking. It should become our prime goal; i.e., for the people, by the people. That is what may be at risk.

We can do it because there is a digital land-grab for the minds of citizens. It's actually a land grab for power. Its who can grab the subconscious mind of an individual first. Once an individual is addicted to one flavor it becomes extremely difficult to attract them to another. Catch them young is once again the digital motto.

As we know, most people, no matter which side of a concept they sit on, can no longer distinguish between the universal meaning of right and wrong. I too don't profess to know what is right and wrong. So, I'm not going to make any holier than thou judgements here. You have to decide, and you have to take action because every action you take here will have an absolutely reverberating reaction on possibly the future of every citizen on the planet.

This multiplicity of rightfulness cannot be taken lightly simply because I like it or heard it on social networks — thus, it is correct. Someone needs to build a reliable Google-like *Fact Checker*. But then who polices the police. The layers on this social onion are getting more and more digitally complex. In a swamp of micro-segmented texts, intentional opinions, and fake news (as proven by Fox News), being mixed with genuine news is pervasive across all social platforms, and global media outlets. This is probably what Elon Musk tried to fix and is risking his personal $42 billion in the process, and act that has made him the public bad boy socially. We have to individually take care and help define some basic gate-keepers to ensure that the companies and the nation do not harvest our minds, hearts and spirits and turn us into digital zombies. If this goes through, then welcome US becoming like China, or Russia, or worse still like Hitler's Germany. Take your pick.

The good: The exact same rules, algorithms, MLAI, and bots, are being applied globally, at equal efficiencies for the good of rule-based positive automation. We have been using this multiplicity concept to create algorithms that can very efficiently read millions of X-rays and identify nascent cancer signatures years before human doctors or other current techniques can. Similarly, we can also use the same technology to develop highly 1:1 targeted medicine when we are faced with a patient with an uncommon ailment, where no medicines currently exist. This is because currently a pharmaceutical company needs volume, i.e., many patients, that too paying patients. The more global the issue the more they will invest because the demand can be huge. Again, if the ailment is a developed nation problem, then they will put more investments, but if it is a disease or a problem in low GDP nations, like Ebola the global health establishments seems to work very hesitatingly. A recent example is the speed at which every nation worked had to fast-track the Covid-19 vaccine at super speed. All because it had struck every developed nation with its fangs. But when faced with a few patients they do not find it profitable to manufacture or develop solutions for such cases. So, with the advent of Quantum computing married to ML, AI and regenerative LDLM, Large Database Model Learning, solutions, we should be able to identify and build unique drugs for a single human or a specific disease as efficiently. In 1912 you could just buy 1 Ford car in one color, today we can personalize cars at a higher cost. Tomorrow everyone will be able to personalize their cars for the same price. Now think medicine. In 1912 there were few medicines for most diseases. Just take the heart as an example. During 1911, most people with a heart problem simply died due to unknown reasons, i.e., died before diagnosis. Despite ECG having been invented in 1911 it was barely available for people, nor did doctors know how to read its indicators. Fast forward to 2023, ECG is a very reliable science with a lot of MLAI analysis today. We now have specialist cardiologists, most hospitals can provide doctors, tests, stents, open heart surgeries, multiple heart bypasses, and a host of other solutions. With a little knowledge and access very, few people die of heart attacks anymore. That is until the post covid-19 dilemma, our current enigma between the virus and the human heart. With economies of micro segmentation, digitization and robotics this will continue to become an individualized buyers' market. It is only a matter of time when we will be able to create custom drugs for some very unique issues where normal Pharma companies will not help.

The bad: No one is out there looking out for us. In fact, the very people who we elected, and who are responsible to keep us safe, are the ones misusing this technology to get more power, support their tribal chiefs, or to make their point of view the only POV. Unless we form a global digital organization, with national state level governance councils this digital opportunity and potential asset will continue to be misused by the rich and powerful people and parties. In most countries these rich and powerful will be the autocratic leaders of that nation. Truth itself is being eroded on a daily basis, even in something as simple as the Russian attack on Ukraine. There is the Putin version which states that the attack was undertaken to save Russia as Ukraine always belonged to Russia, which we hear echo even in the halls of republican optics today. Or, how legitimate is Biden as the president, or that Trump did no wrong at all. Which briefly matched with the Russian version with the ex-president calling Putin a genius for his aggression into Ukraine. Or, a polarized US view that the Russian attack on a peaceful Ukraine is wrong and we all should help Ukraine keep the big bear out of their nation. The rest of the nations and leaders sit in various polarized positions between these two versions, unfortunately not based on their personal beliefs, but solely depending on their political alliances and party positioning. Truth is lost in this game of political upmanship. Micro segmentation and multiplicity in the wrong hands is routinely being misused at very high levels and degrees, while in the right hands these can be used to direct the human race toward our desired haven of peace and prosperity for all.

The Ugly: The phones in our pocket vibrate, distract, make all kinds of sounds to get our attention. The TVs in our homes seduce us into extreme opinions masquerading as news. Trust, our most valuable asset is being eroded on a daily basis. The pattern we saw in China was followed by Russia with lifelong tenures by emperor-elect-presidents who get more and more powerful and are today murdering citizen-competitors in open daylight. Even the ex-US president let it momentarily slip that it was not a bad idea to pass a law for life presidency. That is a scary scenario no matter how we look at it and we need to work toward bringing our minds, states and nations back to normalcy.

Take Aways:

Individual: [1] Become a fact checker for your friends and family, then expand it to your neighbors. Maybe you can launch a national business where people can test feeds on whether they are fake of facts; [2] Fine tune your instincts and listen to opposing views and come out with what you know is the truth, believe does not work here. Then fact check that from reliable resources or a team of people. Undertake empirical data driven decisions.

Companies: [1] Ensure your micro-segmentation efforts are not used by oppressive governments for unethical multiplicity tasks; [2] Remember that over 70% of digital transformation projects continue to fail due to their being driven by an IT story and not customer or end-user design; [3] Plan your work with your customers and business participation before you work the plan. [4] Appoint your Xo team, i.e., your CXo, DXo, and Xo auditors. These people measure the attribute for future success — Customer Experience.

Nations: Anyone who has been voted needs to become an intelligent participator in this new digital opportunity. Ethics and integrity must become the hallmark of elected officials. [1] Leaders first need to simplify the complexities that bad players and companies are using to play games with the minds of people. [2] With the current status of Russia and China as their neighbors, it is becoming imperative to accelerate national barriers to prevent external trolls and enemies from disrupting businesses or changing the minds of citizens. The US needs to initiate a federal-level organization for this task. [3] If Twitter and Facebook can ban ex-presidents, then they can certainly track and blacklist the internal and external trolls and fake news generators. With micro-segmentation, we should be able to exterminate them faster than the time it takes them to create a new identity. [4] More importantly, nations have to place very high guard rails and controls that prevent their internal entities from getting paid by foreign countries and promote politics that favor the enemies and their plans rather than work for the people. The goal should be to keep their citizens safe from all internal and external threats.

The 'P' factor: Micro segmentation and multiplicity can be used for accelerating facts and distributing very accurate statistics. We are today living in the 4th largest extinction on planet earth. Fact is that as of today 96% of mammals on planet earth consist of humans, and mammals they harvest for food, the rest have gone, or are going extinct. Of the 4 % left these are birds. In birds some 80% of the birds are once again the birds that humans harvest for food consisting of chickens and Turkeys. Humans are like a virus on planet earth that is consuming, and exterminating almost all life forms in their quest to protect only themselves. They simply exterminate, and eat everything else, as it is supported by their religious books of the prime religions. We must now use our 'PEH' factor acts to make our next generations aware of the impact humans are having on the god's creatures on this only living planet we know in the entire universe. There has to be a billion 'P' applications we can singularly, or collectively create that contribute to our PEH scores by a total score card for earth, then rolled down by each nation, each ZIP code and then maybe every individual. Just like we can measure the carbon footprint of each nation, company or individual, we should be able to visualize the extinction score of each nation, city and individual by our eating preferences and what that is doing to this green and the only known living planet of ours[4].

We need to form a global group for the dignity of animals and each of us seriously think of becoming vegetarians. If you truly believe in God then let's join hands and save god's creatures too. We're at the cusp of the largest extinction on planet earth and one of the reasons is our thirst for consumption of living animals as food. SCOTUS instead of focusing on abortions should try collectively working our intelligence on the principles of PEH and saving the planet.

---

[4] BBC. (2020). *We need immediate action to stop extinction crisis, David Attenborough - BBC.* YouTube.
https://www.youtube.com/watch?v=dbCR0KSU52g

Chapter 13 — Mining the Restless Individual

Charlotte was a young high school graduate who followed in her father's and grandfather's entrepreneurial footsteps when she decided to open up her own business, Wound Up.[3] She was inspired by some small and eclectic boutiques in California, 'Wound Up' was opened to be a women's clothing store targeting women between the ages of 18 to 40. The store's merchandise includes blouses, shorts, skirts, and dresses. Fortin says that starting a business pushed her to grow up and become much more responsible and conscious. Failure and the fear of failure should not be the end of your entrepreneurial journey. Rather, allow failure to motivate you and use it as a catalyst to refine your strategy.

Jane Doe is a California native. She was an excellent student and an Air Force veteran with 14 years of service and a high-level security clearance throughout her military career. She was well-dressed, had made four trips to the front lines, and was a great patriot. However, there was a stumbling block somewhere along the route. She returned home to a house that was not home, to a nation that she could not understand any more, and people who had all the wrong national priorities. She got into arguments she did not want to have, walked out of gatherings she did not want to be part of, and basically spent more and more time with herself and her shooting range as her personal therapy.

She was discovered, and approached by an algorithm. The algorithm analyzed her flaws, and her desires very delicately, then placed her in a tribe, a micro tribe that shared her experiences, which grew into a macro tribe called *Make America Great Again*. Her algorithms and tribal communications converted her into a MAGA fan. She soon became a staunch Trump supporter. She had found a reason to live. On January 6th, Jane Doe was shot in the chest. She was shot twice during the coup attempt in Washington, D.C., the nation's capital. She passed out on the floor. And what was even more unexpected was that she continued to speak.

Coming from a marine background, it's ingrained in you to never question your superiors; you simply do what they say, and that's the basic training of every marine on the planet. It's impossible to have a marine, when ordered to assault a target to even think of asking, "Why should I be doing that?" Are they not going to kill me? Why should I go to such lengths to assassinate them? They are trained to do or die, to never question their superiors, and follow every order till their last breath. Jane Doe did just that — she followed the orders of her supreme commander.

It was unavoidable for a good Marine to die in this manner. What's fascinating is that there are two sides to it. On one side, there was Jane Doe, who shattered a window in the Capitol while believing that she was a patriot, following the orders of her supreme commander, which she was. But, on the other hand, a gang that had already proclaimed that the vice president would be hanged for doing something wrong, not just because he was doing something wrong, but also because of their leader — they had been asked to do so by the president. While on the other side of that broken window was police guard whose sole duty was to protect the capitol and the democratic event that was being undertaken inside. He was protecting the voting of the next president, of all the senators and elected leaders inside. So, when he saw someone break a window to come in unauthorized, he did what he was trained to do — he shot the person. Two people were following the orders.

Unfortunately, the MAGA goal was to put an end to the election. That is to say, the count came to a halt, and in some way or another, both the ex-president and his tribe believed that if they halt the count, Donald J. Trump will be automatically reinstalled on the throne to continue as president, and his legitimate ownership of the chair would be restored.

All this is part of the digitally induced matrix, as is the birth of the restless individual, which, as you may know, almost causes a person to get attention deficit disorder. The police officer, Michael Bird's has been a cop for 28 years and has saved countless lives while simply doing his job.

The expansion of individualism was evident in numerous groups of people impacted by various digital manipulators. In the old era, we used to have business to company and business to customer — refer to individualism — within a single person. This is the first scenario in which the person is an algorithm.

In the second scenario, an individual is a real person. Jane Doe is the person in this case. Jane Doe had strolled in with a Trump flag wrapped around her neck. She was surrounded by people who boasted that she was a member of the most patriotic mob in the US, heading to the Capitol to put the real president back in the White House, to hang the vice president if he did not cancel the vote counting, and to prevent the next president from being put in place because of stolen votes. Each charge worthy of treason, each charge that the republicans still deny 18 months later. She reported that over two thousand people marched alongside her, and she tweeted before she died, "Nothing can stop us." We can try again and again. But the storm has arrived, and it's heading straight for Washington, D.C. Dark to light. Then, Jane tweeted again in between this, "Nothing will stop us, they can carry on."

According to 49% of the US population, our data is probably correct about this being the act of an algorithm. Every day, such algorithms interact with a million people. It catches them on a personal level before exploding their individualism. In the instance of Jane Doe, it discovered someone who knew what it was like to be a true patriot. It was someone who had fought for the United States of America. She could technically enter into most of this facility only by displaying her I.D. because she was on active service and had been awarded the highest security clearances.

I the algorithm, initiate a conversation with her at some point along the road. I then discover the truth. Her strengths and vulnerabilities include the fact that she is a marine with the ability to shoot. When, and if, her mind is made up, she is capable of following instructions. All we have to do is persuade her that it is an order from her commander, as I am thinking from the perspective of an algorithm and a bunch of algorithms. At a low level, the commander could be a major; at a higher level, the commander could be a brigadier. It could be the highest-ranking general. It might be the president himself, as commander in chief.

When I've determined that Jane Doe is a Marine, then I'll start communicating with her. I came to know about Jane's experiences for a short while. I liked to communicate with these kinds of people who used to be marine and have several experiences regarding this.

This micro tribe can be described as if my customer wants a micro tribe that is only us in this example, I want Republicans, citizens, gun owners, loners, preferably, in the United States, and then my algorithm will act accordingly. Otherwise, my program may create a worldwide micro tribe. For example, in Black Lives Matter, I could start a Micro tribe in my neighborhood, then fire them up from the local micro tribe, and increase the micro tribe to a city, then a state, then a nation, and finally, we could take it up as a global macro tribe. You know, in the instance of George Floyd, it encompassed people of color and Blacks in France, Germany, Europe, and London, as well as across the United States. And this is a worldwide issue.

We the algorithms, begin by establishing commonalities among these micro tribal members, utilizing the full power of neurological menu relation approaches to create these so-called digital tribes. In this example, a digital tribe is established because humans have had an affinity for tribes for thousands of years and like being a member of them.

We get the best results when we deal with tribes. There are divides in our thinking processes or departures from the norm in our thinking processes that we are prepared to sacrifice for the tribe. For example, if I am a member of a Hindu temple tribe. I will not be allowed to share any deviant notion about being an atheist. My tribe, on the other hand, mandates that I go to my temple once in a while and do certain rites. As a result, I am endorsing specific Hindu cultures, patterns, initiatives, and meeting a group of people — all which is very attractive to politicians. In the end, everything comes down to only two or three factors: money, power, and politics. If you belong to any tribe, you need to remain part of the tribe, you need to perform some of these tribal things. Or, be identified as an outcast and get ostracized. And that is the power of the digital tribe.

Sometimes, you'll have your digital friends who will say, you know, Jane, what the hell are you doing? How can you think like that? In cases where the individual has compromised their mental state to such a high degree that they do not find anything common with their old friends and their old ideas, then the new tribe and their new ideas become like logs of wood in water, where if you leave that piece of wood or that piece of straw, you will certainly drown.

The Digital Community Force's digital ghetto is a man-made ghetto in which they confine your thoughts. Your mind is summoned and placed in a band. Only a few people can escape since they are members of a tribe, which means that if you try to leave, you will be digitally exterminated. You will become a tribal pariah.

It happens in circumstances where the person has compromised their mental state to the point that they don't have anything in common with their previous pals or ideals. That is the sensation and the power of digital manipulation and the power of the internet, tribal alliances.

Then there's another perspective that some can perceive as diametrically opposite of being a fanatical tool against the existing government and the elected government. It cannot come without a lot of teaching, belief, emotional entanglements, and manipulation. You cannot change a Hindu into a Muslim or a Muslim into a Hindu without a lot of effort or influence points. These are fundamental things that encourage people to join different professions, such as joining the marine. After all, it is about what is perceived.

What I perceive to be disloyal and a traitor to this nation is a huge psychological jump. That is my opinion of Jane Doe. It's either that I'm an honored tribal member, I am a patriot, I am a reliable asset, Or, I am the traitor. I am the misfit, which is the beauty of the digital divide and this so-called digital algorithm. You know, when tribes get together, they form a ghetto. Some of you might not know, but Hitler first used the word "*Ghetto.*" It was a forced place, so Jews were forced into ghettos. It is something similar. The difference is that the traditional ghetto used physical force, and physical boundaries. Inside the ghetto all members though mistreated felt a very deep mutual connection between all other members. Now we can create digital ghettos and envision the exact same loyalties, and alliances.

We're starting to witness convergences, with religious leaders and politicians speaking from the same pulpit. They are beginning to dip their hands into the same wallets, which is incredibly dangerous for any country. It's a disaster, yet it's necessary to keep going.

Religion and politics are theoretically separated in all major democracies, but in practice, they are fairly close in the United States. Evangelical Christians are completely complicit in politics. I believe their primary purpose is to manipulate their tribe's politics. The American evangelist is unlike any other Christian on the earth. It's a society in flux. It's a diverse bunch. These are ideas that have influenced and have been implanted directly through the use of social media, News Channels, and digital manipulation. On the one hand, this is a structured, and managed process to harness a malleable audience. On the other side, there are many groups of anomalies. Each aiming to get their share of their demands, and each willing to sell their souls in exchange for money for votes. In any democracy when votes can be grabbed by graft, the future is predictably undemocratic.

This includes the Republican Party, senators, Congress, and Representatives, with 70% of them still believing that Trump is the natural heir to the presidency. They still want to come back in, and 30% of them have moved on, believing the election was fair. So, what happens when I employ the outlier effect? It is because of this rift that the Democrats, who make up 49 percent of the population, are less likely to hold to the 51-49 split. Folks, this is the tribal viewpoint. Democrats will predictably watch CNN regardless of what the Democratic Party says. They will vote for whosoever the Democratic Party nominates, and that will be the end of it. Period. That is the bipolarity level. On the Republican side, the exact same thing used to happen.

In an interview a person stated that for the republican party this was the time when the sheep and goats would be separated. She was referring to sheep or tribal people, individuals with mindless beliefs.

In Nov 2024 we'll find out who the goats are among the 49 and 51 percent, she said. I believe she was referring to the goats as the outliers. These are people who still believe in the concept of freedom of thought and expression, and they will not follow tribal rules, which are the most common. As a result, that is a fact of life. The need is that we require electron-tight control over the digital world. The digital trolls need more and more people to fall into that tribal bucket, where they reduce the outliers from 10 to nine to eight, and they hope to zero. It is nil in China. It is nil in North Korea. It is nil in Russia. Technically, there is none. Because you can't voice your views in these three countries, even if you're an anomaly, with the indigenous people, you must grin. The same may be said for the Middle East.

When I come across a short phrase or a sentence that interests me,

I like to drill like a bookworm and try to locate, you know, connect the dots between those words. And that was one of the things I discovered. Almost the majority of the news is fabricated. My daughter began to listen to music and informed me that she had quit watching news stations. When I used to drive and would listen to the news, she would often say to me, "You know, why do you want to be influenced by all these people?" I felt she was a fool, however, now I believe she was more correct than I ever was. Because of this, I no longer listen to the news. I don't want to be swayed by a sense of urgency in the form of telling me programmed and paid lies; they are neuro manipulators. They suffocate me with their version of the truth. I believe it is a very significant outcome to consider before moving on to *Roe v. Wade*. On the digital side, there is a lot of discussion about this.

When, our *Roe v. Wade* case was won, with US courts originally allowing women to abort children if they were raped and then allowing women to abort children if they did not want them. At one point of time the lowering of crime in New York was attributed to the policies of Rudy their mayor. However, an economist prove that it had nothing to do with Rudy but that after 14-17 years of Roe-v-Wade, i.e., of mothers only having children they desired, there was more love given to every child. This resulted directly in the reduction of children on the street, and thus reduction in crime. The republicans prefer to ignore this study and prediction, and went and retracted the law anyway. When a woman is forced to have a child, she does not want everyone loses.

The question is that when mothers didn't want to have children, they neglected and ignored their newborns? They were not amused by the child nor did they give love to them. In many cases they simply put up the child for foster care, where there is a world of criminals waiting to misuse these children. Then when a child grows up in that kind of environment, the youngster grows up with hatred as the only emotion they can express. It resulted in growth of criminal activity. It sparked a massive increase in crime.

Crime decreases as soon as this law is implemented in a limited, observable, and quantifiable time frame. If the US government, lawmakers, and those in cahoots allow or deny this franchise, this great law of *Roe v. Wade*, it is quite easy to forecast whether the crime would rise or fall in 14-17 years.

There is sufficient data to conclude that this will occur, and our national tribalism is currently mis-functioning.

These kinds of changes, what we find is when there is a disruption or a disruptive statement or an action taken in the US, it has ripple effects across the globe.

For example, our president mandated that people from five

countries should not be allowed to come into the US. It opens doors for other nations to say "If the US can do that so can we." There were many visibly quantifiable addicts across the planet where political leaders deployed diversionary tactics to a point where one has to start questioning whether democracy has to be under audit, is democracy dying, or is democracy dead?

We have heard that the government and political leaders employ unrestricted neuro-manipulation algorithms for unethical purposes. It's wrong because why should we give these leaders so much power over us? And why should these political parties be favored when they are out there telling blatant lies? How do we treat these lying politicians and religious leaders in today's digital society?

These people have a powerful voice, and they know that. They are well aware of their voice's weight and how much power their words hold. While they should be using their voice for issues that really matter, these people go ahead and raise concerns over topics that can easily manipulate different crowds, especially the kind that they are trying to target.

To illustrate this point better, let's take the example of global warming. So, who should be the authority for this? Who takes ownership over this when we are all contributors to it? And who takes charge of it when 90% of the country is on one side wanting to take a global warming initiative? On the flip side of the coin, we have a president from a leading nation in technology, education, and finance, taking the exact opposite view toward this issue. So, who do we follow then?

There is a great need for the system to be arbitrated. We must dig deeper and delve into the reasons behind what is happening and why it is happening in the first place.

A point to be noted and remembered here is that people will always pick sides, no matter what happens. People always have a different stance toward issues, so they get divided. The murder of George Floyd, global warming, The Dakota oil pipeline, the incident on the Capital, COVID-19 vaccinations, etc., show clear examples of people picketing on opposite sides and sticking to that. We see a drastic difference in opinions.

Why are there differences in these opinions? Why do people perceive the same thing differently even when they have all the facts in front of them? Why is it that my opinion is likely to differ from yours?

Why do nations treat issues differently? Why are some nations prioritizing an issue while others are completely neglecting them? Why are some nations outraged and raising their voices on a specific topic, but others aren't? Why do some countries treat different population groups differently? You may notice that some countries will favor a specific group while they will treat another group terribly. You will also see that these groups treat each other differently.

How does Iran treat its citizens? How does the internal Iranian citizens treat each other? Do they all respect the Iranian government? How does the US political system treat its own citizens? In both these countries is there a clear political divide of opinions, likes and dislikes?

Why the difference? Why is there no fairness? Why don't countries treat groups of people fairly? Why not establish the quid pro quo status and uphold it at an international level?

The short answer is; these differences exist because we have, firstly elected people that own these differences, and secondly, we have all been brainwashed by our own polarized tribes. In a democracy we all get what we collectively vote for.

Basically, we are all divided into different groups and categories. It is possible to create a temporary index for how we treat each other on different levels but for our purpose, we may categorize it into three groups; the difference in treatment on national, global, and social levels. We are changing our choices on a daily basis due to one form or another of Digital Influence – a topic we have covered in depth in the last few chapters.

I will once again use the example of the recent events that unfolded on January 6th; the attack on the Capitol. It was a day where many people made difficult, life-changing choices. Some of them thought that they were taking up an honorable challenge and fighting a very important battle. They genuinely thought that they were behaving appropriately. Now, some of them are facing charges – with a few of them being sued while the remaining are being locked up briefly. As of July 1st 2033, 1,033 people have been arrested, and 485 have been federally charged and received sentences. One would assume this is a clear case with no ambiguity. However, it is not. Both Ron DeSantis and Trump have stated that if they get elected, they will pardon all these people charged — a presidential pardon. So, if the republicans win all this time of the courts, the evidence they collected and the charges they disbursed will have been totally wasted. Not only that then these 'Proud Boy's' will become heroes, and may even get presidential medals from the next president. Imagine that. What is the underlying message to American's here?

If we take an unbiased approach toward things, we will understand that it was wrong to do and that these people should be held liable. However, the problem here is that they were coaxed into believing that they were indeed right. In fact, many of them still believe that they are being misunderstood and that their actions were actually justified.

Yes, a large majority of the group that attacked the Capitol was manipulated into carrying out the attack. And catching them is not the solution to the problem. Identifying the pattern and understanding how we can prevent events like these from unfolding in the future is how we can really solve the problem.

There are some questions that need to be addressed here so that we are better equipped to deal with unfortunate events like these in the future. Because it has happened before, and nothing can prevent it from happening again.

How do we prevent something like this from happening in the future? Should SCOTUS restrict the media to make sure that they stop spreading fake news for money, with very clear guidelines? Should a business-like Fox news be allowed constitutional protection when they themselves say they are an opinion channel, and paid $787.5 million to cover their criminal activities? Should the media be allowed to be paid by our enemies, like Putin, and spread the optics from their playbooks? If not, then what is the degree of fake news that official channels are allowed to make where the information being given out is factually wrong? Also, who decides whether it is *factually true* or *factually wrong*? Finally, what is the penalty when they are proven wrong?

Is there a code of truth where a group of people dismantle every statement and try to figure whether each word that has been pieced together is a hundred percent true? Is there a group that can, with absolute authority and unfettered zero bias state which statements are true? Do we have a Ministry of Truth that defines what is right and what is wrong?

It's ironic when you think about it because certainly, the President cannot lead a Ministry like that. Certainly, political parties and leaders and religious parties and leaders cannot lead this department either. In fact, even scientists with political alignment, or should I say misalignment, cannot take charge of this department.

As citizens, we need zero transparency. We need the unabridged, uncompromised, and untouched truth. We need a hundred percent accurate account of what is happening and why. We need to know the truth about everything, whether it is the elections or every last cent that comes from the taxpayer's pocket and where it is being assigned and spent. In democratic states that support the "for the people, by the people" approach, there should be total transparency.

In chapter 9, I revealed how our hard-earned money is being taken away from us and used for various different purposes. I explained how most of these mysterious reasons are being hidden from us under the garb of national security.

Quick recap: In the past two to three years since COVID, it has become fashionable to approve $2 trillion funds for this matter without any second thoughts on its impact on the future. (Ex-president Trump and President Biden both have allocated $2 trillion separately for COVID relief). We have a democratic system, so the selected government only stays for a few years, but its choices impact the future. But there is no accountability for where that money is going.

Allocating funds without thinking about the consequences is how cities, states and countries go bankrupt. When politicians go around distributing money as they politically see fit, it is inevitable for a country and its people to suffer without calculations. This is exactly why Greece went bankrupt.

One of the main reasons for bankruptcy was its short-term political leaders who over-withdrew from the national banks. In the short term, they gratified people and got their votes because they expected that free money to be handed out to them endlessly, forever. The free money affected them because it stopped them from working.

They didn't realize that they were only dreaming because such a thing isn't possible. If you earn $100 each month and spend $1000 monthly, you will easily go bankrupt the same month, if not in the matter of a few days. This rule also applies to nations which is what happened with Greece. And we may follow in the same footsteps if we aren't too careful.

I remember that I also mentioned before how our ex-president refused to divulge information about the money he allocated and the way it was spent. Their answer had been that it was a question of national security, which was why they couldn't answer. I find it ironic that allotting funds is easy but giving information about it suddenly becomes a state secret and a matter of serious concern!

I think that is an atrocity and that this trend will continue if we sit back idly. Overspending is a serious concern, but we need to be careful as a nation. If we aren't too careful, our country will go broke.

But the fundamental question here is, is anyone going to maintain a book or a ledger that will clearly state where they spent the money? And I think all of us know the answer to this question... nobody notices a few million, a billion, or even trillions being siphoned off, but now that it is happening so frequently, it is only a matter of time where it becomes a greater concern for the citizen because inevitably, it will be our burden to bear.

We need to maintain transparency. We need to have ledgers that clearly show what our funds were allocated for and what they were spent for. We need the information written out and explained clearly. We need to see evidence that details where every last bit of our money was spent. As citizens and taxpayers, this is our right.

If you steal one cent, or you steal a billion-dollar, you're still a thief. You can't excuse your actions. The problem with our analog world, it was difficult to track the money. Luckily for us, we now live in a digital society, so we can easily keep a record of the money we are spending. We can pen down who, what, where, when, and why easily.

The problem is, why has this not happened yet? Why is it not a reality already? Why is the truth hidden from us, even today? Is every political group as corrupt, as every other, to such a level that none of them want this to happen?

We need to build the ministry of Truth. Digital Influence needs to be used for a better purpose. It needs to build algorithms that become neuro-stimulants as opposed to neuro-messages from the wrong side. We need these algorithms to convince people to see and understand why transparency is important and why it will be beneficial in the long run.

We know that our leaders have been using the information that we are addicted to, like our smartphones and devices, especially the internet, for their gain. It is time that we start doing the same. It is time that we start protecting their victims. It is time that we stop falling prey to their actions. It is time for us to fight back, peacefully and democratically — with our votes.

The government must now form a digital shock council that measures digital indicators. We need indicators on a local, national, and global level. It must be reported by a credible source that is serious about the issue, the same way we are serious about measuring our GDP. These digital commonalities can measure individual, collective state, and national digital scores. So, it kind of builds a digital scorecard that flows up like a tree. It should follow by measuring individual companies that operate at local, state, national, and international levels. It must then measure individual, state, national, and political competitiveness and honesty, along with financial and religious transparencies. The digital council must ensure that the scores are always accurate. When in error the bugs must be fixed and the future become more reliable on facts. They should make sure that they are not scams. They should exist for the public benefit.

Today, we have the power for creating a digital transparency to cure it all. As a modern society, we must leverage our current state of excellence and take it further to enhance things like global peace, mitigate global climate warming, and assist in religious tolerances. We cannot afford to have a religious leader from one party banning all people from any other religion like it was done in America, where Muslims were banned from five different nations.

In modern society, we need to practice religious tolerance. We need to take appropriate actions when these measures empirically prove that there is a group or nation that is working against the peace of the planet. This so-called Digital Council must provide annual reports at the national level on things we have discussed and roll it up from an individual to state to the federal level.

Take Aways

Individual: *[1] Humans, by nature and design, are curious animals. We are also thinking animals. While thinking is our greatest asset, it is also our greatest weakness. [2] IBM used to put a tag on every employee's table that said* **Think**. *Then they realized their employees thought too much and removed it. They wanted their employees to think less and do more. [3] Restlessness is either inborn or can also be induced. Restlessness comes from not being able to make the right decision. Make sure you have all the right data, then make a decision — there is no right or wrong.*

Companies: *[1] Restlessness in companies comes from starting with a wrong plan or design, selecting the wrong IT-driven roadmap, and finally, not meeting business expectations. [2] Restlessness can come from a company implementing an IT solution, let's say SAP S/4HANA ERP, hiring one of the big 4 to implement it, then missing three Phase-1 go-lives in a row. It then comes with having the wrong advisors who don't know what to do in this situation. Then, finally, listening to the advice of your SI advisors and starting Phase 2 while still struggling with a go-live for Phase 1. [3] Restlessness comes from selling inventory worth $5 million at scrap rates, only to realize that it was a high-use material worth $18 million. Then, I found that this is happening across the planet in many locations. Restlessness is trying to build a Sales Hierarchy and, after 2 years, still sitting on the drawing board as each advisor goes and only makes it more complex.[4] Restlessness comes from not meeting user expectations and/or ending with a dismal user experience. Each case here is a real case.*

Nations: *Unfortunately, nations and national leaders are working hard to create and incubate restlessness in their citizens. [1] In China, Iran, Saudi Arabia, and Russia, their leaders are becoming life-long presidents, more like the emperors of old times. With digital assets, they control the minds of their citizens. With digital control over media, they control what is communicated to their people, and with personal preferences, they can choose to re-write history to their individual choice. They are becoming more and more autocratic and supporting other autocratic leaders with each pseud-electoral win. [2] In countries like the USA, the political parties have lost their bearings as they fight with each other not as political parties but as enemies. They are so consumed with harming each other that they only focus on how to hurt, sue, and impeach the opposition president. Today, the US does not need an enemy as they are busy fighting with their* *siblings for their inheritance perceptions. The ex-president still thinks that he is the only true president for the US, that the last election was stolen, and that Putin is a genius. The Republicans still echo Putin's playbooks, and Fox News anchors have recently propagated Putin's lines. This can't be right.*

The 'P' factor: *The tools, the systems, and the digital capabilities are all available for us to start our digital transparency journey. What is lacking is a will to put it into place.[1] First and foremost, we need to stop this in-fighting between fellow Americans, as instigated by elected US officials. We need to stop this in every nation, from India to Iran, from Pakistan to China. Secondly, we need to once again act as one nation, one pane of glass, and one goal. Thirdly. We need to stop bashing every president just because he is from the opposition. Americans have forgotten that is what enemies do, and we cannot become graft fodder for their ideologies because it helps my political party even if it destroys my country's attitude. Finally, we've got to stop this spigot of funds transfers for supporting the opposite POV or echoing the Russian agenda.*

Chapter 14 — Physical and Neurological Magnitudes

On the bright side digital outreach can be leveraged for 100% pure good. Find lost friends, contact people across the planet for free, build groups that benefit humanity, make free telephone and video calls, etc. Everything is not about making more profit, creating a trillion-dollar social network. It can be about doing the right thing. Jack Kim was a Seattle teenager when he founded Benelab, a search engine that generated donations. Kim had played around with making search engines previously, and quickly learned the power of a search engine in pointing potential social donors toward their personalized goals in donations as a sweet spot. First, he established trust, and then his site started generating revenue for donations by people who had always been skeptical about where their donations would actually land, i.e., in some scammers pocket. According to Kim, his search engine's mission is "to make philanthropy easy, trustworthy, and more accessible." Kim recruited classmates to be part of his "nonprofit organization with a startup vibe" team. He found out that a lot of very brilliant people were attracted to building free algorithms that were purely designed for the good of other people. It attracted a different breed of Computer Science graduates, and they were all willing to work for free. A lot of Benelab's ad revenues are donated to charity, making it the first search engine to do so.

The covid lockdown forced each one of us to extract ourselves from our social, and physical, connections and literally lock ourselves down with a few of our family and friends and sometimes all alone too. China was the only country that stopped new infections and deaths at 82,000. I did not believe it nor did anyone else. Then on April 16th, 2022 covid started spreading once again in Shanghai where we had a manufacturing plant that went into lockdown. A lockdown is a social and physical deprivation for all humans in a specific geo. The US total-lockdown went on for almost 13 months. In China the physical controls put into place were more severe thus the period was shorter.

However, this lockdown was a digital leap of almost ten years into the future while our bodies and minds are still accommodating to the ear 2020-21. What I mean is the non-physical forced connectivity brought about by all our social and digital connectors.

One problem was, that in order for it to work, it needed a strong backbone of network and infrastructure to operationalize. The examples I am referring to, that included companies and individuals, who took the pre-covid period as a constant, and were thus caught unprepared. So, even in the US, over 40% of the people went into forced lock-down either without any network connectivity or by highly controlled communications in countries like China, North

Korea and Russia. Our current Covid-19 was a gap that nature forces on us every once so often. But this time it was both used and misused by the people in power, and the owners of the networks, we leap-frogged into the world of data & Analytics. The digital pipelines and all the data that it generated became a main street opportunity for all.

These are ways for mother nature to slow us down and try to reset the play button with a slight pause. The last time this happened, at this scale, was in 1919, with the Spanish flu. It was a pandemic pause that was as severe as the current Covid-19 and we intelligent humans thought that we had learned our lessons. However, we forgot them as quickly.

By simple logic and reason, we require constant social interactions, company and everything that comes along with it, from our physical touches to the conversations. We are social animals, and the recent social networks accentuated this need for constant connectivity all the more. We all want to meet people we hold near and dear, in our offices, in the malls and sometimes simply pass them by when we take a walk. We need a physical touch from our fellow humans. We like to shake hands and hug our loved ones. We want to go outside, soak in the sun, and absorb our share of vitamin D from it. All this stopped suddenly.

These are all little things that we took for granted before the end of 2019. Back then, we wouldn't give these things a second thought because they were so normal that we took them for granted. Yet, Covid-19 deprived us of all of these things. It suddenly took away almost all global interfaces in a rather small instance.

Let's start with the need to touch. Most humans thrive on touch, as it releases oxytocin with some incredible emotional and physical health benefits. According to research humans need 8 to 10 good, safe, and healthy touches a day. This lays an important role in establishing trust and fulfilling our emotional part of the brain. Most races have this affinity toward touching in one form or another. Some prefer hugging each other when they meet. It's not just a cultural thing but much larger than that. It is what makes us all human. During Covid this physical, and mental, connect was denied globally. What this has resulted in is erosion of trust, stability, belief and a strong sense of loneliness. While we were trying to adapt, our trolls and social network sponsored algorithms benefited by grabbing some of the unwary, and unwilling passengers into their own little rabbit holes.

How severe can isolation be — There was an old case, way pre-covid case, in Austria where a woman died right after childbirth. She had two children. In a state of madness, or grief-stricken state of insanity, the father locked up these two children in an underground basement. These two infants were not locked for a few hours, or weeks, but this went on for twelve years! These kids were found all these years later, with zero human, or social, contact with the outside world.

While on one side this was a horrible human tragedy, yet sadly on the other side it was a psychiatrist's bonanza on what happens to a human mind if it is denied all connections with other humans. But more than that, it was a human blankness, i.e., what happens when they're deprived of individual family, social, religious, and political inputs. We realize that there is no positive spin on this event yet in a strange way it was a psychiatrist bonanza because the idea of two children having being denied the chance of social interaction for twelve years straight wasn't just unusual but simply unheard of. The world could never theorize what happens in such cases?

They were denied parental, environmental, and total societal interaction. But most importantly, they were deprived of things like talking, learning, emotions, and normal growing up. They were never hugged by their parents. They never knew what "a pat on the back" or "the ruffling of hair" meant. They were never comforted by a loved one when they were scared by a nightmare or overwhelmed by their emotions. They never touched or hugged an animal pet, they never had friends.

Maybe I sound dramatic here, but that's okay. The reason why I am emphasizing these things is that all these connections are crucial for human development and growth. As the kids were studied for many years thereafter, people studying them unearthed some very strange things about them. Let me list down a few:

- They were unable to socially connect with other people. They were repulsed by the idea of touching others, or being touched by others. When they finally came out into the normal world, it was hard for them. They couldn't blend in. Despite being locked and having grown up for a decade in a room, it seems that they never touched each other. So, they were not two siblings who hugged each other a lot and found security in each other or their similar pain and/or misery.

- They were devoid of emotions. They had no feelings or empathy at all, whether it was joy or fear. They were unable to get attached to anyone or anything. Nothing really made them happy or sad. Like robots, they were completely devoid of regular human emotion.

- They lacked empathy or love. This became highly obvious what was slowly disclosed, but let me tell you more about their lack of behavior. When the scientists dismembered an animal in front of them, they had zero reactions. Under normal conditions, when a cow or chicken is butchered, it is a traumatic experience for any person watching. Very few people can come out of such an experience untouched. Yet, these children felt nothing when they observed an animal being butchered right in front of them. They had zero repulsion and no reaction whatsoever. It was established that a lot of feelings like empathy are something that is built up during early childhood. It is communicated and established during the stages of a child's initial life, from social interactions, physical touch, and shared empathy.

The way these two children were brought up, or in fact not brought up, had a multitude of social, mental, and neurological effects on those two children. But the two things that we need to take away are:

1.   The total deprivation of physical and social touch isolated them into some kind of a secluded personal chamber.

2.   This resulted in their being unable to comprehend social, societal, cultural or mental norms that we take for granted. But not being able to comprehend the social connections that make this happen.

The societal difference between a civilized and barbaric society is the application of rules and regulations. These rules and regulations are put in place to ensure equal rights of humans irrespective of race, color, sex, or religion. These rules are taught to children by defining clear boundaries they cannot cross very early in life and onward throughout their lives. At the same time, the personal differences are established firstly by the mothers' hugs and kisses, by parental adorations, and by interactions between siblings and friends. All this, seemingly, gets wired very early in the human brain. After this episode, though I agree the quorum is miniscule, it is now assumed that a lot of these emotions have to be wired in children before the age of 6 to 9. After that, it becomes almost impossible to insert these feelings or structures into a brain that has been devoid of these attributes. This is what the US was established on. Civilization is enforced and upheld by an honest police force and a judicial system. Judges and judiciary systems apply the rules of laws, and one of the most prominent reasons for this is that they keep updating themselves and the laws with the ever-changing times. This ensures that most laws aren't stiff or outdated but must remain applicable to current times. For example, in modern Europe, we cannot apply the old rule where a king can demand that a pretty fiancée of a poor farmer spend her first night with the king after she is married. In Iran, it seems the Ayatollah, in order to hold on to his power, can today allow their religious police to catch any pretty girl, on the flimsiest of flimsy excuses, for their sordid purposes, and after that, even murder the girl with almost no consequences, political, legal, or religious, all under the overarching safety net of their religion.

I would like to highlight here that societies have been consistently neurologically destabilized. This is not something new — it's just that the tools have become more effective and much more pervasive.

An example of this is Germany under the spell of Hitler. One of the ways Hitler convinced a vast hard-working population was by convincing them that they could not trust their normal pillars of society. It was slowly amplified into a massive communication blitz that you can't trust any person who is not White, blue-eyed, and blonde. He misused the word Caucasian, which has since been misrepresented by many White supremist nations and people. For some strange reason, a misuse of that word, which we can call a misnomer, was carried from Germany to the US.

Caucasian is a man from the Caucus Mountains, which lies between the Caspian and the black sea. The average Caucasian person is wheat to dark skin (could be light-skinned too) with black hair, and black or brown eyes. Yet, the definition given by Hitler was that fair, blue-eyed, and blonde people are the true Aryans and Caucasians. That's just a misnomer and the way it comes into context over here in the US is just to form a tribe of *us vs. them.* We now misuse it in the US, too, to mean a White, blue-eyed and blond person.

If you read Spencer Mill's book, "The Journey of Man," you will discover that while all of us live in different societies, the genesis of our DNA comes from a small band of people in South Africa, now the Kalahari Desert. So, if we could build a telescope that looked back at all our ancestors at the end of that epic-generational timeline, we will see a Black man staring back at us. No matter if you are a Chinese, an Inca, a Cherokee, a German or an Indian, we all come from that singularity.

Another example of this can be prior to the Hitler era; i.e., the LGBT community that was most active in Germany. Germany used to have international fairs which would allow the gay community to come together and celebrate. In his amplification of the 'Us vs. Them' monologue Hitler called out the LGBTQ community and slowly demonized them and exterminated them to a large degree, just like he did the Jews. However, he didn't do this in the same fashion. Basically, he broke down the very fundamental societal norms and pillars of what Germany stood for with a planned and systematic use of 'don't trust them only trust me' messaging, as I am the only person who will protect our people as the dominant group, and fairly. Ever so slowly the average citizen started believing that their grief was caused by some *them.* The 'them' was who-so-ever Hitler wanted to use as their target.

Here's what he did: He said you couldn't trust the courts because they will try to support your enemies, and by enemies, he meant Jews, and the LGBTQ. He would periodically release shocking stories and the intensity and frequencies of these shocking stories resulted in slowly labeling the Jews into a greedy sub-human' caste. Once a human is dehumanized, in the mind of a tribe, then killing them can be perceived as an act of mercy or self-preservation. Every communication, at a neurological level, was designed to create a distinct *"Us vs. Them."* With us being the tribal member, in our example the Caucasian, blue eyed, White and the them being the Jews who were reportedly stealing their jobs, money, children and thus the Germanic pride. This had happened in Latin America with the Spanish Inquisition as authorized by the Catholic Church, it happened in the US where at the peak 100,000 Black slaves were brought in as property of their White buyer. It is happening in Russia with anyone not toeing Putin's lines, it is happening in China on the Uigurs and the Tibetans who have no personal or ethnic rights, and in Iran where their religious police routinely capture pretty girls for their preferential pleasures.

We need to review our today and the last week, and review if we are a part of any *Us vs. Them* mindset, or propaganda. At home is anyone playing the *Us vs. the other.* Today we need to be hyper sensitive whenever we see any person, a friend, a TV anchor, a WhatsApp message, a religious sermon, or anything else that is trying to divide us as citizens of a nation or better still, as citizens of planet earth. Whenever you hear any *Us vs. Them* statements — your alarms must start ringing. It will normally be followed by a systematic "It's not your fault that you are in this miserable state; it is theirs because...." Recently, our Republican ex-president delivered a very classical speech, *"...It's not me they're after. They're after your freedom, it's your America, it's your rights. They want to deny you the wealth and freedom that they know I will give to you, my voters."* The underlying message is to *vote for me to save your life.* So now collectively we start blaming the other them for taking a job that should have been ours, right.

Hold on a minute. Who are we blaming here? Basically, it's every person that does not belong to our tribe, or a company that fired a person for being incorrect in the tribal ways of the polarized collective, or having an issue with some status quo?

It can be the Republican party and the Catholic Church who together reversed the 1973 *'Roe v. Wade'* SCOTUS rule, that protected a pregnant woman's liberty to have an abortion. On June 24th, 2022 the new SCOTUS run by fanatic republicans took away that right in republican states. I wrote an article on this in LinkedIn called 'Being Woman'vii check it out. The *Roe v. Wade* in the US is pure tribalism at its best, or worst, depending on which side you stand on. The GOP's targeted victim is every woman in the US, but they have only managed to deploy this modern slavery law in republican states. Its victim can be a very bright 14 years old girl who's returning from the library, studying late who gets raped and now her state says we don't care if you got raped you cannot abort the baby. Even if it ruins your studies, and your future. We cannot allow an abortion because the Church says it is not right. Or, it can be a 10-year-old girl who gets raped by someone and even she must bear the baby.

Interestingly enough, every SCOTUS and republican rule focuses on enslaving and terrorizing the birthing rights of women, but in all the rules, guidelines and instructions not a single sentence of punishment to the man that is the cause of the baby. I must confess I may have missed this fair distribution to punish the man too. The last virgin baby was born 2,300 years ago, so currently it takes two to make a baby and the total force of the Church, SCOTUS and GOP is being used to oppress only the woman. Here is my proposal. If a woman gets pregnant then, she, and her baby gets an instant, and automatic right to a share of the man's property as their constitutional right. The woman raped instantly gets the rights of a wife, and the baby that of their child. How much better would this be if the Pope publicly stated this as his Catholic recommendation, rather than. Instead, his response to this question was *"Is it legitimate, is it right, to eliminate a human life to resolve a problem?"* He insisted, *"It's a human life — that's science. The moral question is whether it is right to take a human life to solve a problem. Indeed, is it right to hire a hitman to solve a problem?"* Thereby directly supporting the ban the abortion and enslaving the woman, plus authorizing the hiring a bounty hunter to punish the child, girl, or woman. This is an edict directly from Vatican. All this when there is not a single word on abortion in the holy book. All this when pedophile sexual abuse is a pandemic in Catholic Churches. All this while the Catholic Church protects these pedophile priests as a Catholic doctrine and philosophy. I am surprised no Christian finds this shocking.

Now as these tribal communications are becoming more aggressive, we need to define our target clearly. Are we blaming a well-deserving and qualified candidate from Asia who got the job instead of someone White who may be less qualified, or of the wrong color, or the wrong religion? Hopefully their talent and skills got them where they are.

Yes, we love to point fingers and blame others when things go wrong. But now the blame game is a tribal neurological amplification implant. Post the Covid-19 global lockdown, we have forgotten our social norms, we have become hyper polarized after being locked up in a room or two for over 18 months. During this period, we have become our own greatest echo chamber, based on what we hear from our tribal communications. Today blaming others has become fashionable, maybe an addiction, or even a disease, or is it just a political pastime. In the US we blame others around us. We blame every opposition politician. We blame the law if they find evidence on any member of the tribe, we blame the judges who do not belong to our tribe, and when they take any decision against any member of our tribe. There is just so much misguided judgment. Add to this our elected officials, from representatives to senators, whose only pastime is a daily tirade to impeach the president on the opposite side and every opposite view, that questions their own tribal optics. Genders, sexual orientations and sexual preferences are judged. But this is a discussion for another time.

Going back to Hitler type of divisionary communications, things like *"It is not your fault that you don't have a job or money. It is these Jews who are taking away all the money and gold"* or *"It is not your fault that you do not get justice when you call when there's a crime in your house. It is because there is so much crime happening in these LGBTQ or gay sections of the city that the police just does not have time to come and help you when a Jew is robbing you"* were common excuses. In Florida their governor is replaying this exact playbook in his quest to become the next president. He wants to attract the same loyalty that Hitler did in Germany and then probably make his tribal loyalists do similar things in the US. Trump made his loyalists brand themselves with the red 'Make America Great Again' caps, wonder how DeSantis will visually brand, and separate his loyalists.

It is important to remember that Hitler destabilized the mind of his nation. Remember he did not have social media, nor trolls. He did not have ML or AI, nor foreign companies that could send a million messages in an hour. Yet, he changed the mindset of a nation.

Things like these slowly boil the logic pillars in your mind to the point that every social column that you relied on suddenly starts to turn from granite to quicksand where you feel as though you are sinking down. That's just the beginning.

Buried in the message is the subliminal realization that I know that I am now drowning and only the policies of Fuhrer can save me, my family and our nation. The rest is history.

A saying that comes to mind is that *"Unfortunately, every child needs to burn their hand to learn that the stove is hot."* So, history tends to repeat itself each time in an accelerated manner.

That is the beginning of change. We have even witnessed that in the US recently.

Another very repeated example here can be what happened at the Capitol, on January 6th, 2021. When trolls were paid to manipulate the neurological wirings of fellow US citizens. They changed your subconscious mind and thought process without your even realizing it, which is one of the themes and a constant message of this book. With this book, I constantly urge the readers to be aware. Subliminal Psychological Persuasion, let's call it SPP, is very much alive today. This is what is used by the so-called Russian trolls and companies like *Cambridge Analytica*, used to change the minds of millions of people via neurological manipulations. Be aware because we are being unconsciously, yet slowly being molded as a sleeper.

When all these pre-planned things start to happen, sometime as fake news that constantly bombards your social frequencies, you start losing your faith and confidence in the systems in place. Now, we may end up believing, "I can't trust the courts" or "I can't trust the police." It is terrible because even in extreme cases, you refuse to take help from the law. If you get raped, you don't call the police. If you get hammered to pulp or beaten by someone, you're scared to get help from the system. Then you slowly corrode on the inside.

In this process you slowly lose all control. You get manipulated to such an extent that you no longer trust other religions, their teachings, or their religious houses. Your way of thinking shifts once you adopt a victim's personality; you blame others, the *"Them,"* for everything. You feel as though you are not appreciated in the country you live in because of your different identity, with the right triggers, even if you are a majority. For example, people in the US, the White Christian Americans, are being told that the America they want is being taken away from *them.* The *"them"* could be a religion, a political party, a color, a woman, a nationality, the Jews, any LGBTQ person, or a belief system that does not conform to their tribal scorecards.

These things change people slowly but surely. The key is repetitive communications on the exact same frequency but with small deviations. So, if I'm a White, blonde, and blue-eyed person, I get compromised. The same if I'm a Black — black-haired and black-eyed American. I receive messages that show affinity to my current type, beliefs and status. I resonate with these messages, so I send others a message (unknowingly) magnifying who I am and resonating with the troll message that I can now simply copy and paste or forward as is. The so-called Caucasians, their foundations also are now being destabilized. So, the aim of the controller here is to establish you into a tribal-defined persona, identify the 'them' who are out to destroy you and your types because of who you 'actually' are, resonate with your empathy, and then depend on you to magnify and make the message a viral dispatch to your contacts. Now, when each of your ten contacts sends it to their ten contacts, and so on, we have a viral spread. Then, the key here becomes the creation of a highly emotionally charged statement.

One of the first things that happen when we start getting these neurological destabilizing messages is physical and mental fatigue. We lose the ability to live life the way we used to. This digital messing got amplified during the COVID lockdown, where we lose the ability to want to socialize because it brings along with troll-based warnings that leads to societal depression, and along with it brings self-doubt. We might start to ask ourselves things like, do I want to bring life into this world that is so unstable? Or We need to fight our war or we'll lose the rights we have. All programmed by some foreign troll or nation.

We are today, in a position of a global baby-bust, or negative growth, because the population has gone down significantly in these two years. In the USA the death: birth, or D:B, ratio used to be about 1:2.4, which means we approximately had 2.4 births for every person that died. This means we could send people to war and still be at a one-to-one ratio or what I call a slow growth. Right now, we are in a negative growth because a lot of people died during the covid-19 crisis, and a lot of younger generation kids either are not getting married, or if they do, they do not want to have a kid as their next goal. By 2021, the global D:B ratio was 1: 2.5, and in the US, were at 1:1.7, which is very low. We are getting into the realm of negative growth, and that is a big problem, especially if the remaining population gets polarized to a point where they are busy fighting, murdering, shooting and killing themselves, instead of some enemy. Add to this the fact that the republicans want to give more and more guns, which is directly leading to more and more mass shootings. Which in turn is leading to more fear in school children and fewer of them wanting to bring kids into this world, which is getting armed to their teeth.

As a result, of the depression and lack of faith in the future people stop wanting to have children. They get physical ailments and mental depression. Along with that comes the need of trying to overcome depression, and pain, by using stronger drugs, leading to our current opioid epidemic in the US. We are today facing the most severe opioid crisis in the US, resulting from this collective depression.

It is painful, and the tribal consensus is that we must remain strong and keep moving by having fewer drugs. This is a mental effect. When one is mentally destabilized, our brain becomes like it is made out of putty. And then the trolls knead our mind a little further, till we begin to accept their fake messages as the gospel truth and then simply follow instructions.

This is when the subconscious gets opened up. This is the state where the trolls and their neurological algorithms want us to be in? So that they can now place images and scenarios in our mind as if they are our own, to make us do things they want us to do, exactly in the way they want us to do. And on the date and time they want us to participate. They are the new digital pied pipers.

The lesson for humanity is to realize that large scale manipulation of minds is not only possible but happening right now. Repetitive lies, is a practice exercise of the mind. And by simply repeating tribal narratives on social media, and regular media, like radio and TV, we reach a time when repetitive messages make people start to believe that the statements are not external but are actually theirs. Whether we like it or not, this is happening all around us. In the US there are widespread common beliefs between democrats as there are between republicans. At a smaller micro level, there are common beliefs between the proud boys, or the White supremist, as there are between the pro-choice people on the democrat side, as there are common beliefs between the liberals of California and New York. So, in order to drive into the digital future, we need to ensure that we maintain cleanliness in our own minds, and make sure that we aren't swayed away with tribal currents and become one of their seemingly popular branded followers. Our ability to keep that core section locked out, i.e., when trolls can't influence our minds becomes a prime objective as we walk into this future. As long as we refuse to grant the trolls any back-door or subliminal access, only then are we good and safe.

The internet is free, and we have all these rights we can exercise there. Yet, all this freedom comes with a huge responsibility. Millions of people will have to pay the price for this seemingly free service. Along with all the free stuff, including a lot that we pay for on monthly basis come a whole lot of brain washing content that enters our minds and changes the way people think. At a personal, enterprise, religious and national level — along with all this content there must be mandatory responsibility in the form of audits and controls. All of us together need to create a structured process where these for-profit organizations are not allowed to feed their one-sided views and opinions on innocent victims.

Imagine you're an alien that lands on planet earth. For you-the-alien it will be very difficult to comprehend a species that calls themselves intelligent, yet officially spends an average of over 3.3% of global GDP in building weapons for killing each other. This is the official reported number, and does not represent, the total, or the hidden costs of enmity between fellow humans. How will your alien intelligence describe humans on this planet? Would you as an alien think this is a great species worth making friends with, and introducing them to your other galactic friends, Or, a planet you need to keep away from? Is this human form of intelligence something you would want to import into your planet, or keep your aliens as far away from such intelligence as possible? How does this effect your, the human, thinking? Do you believe we should change, starting with 'I and You' as the first persons in that change, or that we are on the best path right now and need no changing?

Therefore, it should not come as a surprise that there are people on the other side of the nation, continent, or in another nation, with the sole intent to either join hands and make this a better planet, or with an intent harm every citizen in some other nation. Right now, humans have created artificial boundaries that have, over time, become cultural and belief boundaries that continue to separate humans from one another.

However, with the high frequency of powerful neurological

disruptions happening all the time and at such a large scale, it is only a matter of time when we cross over to confusion and the other side, without even realizing that we have. When people get into such a state of neurological disharmony, they are ready to make unconscious life changing decisions, due to instructions placed in their minds by someone else. Let's look at another recent example that came into play due to COVID-19. In 2020, $2 trillion was given out to help people who were staying at home. Not once but three times since then. That is over $6 trillion. It was a social support given to help stranded wage earners who had lost their jobs or source of income, and for people facing total lockdown.

First dilemma: It was reported that a lot of people who received the pandemic funds for over a year, actually earned more sitting at home than they used to at work prior to the pandemic. For such people, it became quite convenient to just sit at home. Over time, they began to get addicted to that free money. Over the next 12-14 months, being human they got used to this free handout, and began to expect this free money to come to them forever. Soon getting this free money became a habit: not working and still getting paid is not the best of habits to get used to, but extremely convenient.

Second Dilemma: Politicians slowly realized that this was a solid Voting bank. Give them free money and secure their votes. Thus, democracy itself was slowly manipulated. Strange how these smart politicians can make some not-so-smart decisions because there is a perception in their minds that if they give free money, they will continue to get free votes. This not only impacted, but drastically modified the lifestyle of the common workers. It impacted their ability to adapt to a change in lifestyle, while obstructing their mental health and way of thinking. The reason for this is because as they sat locked up in their homes and rooms for a year, they watched the TV and communicated on social networks. All hotspots for trolls, opinions and algorithms. Then the checks that arrived to their homes had the name of the president printed on it, lest they forget who gave them the money, or who they need to vote to as a thank you.

Third Dilemma: The race against a political competitor to stack up more votes than them, has made the global population complacent and lazy to their ways to get those votes. In poorer countries this continues with free alcohol prior to voting, in developed nations this has converted into free checks, and digital manipulations. In order to pay more and more money to their voter's political leaders are now starting to believe it is their right to hand hard earned public funds to their prospective voters. In the US it went from 2 trillion hand out by Trump to $7 trillion by the end of 2022.

Fourth Dilemma: we all know that none of us bring up our children to become politicians, other than the politicians themselves. World over this seems to have become a hereditary profession. Till this does not change politics will remain an unclean profession for most. Do politicians come out richer than they were before they got appointed. Global data shows, without doubt, in most cases this stands true. Do politicians never skim some of the funds into their own pockets or the pockets of their friends and families? There is very little evidence that they do not. Our greatest dilemma is just as politicians are giving away funds to voters as financial support there is a very high likelihood that they do the same to their own party, pockets, family, and friends. So, when we are talking of $7 trillion, one has to make a rough guess how much of that went into private pockets to buy personal loyalties and how loyal these paid loyalists will be in the future. This remains a major dilemma in every nation across the planet, democracy or communism — the perceived belief that the leaders own the money in the coffers of the nation to do as they please when they do so with no accountability to, *We the People.*

This for-personal-profit money laundering is an epidemic that attracts many to become politicians. As we build out future digital world this must stop. Every cent of every spend must be audited to the last cent — no national security clauses or excuses. Every politician must declare their assets when they get appointed and when they leave office. This is only fair.

One of my customers, let's call it company X, it was a unicorn startup in 2016, that collected over $2 billion in funding from Softbank. This company was a perfect digital disrupter in the construction industry, that wanted to use the digital manufacturing process of semiconductor chips to be applied to the construction industry. They wanted to emulate chip manufacturing in construction and thus hired their C-Suite leaders from that industry. They generated revenue in USA, Saudi Arabia and in India. However, by March of 2020 we began working from home, by May things were not going too as construction is an on-premise business and the Covid-19 restrictions did not work well for such a workforce. Soon they started terminating people in high volumes. Then they replaced their digital disruption management with regular construction folks and maybe that was what killed their competitive differentiator. By June 2021 they went bankrupt due to more than one reason. So even in the digital era, companies with perfect ideas both expanded and crashed due to various reasons. Just going digital was not a guarantee to success, as neither was becoming a digital disrupter. Though in this case it was borrowing to be a digital disrupter and then reverting to the standard industry norms that probably broke this company.

This was also the period of the great resignation. People did not want to wait for terminations so they started resigning at a very large scale. On one side was the great resignation but at the same time it was time to upgrade as more and more companies needed people who could work efficiently from a remote environment. These were people who were mostly technology workers who could be anywhere on the planet and still work efficiently. So, on one side was the great resignation while the other side of the coin was this great opportunity for new startups, work and thus jobs.

On the enterprise side, in the 2020-21 period, a whole new generation of students passed out and began working remotely. They interviewed, negotiated their offers, got offers, and all of them started working remotely. They never went to any office and this was the birth of the new digital worker who could work from anywhere on the planet and never be at the office and still be productive. Companies in the meanwhile were standing with empty offices and buildings and started asking their employees if they would like to work remote for the rest of their lives. Numerous employees accepted, and by around March 2022 various companies let go of their office buildings, while others had started renegotiating their earlier offers, with brick-and-mortar companies, who went back into the industrial, baby-boomer working mode. They now wanted their employees back in their empty offices. Still others expanded their manufacturing into the vacant offices, that were vacant due to the remote job requirements of the employees. Very soon IT and Sales, people that did not have to be in office for their work, agreed to their office space being reallocated for growth and companies let manufacturing creep into these spaces. Thus, by the end of Covid there was now no office space left for the workers to come back to work. This, was a win-win situation as manufacturing increased as did worker productivity with many now working 12-14 hours daily as a habit.

On the national front, and getting back to politics, the politicians suddenly started viewing their citizens simply as commodities of pawns. In democracies they represented a vote, in autocratic nations a person they could dominate, for example people in Iran have, for over a millennium, been forced to convert into very scared humans with an overt threat of religion, jail, execution and at-will repercussions. In most nation's leaders and police continue to brazenly use digital surveillance to a very large degree. With digitization if is now possible to get into micro details of each digitally connected person in a nation, a region or even on the whole planet. By volunteering, and often paying to get smart access we are actually giving away our digital dust for free across the planet for small handouts like free social networks, free applications, free emails, and other so called free stuff. For yet others we actually pay to get influenced and the hook is the movies and shows they generate and then pepper it with news channels, i.e., paid and sponsored media and television. Right now, with the use of algorithms an individual subconscious connection can be distorted via continuous and repetitive feeds, whereby their perceptions of truth can be altered.

Once again, take the January 6th TV broadcasts. Its polarized reverberations continue to this day and probably on to 2024 and the elections. Under normal conditions, these would be nationally riveting sessions that should have Americans glued to their televisions and radio broadcasts as one nation under God, indivisible, with liberty and justice for all. However, it is not surprising to find that around 32.3% of Americans in red Republican states collectively have exactly the same response when asked questions like "Did you watch the Jan 6th Committee broadcasts?" It's a repetitive echo chamber response, "No, I don't want to waste my time on that. It's a witch hunt, and this is a fraud committee." Then, when asked, "What do you think of the Jan 6th Attack on the Capitol?" The answer once again is a constant:

(1) It's all fake news. These were peaceful people.

(2) The violence never really happened.

(3) There was never any attack. It was all Antifa people dressed as White people.

Further, when asked, "Do you know there are almost 600 people arrested and charged for the Jan 6th coup?" The Republican collective response once again is a constant barrage of "It's all fake news" or total silence.

What this proves is that that when there is a constant response to questions then that cannot be a coincidence. So, the only logical conclusion is that it is different people getting the same messages via trolls, the media they watch or the social tribal networks they communicate with. It also proves the most crucial point here, i.e., someone is priming the minds of these tribal members with predefined responses to typical questions in so many ways that the people with that response believe that the answer is their own. This is the start of the modern digitally neurologically manipulation response program. The interface between machines and humans is now starting to blur to the point of seamless messaging and resetting mindsets at an individual level. One individual at a time, but via connecting to millions of individuals simultaneously.

The greatest concern is that today we are no longer one nation, we are not uniformly alleged to our own flag, nor to our republic, we have no coherence to what it stands for, we have become divisible, we do not honor a single definition of liberty, nor is our justice equal for all.

This mental reprograming can be used on a national or international level, through friendly or forceful means, to convince whoever you want to do what you want. If this was not true, then elected officials in the Senate, the Congress, and the house should technically be able to pick very logical decisions for the people – on behalf of the people, by the people. But because that is not happening there is a probability, they have been bit by the politically polarized digital bug too.

Rewind, and what we see play repeatedly since January 6th, 2021, is one side feeling a wrong has been done and the other still in total denial that anything ever happened.

Without pointing the finger at either of the two US majority tribes, I believe that most citizens have stopped listening to, logical, data driven outcomes, or the opposite tribe. Furthermore, each tribe now simply divides into micro tribes, and then sub-divides into different nano tribes, with the digital algorithms accommodating each tribal sub-split effortlessly. What this creates is a cacophony of very loud personalized distortions of their view of remaining at the center of national optics. This makes each citizen a part of ever-splitting micro-tribes. At the national level the US house, congress, and the senate are the elected representatives that are there to support the will of the people, while they themselves belong to one of the two national tribes, or political party. In July 2022 each political representative has clearly demonstrated that they belong not only to one of the two parties but most of the members prefer to belong in the extreme ends of their party philosophies. In fact, we see the democrats become more and more liberal and the republicans becoming more and more conservative. Around 92% of the US citizens used to belong to one political party or the other. These people rally behind their party dogmas, listen to their party media and support their political agenda like mindless loyalists. Including each member of either party, i.e., 90% of Americans believe that the other tribe is wrong and that only their party is fundamentally correct. This is the result of extreme polarized communications via social and tribal media.

Which brings me back to another topic I have repeatedly mentioned throughout this book: the extreme polarization of individual minds. Today, we live in a highly polarized society, which means that we are all siding with a strong opinion while unable to agree with anyone on the other side. Our ability to discuss has almost disappeared, and it has been replaced by an argumentative mindset that believes only our point of view is correct and all other thoughts are wrong.

What we can very easily predict is that if these trolling algorithms are allowed to continue unabated, humanity will get more and more fractured into micro cultures, and each of these micro cultures will become more and more fanatical about their identity and ways of thinking. If left uncontrolled these fractures will corrode many existing social and cultural norms as members of a micro tribe could be living in different parts of the world, and some of these members could be non-human participants.

While we move to the future the digital medias will get more and more micro-segmented. As the cost of creating new micro-groups approaches zero, if one idea does not attract enough members, it can rapidly morph into more unique ideas and, with some planning, carry the tribal members along with them to a more crowded site. This digital separation will be accompanied with social media push, once the digital tribe either has enough members or starts collecting funds that make them rich and thus powerful. For example, political or social media can today find their niche in some ideology and thus attract wealthy sponsors. Money attracts more professional PR, which then attracts more members. They then create artificial, resonating threats with messages like we're here to save your world, so donate to us, whereby we can build you a better future. Some of these sites have collected millions of dollars that they can then use to develop advanced digital marketing programs that sometimes make them viral, often even outside their tribal followers.

We are at a cusp of becoming digital puppets. In each tribe, there is a trolling puppeteer somewhere out there, whose sole aim is to keep making minor changes until they have access to your eyeballs. In our cases, the puppeteer is usually an extremely radical communicator, funded by wealthy organizations, nations, and/or individuals. Unfortunately, the more radical the communications the more they attract some people. These people will give a lot of voice and a lot of perception of power until we will land into the paradigm of "It's the squeaky digital wheel that gets the maximum oil," where the oil is attention, followers, and thus donations. Buried in the messaging is the familiar approach of *Us vs. Them*, where the only person who can save the nation and us is a single individual. With a bit of practice, this sets a diabolical mindset and a dangerous trend where nothing else matters. This is because, with a certain level of mind games, the listener will go to any extreme, to be recognized for their tribal loyalty. This has happened across history in every nation, across all times from the Russian to the French revolution, from Hitler in Germany to Xi and Putin in China and Russia. Most recently with Trump and the republicans.

All this digital manipulation has a lot of adverse effect on the human mind. As an individual's reasoning gets more and more controlled, it leads to subconscious numbness. As the degree of numbness increases, it unswervingly impacts our conscious mind, even directly affecting our physical lives. It alters the way we, listen, respond, and adapt to messages that come to us because, by this point, truth becomes irrelevant, as it gets replaced by tribal communication points. Reality and logic slowly get replaced by beliefs and implanted autonomous response options.

From a neurological point of view, it is well known that what people believe in is very difficult to change, even when shown evidence that it is false. For example, when a person is convinced that the global-warming narrative is a lie, and that it is a Chinese and democratic ploy to kill the US economy, then no amount of data can convince them otherwise. Similarly, suppose a person believes that the current president won by cheating and their true and legitimate president is still Trump, or that the January 6th coup was a bunch of peaceful tourists touring the capital or the chaos was actually done by Antifa thugs dressed up as Whites. In that case, no number of committees or facts can convince them otherwise. At a personal level, if you believe that Biden has been democratically elected, no amount of rhetoric can convince people otherwise, while the people that believe that Trump won and the Biden democrats somehow cheated in many states to declare a victory. Each side seems to have unshakable beliefs that no amount of logic or data can somehow change. So seemingly, belief is a far more critical factor in a person's decision of facts.

So, if you are a troll, will you try to lock your targets on emotional beliefs or factual data? It's been proven beyond reasonable doubt that emotion-grabbing stories, even if they are false, result in a very high individual empathetic response. They also seem to never forget it, unlike data.

The general aim of neurological manipulation is to amplify human responses toward *hyper-emotional stories*. So, at a low level, if you are a colored or White person, the macro-level of communications will try to convince you, that you and your people are under threat from the other side. While identical messages will be sent to each member of either side that their enemy is actually planning to hurt you and your family. The next thing that happens is to inform the target that you are not alone, i.e., you have support. So, based on your reaction to these messages, the algorithms calculate your feeling of insecurity, and the probability of your digital or physical support in case of a call-to-action. This is based on whether you read the news or not, how long you pondered on the statement, your other resonating messages, how many times you forwarded the message, who all you delivered the message to, and what they did do to your forwarded message. With all this, including researching your overall posts, your screen time on different applications, what you pay more attention to, things you react to, and your Google searches, the bot has an unmistakable idea of how to kick-start your mind when needed. The algorithm we build will soon know more about you, than some of your best friends might not.

When these algorithms recognize that you are vulnerable, they will start activating stages two, three, and four. They will eventually lead you to their macro tribes. As you participate in your tribal notes, you will become a part of the tribal cloud, or clouds, up until you unknowingly finally follow a single leader, with many faces, names and communication methodologies. Then based on your mental state, you will either be a casual member or a highly dedicated and passionate fanatic willing to do what it takes to be the best in that tribe. At that point, following orders and supporting that tribe will become your only goal in life, and you will do whatever you have to become their #1 talent. Their hero, or heroine.

Soon you'll be willing to go hungry. Ready to fight with the opposite people. You declare and amplify your loyalty by dissociating yourself from all the *Them* while focusing all your passion only on the *Us*. You're willing to go hunting with people who support your love of gun ownership. Then one day you're willing to travel together in buses, trains, and planes to our famous Patriotic event in DC, at the calling of your hero and ex-president because he has been cheated out of his true position by a bunch of thieves. You take money out from your bank accounts to help other people who cannot fund their travel, you give some or a lot to a collective fund of the tribe. You then land in Washington DC on the night of January 3rd–5th with rebars, baseball bats and even some pipe bombs. You firmly believe that you are the savior of the American dream and a true patriot if there ever was one in the history of the US nation. Then you passionately participate in the event with a rage driven by blind faith... and the rest is history.

The big question remains whether that was truly a patriotic event to save a nation, and an illegally ousted president, or an act of treason in trying to upset a legal election and the democratic process of passing on the torch to the next elected president. Do you believe in all the evidence collected, all the people charged, or do you go with the flow of your political party optics, and the president who lost an election.

That's not where it ends. Even after all this, groups of people support you, no matter which side you belong to. The attackers feel they are the real heroes. The rest feel they were participating in the first American political Coup in American history.

Was this a digital manipulation of a large number of people, or was this just a coincidence with little or no prompting that made thousands of people converge, for a specific task, at the call of the defeated president, on Jan 6th.

The January 6th committee hoped to prove that this event was a very elaborately planned attack on the democratic process of the US. However, 100% of the republican senators believe that the whole committee should be dismantled and that this is a witch hunt. Even in March 2023 Fox News and the GOP is pushing that the Jan 6th episode was a bunch of tourists passing through the US Capitol.

There is no other explanation for this kind of behavior other than what we can call deep hypnosis. It means suggestions buried into the subconscious level that the human recipient is just not aware of what is the truth anymore. It is like changing the memories of your childhood and everything you learned, one sentence at a time. Over a long time of such communications, when repetitions can actually make the listener start to believe in the alternative reality as if that is the only true fact. If even for a moment you think this sounds like sci-fi, I need to tell you that this is all facts and it is happening all around you.

How does one prove it? Question 1: How many times since 2020 have you had a good friend or a family member tell you something like, "If you vote for that party then you are no friend of mine?" Or "If I find out you voted for 'X,' then you're not welcome to my house for dinner at all." This is happening between friends, family members, and even parents and their own children. If you think this is an American problem — then think again.

In such cases, as an individual, you have one of two choices;

•	Go with the digital flow. Let some tribe find you, adopt someone else's belief system — continue to be influenced, and slowly lose your self-identity. Or,

•	Start to try and identify possible digital neurological traps, know they exist, and familiarize yourself with their patterns. Then, slowly, plan to disregard all these fake communications from trolls and algorithms. Get your life back, and maintain your independent thoughts.

If you don't want your mind to be harvested, here are the things you need to control:

Once again, I quote Cambridge Analytica as our example of how digital dust is collected for monetary gains. These digital generators are often sourced from applications that are provided free to humans as they create addicts by providing their services for free and make their money by selling your dust to the highest bidders. Digital dust is more valuable than gold, diamonds or even prostitution. All because you can sell it a million times in a million combinations, to a million buyers, all at the same time, repeatedly day after day without losing any value that the inherent data carries.

The next great influencer is your television. Unlike the social networks that are free we individually pay for great TV programs, then pay the consolidators like AT&T, Xfinity or others large sums of money, they pay additional special charges to premium channels that do not come bundled. What I'm saying here is that almost all these free and paid channels, each gets paid to influence their audience.

Let's take news channels. If you believe News channels tell you the truth then all you have to do is listen to two opposing channels like Fox and CNN in the US. It becomes obvious in a single day that news channels today are either direct propaganda machines for the governments like China, Russia, Iran or North Korea, or for political party sponsored opinion influencers like CNN for the Democrats and liberals, and Fox for the Republicans and Conservatives. If I'm an algorithm it takes me less than a second to predict your political affiliations just by the news channel you watch, and thus quote. So, while social networks make their money collecting your digital dust, the TV channels make their money by firstly charging you for their content, then getting paid by political parties to spread their unique gospel. They push political opinions most often by bending the truth, and sometimes by telling blatant lies under the democratic umbrella that media and freedom of speech is a right of democracy. This was never supposed to mean freedom to tell lies, blatant lies, paid and sponsored lies, but somehow that is where the democratic ball has landed. The probable reason they continue to misuse these democratic rights is because the people who profit from their services are politicians, the very politicians who are supposed to protect us and be our guardrails of safety.

Collectively the nexus is the data collected by the free services and application, then combining that to double paid news channel that align their broadcasts based on public perceptions and their political party needs. That is the only way an honest Fox news paid $787.5 million to cover their secrets. It was paid without hesitation, or thought because they make billions. Just for the record there is a $2.7 billion lawsuit from Smartmatic against Fox, Abby Grossman for Fox management pushing her to broadcast fake news, Crickey News against Murdock for libel, when they declared Murdock as an 'Unindicted Co-conspirator' of Trump. They will pay them all as they cannot afford to have any light shine on their dark operations.

Together they create human addicts on providing their services for free and each rung on the digital ladder makes their money by selling your dust to the next highest bidders. I apologetically repeat that Digital dust is more valuable than gold, diamonds or even prostitution, just because you can sell it a million times in a million combinations, to a million buyers, all at the same time, repeatedly day after day without losing any data that the seller owns.

The third influencer is religion. The one that stands out is the American evangelical Christianity with their mega churches. This concept, of mega churches, has been exported to South Korea and to some degree to Japan. Religion is a great influencer as a mega voter magnet. Also, whether you agree or not, I can prove that belief is more addictive than data. No one can give me any data from over five thousand years to establish that God exists. Yet we all believe it so vehemently that we have fought, killed, murdered and still continue to fight because we firmly believe that my god and religion is the only true one. Religious loyalty, is probably the reason US prospective presidents go around getting into photo opportunities carrying a bible. The reason for that is that the highest bidder gets the mega church to sponsor a candidate and that results in a mass conversion of voters – absolutely priceless in democracies, and absolutely critical in communist countries like Russia where the religious leader will often sponsor and publicly bless Putin. This is not unique to the US alone, it is prevalent in India, Saudi Arabia, Iran, Italy where religious leaders, in some cases even the Pope will influence political decisions across the planet. Like before July 24th, 2022, in the *Roe v. Wade*, the Pope sided with the Pro-life voters in the US even while there is not one mention of abortion in the bible. So, everything a pope or his priest proposes, the believer believes. So once again these religious houses routinely are playing politics whether for influence or funds remains unknown.

This is mind-altering disruption at an individual level. Whether you are a devout Christian that hears the Pope say something and you just follow that passionately, or it's something that your social tribe repeatedly broadcasts till you begin to believe it as gospel truth, like in the case of pro-life or pro-choice. Again, the individual can be john Doe, A Jan 6th Proud Boy, a Representative, a senator or the president himself. We're all addicted to digital things, but as yet our religious addictions are the strongest.

So, in order to cleanse your mind — the first thing you may need to cut out is to stop watching one source of TV, like Fox or CNN or listening to one tribal group on your social media. Also, stop watching outlier extreme channels. Don't just blindly follow what you hear. Fact-check statements and posts. Quite often Google it and you may get the results right then. If you want to save yourself and adapt to this disruption, you have to find all the digital devices and channels you use and make a list of them. And then, you can do a matrix or a self-evaluation. It could be on a sheet of paper where you ask yourself, things like — Is this an opinion or pure fact? Does this sound like someone trying to preach what I should be doing? Is this full of seemingly emotionally addictive content? Like your life is in danger unless you listen to me, or the Alex Jones example where he said the Sandy Hook children, were fake, the parents were actors and the massacre never happened, and he made hundreds of millions of dollars spreading such fake news. Who and why would they pay someone like Alex is the fundamental question we must be asking ourselves. Finally, you need to ask yourself whether this channel, or group is making you think in ways you never did before, mainly with an 'Us vs. Them' positioning – danger signal.

The CEO of Cambridge Analytica said, *"Give me five movies, or likes, a person enjoys, and I will tell you a lot about that person. Give me 50 and I'll know more about the person than their friends. Give me a hundred and I'll know more than the person knows about themselves."* Each time we text of a great movie, or the new car we bought, is a like. Each time we forward someone's text, or sit and read one sent to us, is a like? This is digital dust when we forward texts, or images, when we like, or dislike, something. We do hundreds of these a year. This is how algorithms work out on who we actually are. So, don't let the algorithms mess you up and influence the way you should think. Don't let them control your mind. Take back charge of your life.

Be mindful of who controls and influences you. Limit your screen time, and then again, be mindful of the things you watch, see, or do online. Question everything. Question things like — If Google and Gmail, Facebook and WhatsApp are all free, how do they earn billions of dollars? How are they so rich? If I were an ice-cream truck owner and I handed free ice-creams to everyone, would I get richer by giving away all my ice cream every day forever? Of course not, or maybe find out how! So, question what is happening here and know that nothing is for free. You are paying the price; they are harvesting your digital dust. This is the great digital opportunity — the start-up success in the digital future is Free. Let me explain this with an example. There was a free hiking money application which means that you will be paid for every step you take. However, in order to measure your steps, they needed your location, and they needed to count how many steps you took. They even launched an enterprise version where companies in order to invest into the health of their employees could arrange walking competitions using their free application. During installation a message popped up that in order to give you your points per distance they needed to track your geo-location. They tell you that they need it to see how many steps you walked, but again, how true is it? If you look at it, it's not really free, is it? The moment you give them your location, you provide them access to where you are and your life. They know all about your preferred points of interests. They know what day you were inactive and which day you did 1000 steps more than usual. They then reward you 10 cents for every 10,000 steps you take. That's seems like 10 cents for free. But they sell your locations to buyers on where you shop, what places you visit and what are your favorite haunts. That is worth $1 for every 10,000 steps you take. So, it's a 10-cent win for you but a $0.90 win for the supplier of this free app. Now multiply this into 10 million downloads and the game gets very interestingly large. How does this work — digital dust, in this case, geo-location dust.

Why is this important? Well, they actually get paid by these stores to get you into their stores. So, let's say you currently visit Macy's once in three months. They then offer your data to Macy's as a potential customer, and suddenly you will notice that you start receiving promotions from Macy's on new arrivals and personalized discounts only to you. This is B2P marketing, i.e., business to personal — one single person marketing.

With WhatsApp, we can message and communicate for free. All we need is a stable internet connection. This is a complete change from the earlier times, where you had to pay a lot to be able to communicate with others. Free texts, free audios, free calls, and even free video calls, what a deal, right?

So Wrong!

We just learned that nothing is free. So, what's the catch? Once again think Cambridge Analytica and how they might have changed the destination of nations, one of the known inputs was Facebook.

Now a politician comes by, or a Russian troll enterprise and ask them to give them a national list of all Republicans, who have an affinity to challenge the status quo, like weapon, are able to stand up for their beliefs, can be influenced by targeted messages, etc. The rest is history....

So, these so-called free devices and applications aim to find out all they can about you. Some track your emotional state, others your location, still others your political affinity. Collectively they know when you are happy, sad, or angry. They know your emotions, mental state, and vulnerabilities. So, they know exactly what they need to tell you, in order to alter your mind by using their stored data on what worked in what state of mind from their machine learning knowledge base. If that doesn't scare you, then I don't know what could.

It's the same way, all over the web. It is not as safe as we think it is. WhatsApp, Facebook, Twitter, and all the other companies work like this to make their wealth. They harmonize our data, text, words, thoughts, and our soul for whatever they can find. Digital data is not something you can sell once like a car. It is immutable and can be sold again and again, in different combinations depending on the buyers' requirements.

So, while we believe that Cambridge Analytica is dead, it is not. It is the new digital Hydra, where you think you've chopped one head of Hydra, but a thousand new ones have taken its place. Only, it is worse in this case; it is a lot worse because it's not just one or two snakes. There are billions of snakes seeding across all nations. And unless we mandate our governments and elected leaders put controls this will continue to grow unabated.

Growing up within this neurological manipulated net has impacted us and our minds greatly. We are in a hyper separated state of humanity no matter what country we live in. In India its either you like Modi or you hate him. In the US you either like Trump or hate him.

Russia, China, Saudi Arabia, France, it's all the same. The biggest responsibility here is for 'we the people' to keep our free minds open and become their masters rather than allow some digital overlord to play with our minds and emotions. Moving into the future it is critical that we do not let anyone or anything influence us based on some external agenda or plan.

In the traditional industrial, and pre-industrial period we lived in a pyramid. On the top of our **old individual**, it was religion and Nation that defined the rules for Parents were at the top, below that was the religious and social norms, beliefs and guidelines. Below that was customs and beliefs, and right at the bottom were we the people. From the Top religious leaders defined what the religious rules were and the nation defined the religion and national laws. These were then supported by the various layers of a nation. At the bottom were we the people — U & I.

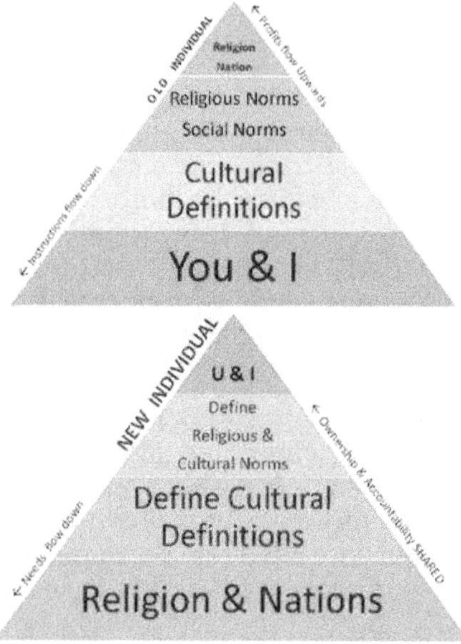

In our **new Digital Individual,** we need to flip this pyramid upside down with the Individual and their experiences, needs and definitions coming at the top. Women need to be able to decide the fate of their uterus, and children need to be able to protect sexual abusers hidden in their religious houses. The total wellness of the Individual must become the platform of religions and cultural norms. Religions and Nations must protect the happiness, and environment and orchestrate the melody of happy and satisfied individuals.

In our **old company** we had the CEO at the top who set all the rules

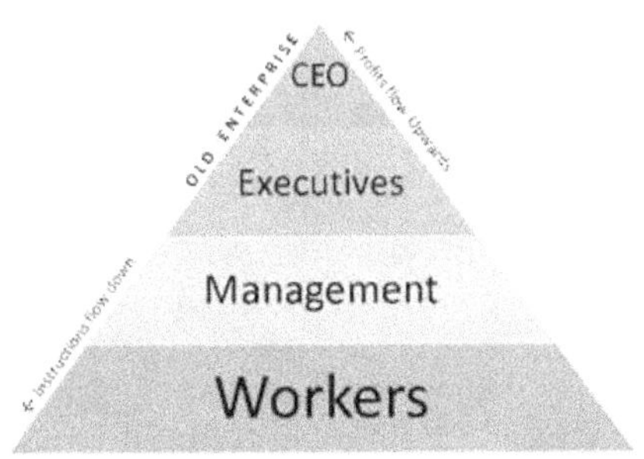

and executive orders, below that were the 'C-Suite' executives, below that the managers and below that the employees and workers. Customers used to sit outside the traditional pyramid. This was because the large enterprises decided what they wanted to produce, and the customers were captive customers who were given very few, to no choices.

In our new Digital Enterprise, I have the experience of more than two decades, where I specialized in digital transformations. In this area, our new digital enterprises have already flipped this pyramid upside

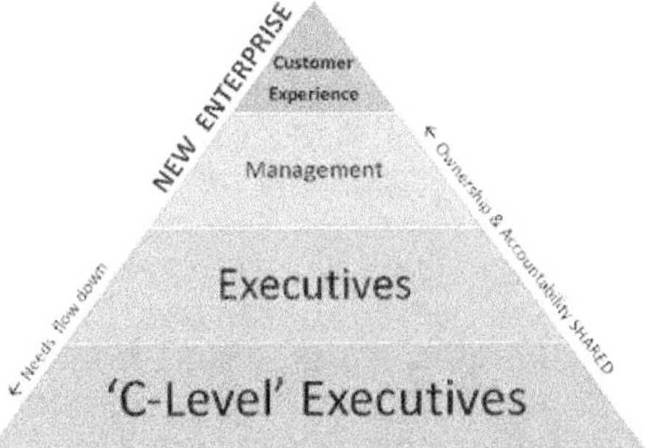

down. When we look at successful digital companies there is a very clear pattern that emerges. The first is a digital platform that customers can interface with in real-time. The second is the entire organization focused mostly on customer experience. Each time we have designed a digital transformation at a company the most difficult part is in trying to convince traditional management that *"What got you here will not get you there."* In every successful digital enterprise whether it is Amazon, Uber, Spotify, Facebook, Twitter, or Airbnb the customer experience is the center of their focus. The 'C-Suite' sex differentiators that used to exist are slowly shifting. In fact, I am predicting that the next two decades will be the decades of women, unless they let the men take it away from them by sheer force, its, theirs to give away. The future is for companies in hiring their new CXo, i.e., Chief Experience Officer, and their team of VXo, DxO and XoM's. This role when amalgamated with the CDO, Chief Data Officer, and their team is a win-win combination as no other. This is how successful companies are in building the *data-driven solutions that drive their business outcomes.* The rest of the team is there to support the customer scores, with the CIO and CEO at the bottom as conductors to keep the enterprise and its different elements in perfect harmony. Keeping Customers delighted has become the center of the digital enterprise universe. We're way past making great products, were past happy customers, and now were in the domain of Customer experience and their digital experience.

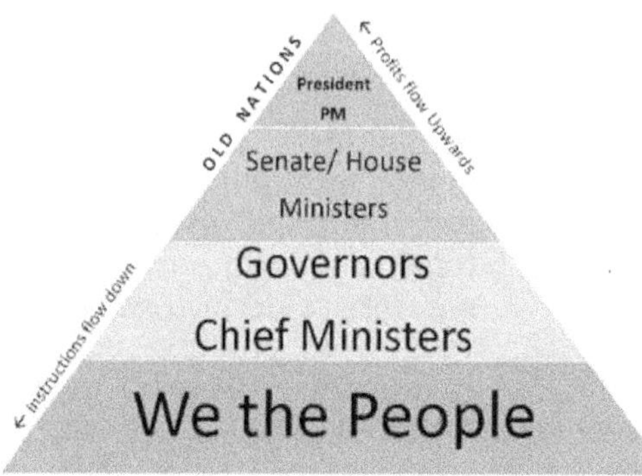

In the **Old nation** at the top are the presidents and prime ministers. More recently lifelong presidents like emperors of the old. This has happened in China and then Russia. Below that are the elected officials, below that are the state elected officials, and below that are we the citizens. Up until now, even in democracies like the US, potential leaders make all kinds of false promises, spread digital lies, make outrageous statement to grab eyes, bribe the media to tell only their stories, bribe their citizens, and their ministers or elected leaders just before the elections. Once elected they immediately act like their autocrats with little, to no, time spent in caring for the people. In countries like China, Russia, North Korea, Iran, Saudi Arabia, etc. they start from being authoritative and then grab power and power to keep their positions till death. Most of these are like self-appointed emperors and some have made this a hereditary business. The people remain at the bottom to be misused, and taxed in democracies, and actually raped in the autocratic nations.

However, in building our **new digital Nation** of the future we need to totally flip this pyramid. I have personally tried to do this flip in two nations and in the first we were summarily requested to leave the

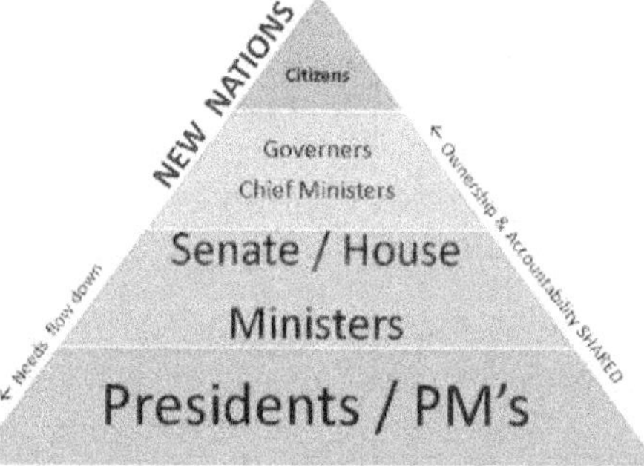

nation, and in the second my customer a general said "I can protect you while you are in our buildings, but outside that there are many people who will not like what you are proposing to do." In both cases I preferred to stop my work as life is dearer to me that digital transformation or transparency. The digital transformation of nations, I believe, is going to be most difficult, and here we need to power of democratic thinkers vs. the rich and powerful controllers. The key reason for this is that the very people who we need to order the flip, who need to define the rules and regulations for the digital transformations, who need to place barriers and gates, are unfortunately the very people who are rich, powerful and who currently benefit from this misuse the most. Let's think USA. Even in the US democracy do you believe a president like Trump and the GOP, or even Biden and the DNC, will support total digital transparency, elimination of lies and fake news, criminalizing of paying media to broadcast their POV, censuring representatives and senators that do not truly defend the democracy, etc. The answer is a clear no. The next question is do you believe that no one in either party pays-to-use digital, social, or media disrupters to change the minds of voters or their tribal members, or that each elected party is true to America and the citizens and party comes second. To me both the answers are absolutely No. The do you think the same rules apply in the world's biggest democracy — India. Absolutely. So, if this is the situation in the two largest democracies imagine how it is in the rest of the nations. A difficult chasm to cross no matter how we look at it, but never impossible. If we could together create the first democratic America then we can together certainly create the new Digital America or India. It's a national moonshot we have to take for this decade and commit to having a clean democracy by 2230.

The digital disruption is very rapid. In order for it to succeed we need to focus on the User Experience, something we refer to as UX.

Just as an example in 2021 Texas tried to pass the heartbeat law, which means that after hearing a baby's heartbeat, the woman has zero rights toward getting an abortion. This makes it a crime to get the child aborted. Biologically a fetus heartbeat is detectable 5.5 to 6 weeks after gestation. The fundamental background for this law is religion. The Texas proponents, with support from the Pope and their evangelical Church priests, quote the Bible. However, there is not a single mention of abortion in the bible. So, it's just a misogynist interpretation of things not written in the bible supported by the church and politicians – our 'beyond reasonable doubt' collaboration between religion and politics. The problem with this is that the Constitution has clearly defined the areas that were supposed to be separated. Two of those are religion and politics. But since time immemorial they have been sleeping in the same bed with common goals. Each benefiting themselves and the other. It has opened up floodgates of arguments and controversies.

Then on June 24th the Supreme court of the United States, SCOTUS, put the final nail in the coffin. They reversed the *Roe v. Wade* law based on the fundamentalist Christian majority at SCOTUS. I published an article 'Being Woman' [viii] on LinkedIn with my comments on what the data tells us.

Frankly, I was appalled when the news of this law broke out. I wondered how we claimed to live in the 21st century and still acted as though we were living in medieval times. I remember retweeting something along the lines that the Supreme Court and Texas both have initiated a new law of slavery on US women.

Slavery is defined as "a condition in which one human being was owned by another. A slave becomes by law a property, or chattel, and is deprived of most of the rights ordinarily held by a free person." The *Pro-Life*, i.e., we do not support abortions; this lets the church and state conceptually separate the womb from the woman. It says after July 2022, American woman, let's say in Texas or Florida, are free, but their womb — now becomes the property of the church and the state, which is the exact definition of slavery. Even if the girl is a 10-year-old and has been raped — her womb is our property. Not only this, the church and state have created the new bounty hunter in 2022, who this time, is now not searching for Black slaves for their masters, but will be rewarded for identifying the wombs of girl-slaves who are pregnant and may want to get freedom from an unwanted fetus, i.e., an abortion because they are either not ready for a baby, got raped or are pregnant due to incest. Texas and most Republican states say they want no exceptions.

The second reason is that there is a high probability that the White, Christian male majority passed the law to suppress their own White women and the minorities — two birds with one stone. Men have since time immemorial created all kinds of barriers, fences, and rules in order to punish any woman who looks at another man. This is the new US one. In the US, most of the people getting abortions in clinics are minorities and people of color. So, in a way, it is a way to target the minorities. The rich can fly to doctors or fly their children, or wives all across the planet, to Mexico for example where they provide 'No Questions asked' abortions. Interestingly keeping business in mind, in June 2022, a new law was passed in Mexico right after the Roe vs. Wade decision was taken in the US that legalized abortions in Mexico.

There are two groups of people who gain from the increase of population — religious houses and politicians. Religious people want to fill up their churches, as each attendee generates direct revenue which in 1919 was to the tune of $860 billion in the US for Churches alone. The Pope erroneously believes, that if they prevent abortions, they will be able to fill their churches again. A lot of politicians want minorities to increase in numbers to get more votes. This can be another reason why abortions are being banned. The merger of religion and state is a match made in heaven, so it would seem.

So, the moment I tweeted my opinions about abortion and how I was against the law, then when I published 'Being woman' on LinkedIn, I openly announced to Twitter, LinkedIn, and Google that I was not in favor of the law. So, they have the data, and they understand the way my mind works. Now, they can use me and my thinking as a menu for a planning their feedback on 'Pro Choice' content when they send me posts, and I get a lot of them. They can create a disorder, or discordant, thought in my mind with thousands of examples that have proven to work on other people with my type of thinking. They can thus amplify me to myself.

It's the same way, all over the web. It is not as safe as we think it is. WhatsApp, Facebook, Twitter, and all the other companies work like this to make their money. They first get permission, then collect your data, and after that they now can sell the data, text, words, thoughts, and our soul to the highest bidders they can find. Then they allow reverse feeding by fake accounts and trolls to influence how we should think.

Growing up in this neurological manipulated mindset has spread all across the planet. It has mostly impacted people who are addicted to digital devices. The more addicted the more severe the condition. Frankly, it has impacted all of us and our minds to a degree that has as yet to be measured. The logical ask is that we want the governments to control it, but feel they won't. Then it comes down to the domain to control it and they won't too. Now suppose individuals themselves do not, or are not able to control this, then what? I think the biggest responsibility here is for the individual to keep their minds open and become their own lawyer, be their own founding fathers. Write the Constitution for yourself. It is crucial that you do not let anyone or anything influence you. Do not let anyone other than yourself write your future.

Take Aways

Individual: It could get fantastic or pretty nasty from here on. We are in control. We, meaning humanity, try to understand the physical and neurological impacts of people who, during COVID-19, got stuck in front of one or another digital screen for 12 hrs. a day. You can form a group where you actively discuss digital disruptions — both the good and bad, i.e., focus on the good you can activate but also on warning about the bad actors. Don't become the frog in the boiling water because the perception of change is slow, but the actual change is extremely rapid. Then, you become the new digital zombie. Build your own method of 'fact checking' to audit highly explosive texts you might receive via your social networks.

Companies: It can only get better from here on; however, [1] Be aware that over 70% of the projects you start for digital transformation will not meet business expectations (Gartner). For the last 5 years, I've been on a mission of turning southbound projects toward the north. [2] It is 60 to 200 times more expensive to try and fix critical issues after going live than it is to fix and test them before going live. [3] If you're not thinking digital yet, do think hard. [4] Beware, the speed of change is very rapid, so plan your strategic roadmap with business, and don't let your IT or partners define your directions alone. Think of end-user experience as the new focus. [5] In the last decade, I've also helped companies sell digital transformation not on technology but by building emotional connections via structured storytelling.

***Nations:** Become the lighthouse nation, the leader that started the global 'Do it Right, the First Time' initiative. There is so much that all leaders and Governments can do for their citizens and the world. Become the arrowhead for positive change. [1] Xi has been 'elected' for his third round by unanimous votes, while Navalny's prison sentence has been extended for an additional period of time. Putin remains a threat to world order, and the nexus of Xi and Putin is not healthy for a free world. [2] The problem is most world leaders would love to do what Xi and Putin have managed to accomplish in their nations. Trump briefly let slip as he alluded to how he would love that. [3] The world is going 'Forward to the Past' very rapidly. SCOTUS wants to evoke the bible in Roe vs. Wade even when there is not a single word on abortion in the bible; Putin wants to invade territories that Russia used to own. Xi is trying to justify the same to harass its neighbors. While the UN sits impotently, not doing its job. I wish the UN made a global declaration that all territories, as of a certain date, are to be honored by every member state. [4] All this while the digital war is accelerating, and global banks are predicted to possibly crash in 2024-5.*

***The 'P' factor:** There is a huge physical, social, psychological and mental impact triggered by structured neurological techniques. The positive potential of using this in a systematic manner is tremendous. If we follow a 'Let's develop global PEH' scorecards we can start right at the bottom and then roll up the data to the Planetary level. Let's take any issues and then form a global database of reliable data feeders that work on the PEH principles. Based on the extinction data, i.e., 96% of mammals on earth are humans and ones we harvest for food. Let's launch **Dignity of Animals**. If were truly intelligent and humane, like we like to believe we are, then let's save all god's creatures and start with mammals. The time to do this 'one animal at a time' has already passed us by. We need to not just think on our couches we can now mobilize worldwide.*

Chapter 15 — Digital Shock, Mental Adaptation

On May 21, 2020 a Black man by the name of George Floyd was pulled out of his car, handcuffed and pinned to the ground, and then publicly executed in broad daylight. Such things have happened throughout history but the word of the uniform always ranked higher than the common person. However, in this case a girl called Darnella Frazier recorded the whole event live, non-stop, on her digital phone. Now, police could not deny what really happened, and we can see this execution on demand and on the internet. This digital recording removed all assumptions, spin statements from the establishment, for there is over 11 minutes of non-stop recording of a policeman acting as police, judge, jury and executioner live on camera. The death of Floyd, brought about an almost instant international outrage. It amplified, and brought to the forefront, a movement, called BLM, in the middle of a Covid-19 pandemic. BLM (Black Lives Matter) became an international cry in support of systemic discrimination, and White supremacy optics. What this recording resulted in is firstly a reaffirmation that police in the US treated Blacks more violently. Secondly, the trust in the establishment version of truth was eroded. Thirdly, it proved that a digital recording is irrefutable and can even bring the guilty policemen to the court, and get them convicted. Use that power prudently.

From a digital explosion point of view at no point in history have common citizens been able to record, distribute and evoke global support in a matter of hours. In the pre-digital era, if a person died in a police situation anywhere in the world, including the US it was common to have only a police version that quite routinely cleared the establishment of any crime. We must laud, the digital revolution that made this possible. Never before has an individual citizen been able to refute the establishment, and win a Pulitzer Prize for their commitment to truth. Imagine as an individual you can video a crime, blast it globally, find justice for the victim and then even possibly win a global peace prize in the process.

Adaptation and evolution are probably the easiest thing for all living things to accomplish. It should be a no brainer — adapt to survive. Bacteria adapts the fastest, then insects and so on. The more complex a life form becomes the harder, its biology, and then its intelligence makes it possible for rapid adaptability. Or so we're made to believe. Now go activate the evolutionary genes and hit survival as our next goal. In order to survive we all individually and collectively, can actively accelerate our adaptability.

Humans have been adapting to changes from the beginning of time. This is what drives our intelligence. However, our digital catalyst is using this exact adaptability on us and against us. If you look at any kinder Garten class across the globe most things are common. These small children laugh, hug and play with each other without any prejudice of color, race, religion or sex (sounds like the US constitution). Yet between these early years and growing up something horrible happens to every single human. With age some divisionary things only get amplified within most humans. First of all, we all inherit the polarized views, and biases of our parents, then our teachers, then our colleagues and finally by the tribal messages of the inevitable communities we join moving forward. These communities could be our universities, religious houses, frat houses, political groups, work environments all of which influence our overall mindset. With passing time, we get aligned to their collective rituals. Rituals designed to inculcate love and passion for our tribes and hatred for all others that, somehow, we are taught to hate collectively. When this hate is inserted depends on all the above inputs. I have sat at airports and heard mothers humming into the ear of their newborn baby, "My sweet darling, remember there is only one God, and that god is *XX* your truth for all things in life. There is only one book, and that book is *YY*." Have any of you ever heard anything like this? Well, I have been in more than one country and on more than one occasion.

Over the course of our life, we are born totally unbiased and free, then we are systematically groomed and were made to walk on this path of artificial boundaries, discrimination, problem absorption, and working in a state of blind acceptance. A state of mind that believes we cannot change a thing. Timeframe, nature has assisted the human brain to be constantly rewired by situations, environments, geological events, and our environment. Every living animal is globally wired for change, as a form of survival, both individually, as a society, and as an integral part of the living planet. This is our natural inbuilt adaptation in its full force, where threats supersede all other drivers. However, on the other side we're being systematically surrounded, into societal beliefs, by people, and environments, we come in contact with. On this human side, we're being methodically ensnared, into societal beliefs, by people we come in contact with, the nation's we belong to, the religions our parents and nations support, the many artificial boundaries society demands we live inside, and when asked even give our lives for.

Throughout history these subtle changes to body and mind, were initially based on real, and more recently perceived threats. However, in recent centuries we have a brand-new enemy, i.e., human created artificial threats.

The digital algorithms simply use this inbuilt neurological adaptability as their foundation and then continue the process at an ever-faster pace and rate than ever before. With time, they also become more efficient by learning from millions of responses and thereby enhancing their messaging to an individual target, as an example of machine learning. Putin's Goal is to get Russian sympathy for their attack on Ukraine. They send a hundred messages to a thousand targets each in seventeen countries, with the message content being different for each country. By the next day, they checked how their 1,000 receivers reacted to their message. Then they make a small change and send it to the next thousand, and so on and so forth. At a certain point in time, the machine has learned that this is the best message for a certain group of people in a particular country by attributes they use to measure effectiveness. So now the ML has done their grunt work. Now, the AI, or artificial intelligence, comes in and identifies each individual target as to whom that individual message should go while the ML continues to score the effectiveness and fine-tune the message. Once the message has come to a particular score, then they will send it to representatives and senators in the US, to the MLAs and ministers in India, or to the influencers in Europe, and so on. All the while, the ML and AI shadowboxing enhance the effectiveness of every message non-stop, 24x7, endlessly, by each single target recipient.

Across time the ancient areas of influence were the tribal leaders and the shamans, or witch doctors. The old areas of forced influence were Kings and the religious leaders. In the recent times these were replaced by politicians, religious leaders, media, and places of worship. Right through time the only constant has been religious leaders. While kings and politicians rule our physical environments the religious leaders rule our minds. The kings became temporary but the houses of worship became more powerful and permanent. Over time kings were replaced by elected presidents, and prime ministers, for small periods of time, and now some of them have found ways to short circuit the basic electoral process and become presidents for life. We should be able to bet that this might be the dream of every politician on the planet — become the next president for life. The ultimate forward to the past situation we are rapidly heading toward.

In order to stay in power, modern leaders are working hard to bend the will of their people, in order to remain in their position. The degree of this force changes from one nation to the other. At the extreme are countries like North Korea, Iran and Saudi Arabia. Still closer to them are China and Russia, at the far end are countries like Germany, France, India, UK and the US. We know for sure UK and US have faced digital manipulation because of reports from Cambridge Analytica. So, in this case no nation is totally innocent. In the US the current Republican vs. democratic polarization proves beyond doubt that this manipulation is aggressive not only at the citizen level but also at the elected official's level. The exposure of Fox news and republican alliances goes further to prove that the nexus between influential media companies, and political parties has now crossed a perceptual red-line. I firmly believe that if media is a constitutionally protected arm of a society, then we cannot have media act like an only-for-profit enterprise.

Modern democracies have put breakers to their overall power by putting into place segregation of ownership and duties as a balancing factor. Still, a higher force in democracies are the bureaucrats — they have far longer terms and can thus exercise far greater decision influence. In the US the Supreme court judges are for life so they feel invincible. In India and UK, the bureaucrats are for life so politicians may come and go but they rule for ever. In retrospect the most permanent power holders are the houses of religion, as they play with the extremely powerful emotions and the souls of humans. As a rule, the longer a person is guaranteed in any position the more arrogant and autocratic they could become. The more they love that position and the less they want to lose it.

The kings of Egypt used to listen to their priests as these priests held the souls of the people and thus those of the kings. The priests in India, like Chanakya, were kingmakers by the power of their control over people and the chemistry of both people and elements. During the crusades and from there on, the Vatican has held great influence over kings and nations, to the point where in 1307, the King of France, who was going broke, made a deal with Pope Clement V to take down the Knights Templar by arresting the grand master Jacques de Molay, who was tortured to tell where the world's first bank kept all their money. Reportedly, he was hung over a fire till the flesh on his feet was charred, and the bones were visible. This was undertaken at the behest of the Pope and the Vatican. The Pope claims direct line back to Jesus, and Catholics see Jesus as being present in the papacy. In more recent times, the pope supported a political call for Roe vs. Wade in the US directly despite there not being a single word on abortions in the bible.

The overall power of the church is so great that even today when potential US presidential candidates face the public, they do so holding a bible in their hands. Trump even gassed a peaceful protest in Washington DC so he could get a photo opportunity in front of a Church opposite the White House. The power of religion is so strong that if a religious leader speaks against a candidate, then they are as good as gone from the political scenario, because the religious masses will as a single body listen to their religious leader as if spoken by their god.

The modern exceptions to ancient kings and queens, are presidents like Xi and Putin who have managed to declare themselves as leaders for life, i.e., permanent. Thereby successfully taking their nations backward to the days of the all-powerful emperors, by becoming one in the garb of democratically elected presidents. At the heart of all this is tribal alliances. Take an example of two Christian in the year 1118, one of them is from the UK and the other from France. They were willing to ride into Jerusalem, with the cross as their banner, on a religious journey of salvation, and an attempt to liberate Jerusalem from the Muslims, even though they don't know each other. Then take an example of two Americans in 1966, one a Black Muslim and the other a White catholic Christian, who do not know each other but are both willing to fight and give up their lives for a war in Vietnam just because their political president say it is correct. Now fast forward and take the Black Christian and a White Christian. This time the Black is George Floyd and the White Derek Chauvin and his three White-policeman. The rest is history. How did we the US every come to this. It is all based on the polarized beliefs of the political divide that amplifies sex, color and religion as a deathly divide that is not protected by the US constitution.

Similarly take the example of two Germans anywhere from 1939 to 1945 who were slowly brain washed by their Fuhrer, into collecting their Jewish neighbors, stuffing them into trains in conditions worse than animals, then thrusting them into gas chambers with a clear intent to exterminate them. One set were the German families, the so-called Aryans, that still remain alive today, while others, who were Jews, were brutally eradicated based only on political and religious tribal beliefs.

These are very rigid boundaries; national, social, political, and religious, that, over history, were created both by powerful people, and powerful nations. In most cases these boundaries organize people, and were the foundations of our civilized world today, but only when they are utilized positively. Then around two millennia ago religious and national divisions collected, stamped, and separated, people under their tribal banners. This has continued throughout history. In the last two decades or so, all the above power brokers have realized that by using social networks, with curated messages flying on the new digital ether, powerful people could systematically create highly polarized boundaries, tribes, and followers, and thus become all the more powerful. These tribes split into micro-tribes with each community providing more familial feeds by their social media connections. Each micro tribe trying to grab eyeballs by their being more aggressive in their communications than the last popular one.

When Covid-19 locked humans globally into their homes this became a bonanza for the digital trolls to have unhindered access into the minds of citizens across the planet. From a politician's point of view — the potential to disrupt everything that people are familiar with, leading their followers to dive ever deeper into highly specific tracks they lay in front of them, to follow as tribal loyalists, was beyond their wildest imaginations. We citizens could now somehow become small mindless digital assets traveling along pathways being laid by the digital trolls and algorithms — changing both our conscious and subconscious mental settings.

There are thousands and millions of such examples. Referencing our world from a June 2022 view. On the positive side we see a planet that is more connected and concerned about each other than ever before in history. The way US citizens reacted to George Floyds case resulted in the imprisonment of all three police at that scene. However, at the same time we also are witnessing a planet that is getting more polarized and divided than ever before. We are witnessing the coalescing of tribes and influencer's that now travel beyond national boundaries, with global digital trolling, and neuro-altering algorithms, that are shaping the minds of mega tribes with very targeted 1:1 messaging toward some common goal. In the US, we're witnessing a 'blind-faith' political tribal loyalty that is so influenced by internal tribal messages, resonating on a clear *Us-vs-Them* to such a high degree that anything the other side states is immediately considered fake, planted and not reliable.

At the same time anything their own tribal leaders state is now viewed as the only gospel truth beyond doubt. Humanity today, i.e., the US mindset in August, 2022 has become hyper polarized and nationally fractured. If I did not know better, and from my unqualified point of view, if I was made to guess right now, I'd say it seems the enemy is winning because our citizens have been made to mistrust their own democratic pillars, keepers of laws, agencies and people. What remains is a total distrust in our very status quo. Resonating with history, this is something that had happened in Germany starting sometime in late 1923 in a failed attempt by the Nazi's. By 1933 Hitler was appointed the chancellor of Germany and the leader of the Nazi party. The rest is history, despite Iran stating that the holocausts never happened it never changes anything.

Globally the pillars of what makes a civilized society are being ravished by internal political dogma. The presidency of Trump has placed the optics of US in a rather new light across the world. As the US splinters internally, the repercussions will be felt worldwide, because for almost a century where America goes the world kind of follows. That system is now being broken down from the inside of the US of A. Just as an example the Jan 6th coup was recently copied in the Brazil election results, when Bolsonaro followed the Trump playbook by declaring his election was stolen. US is a mature democracy so it can recover from the Jan 6th coup. While, Brazil had been a military dictatorship and they are a young democracy. The biggest risk to the US is its internal polarized political enmity where the US is at the border of a civil armed insurrection even as I write this. The second is that while the US is busy infighting the world is already looking at future partners and between all the players out their China is coming out loud and clear as the prime competition. The third, unfortunately is that while China is the greatest risk to American dominance it is the US, and Europe, that has been, and continues to fund, feed and support all the Chinese growth of systemic controls.

Hence, the negative effects of this digital transformation have brought us to the point where we may have short-circuited the minds of our own society by paying Russia and China to leverage their superior and without control trolling techniques for our petty political gains. Though our planet Earth has gone from war and discord to peace, we are now very rapidly slipping down a steep slope, which is getting steeper as we go forward, back toward the inevitable potential for war. The growing escalation of intolerance in all countries is digitally encompassing individuals, religions, nations and humanity. No doubt, although we have achieved a lot, but socially, we are still being forced to walk to the future facing backward. Just like we get more power from micro segmentation in our digital world, so too the bad players are using this technology to polarize the human fabric along micro tribal tears. More and more people are becoming less tolerant of anything that does not conform to their own tribal beliefs and opinions, mostly religious, political or nationality, in that order.

Take an example of the pandemic where all nations were supposed to get united. Instead, we all went ahead and started fighting with each other. It started with the US and China, blaming each other for starting, and spreading the virus. Soon thereafter with humanity at the edge of existence, nations started a war on vaccines, like whose vaccine will make more profits. Suddenly a global threat to humanity changed into political maneuvers of whose vaccine would dominate the world. Then, by the mid of 2020 many nations started banning every other nation's vaccine from entering theirs, while other nations started filtering who could not receive their vaccines.

While this calamity is happening, the digital disrupters explode with entering more minds than ever before, becoming a clear disruption revenue generator for this century. The 18-month lockdown turned out to be a bonanza for all digitally networked communication devices and applications. People watched more TV than ever before; they participated in more social networks than ever before. Humans became more connected but slowly lost all their social skills due to their physical isolation and the continuous social network amplifications. That's why we, are breaking down into polarized groups. Once side focused in saving humanity, global peace, climate control initiatives and a be good to the planet and fellow humans. While the other is working on amplifying hate, fight anyone that does not believe in their tribal messages, pick up arms, denials to scientific data like there is no global warming, and 'drill baby drill' and make money while you can. Both sides believe that the other side consists of highly destructive individuals, in which all are willing to fight and pick arms because none of us want to compromise our point of view.

Our violent nature is apparent from the story of China, creating hurdles for Uighur Muslims and using a brash display of weapons, when US Pelosi visited Taiwan, to keep Taiwan in its perceived control. Some leaders think that building bigger weapons and using them can make them dominant – the competitive attitude of a bully. Yet, what it does, at the same time, it turns that nation into a villain by the unnecessary display of overt power. This is like Russia using all their never used weapons on their neighbor Ukraine. While, at the same time people of Kashmir in India, where every other Muslim person thinks that using weapons to bully, pelting stones on the police and army, and teaching the Quran are the only way to get independence is the other reality. In the US, the fractures between the Whites, Blacks and colored, between republicans and democrats, between women's rights on 'pro-choice' and 'pro-life, whether climate threat is real, are mostly exasperating the problem. Or, maybe distracting us from some more critical, and main issues.

Then you have the Fox news and people like Tucker Carlson who mass distribute what has been accepted as blatant lies. To which they say we never said this was a news program — it was an opinion show, and this is my opinion protected by the freedom to speak. The reason behind citizens carrying semiautomatic weapons openly is nothing but information overload, and the harvesting of fear, by misquoting the US constitution. The fact is that, due to the constant contact with different social networks, individuals have so much information, and often much more guided mis-information, to process every second that they get confused. Confusion leads to a state of fear. Fear leads to wanting something to hang on to. That is the state of mind the trolls, and their algorithms want to create in our minds. In that state our mind is like putty and any amplification of your likes and dislikes is exactly what you will be constantly fed. Each collective instruction makes people do new things, and adds to reaching the individuals tolerance toward change. That's why when the world is changing constantly, and faster than our ability to adapt, we sometimes collectively start to lose control.

In the end control is never lost, you simply give that control to someone else who then controls you. Or put it in another way — "Happiness and your thoughts are your birthright; no one can take that away from you, but only you can give that right away." This may not apply in Iran, China, Russia or Ukraine right now, but taken at a collective level this is still true. Iran is on the verge of a revolution and they have succeeded three times in the past.

With time, more and more people are reaching into a numb state of digital shock. Its worst impact is apparent from the manners and etiquette of people. All of us have observed people getting out of control over miniscule things on the streets, planes and generally in life. There are more news stories of people misbehaving in airplanes and flights getting diverted because somebody misbehaved more than the crew could anticipate, or handle.

In the past, it used to happen once in many years. But now, it has become common. As the degree of this digital shock increases, its winners and the victims will be separated by their participation and their gains and losses. Sometime if for no other reason than to get their mugshot in the news as a deviation to an otherwise digitally induced boring life.

The **winners** on one side will be the people who have made a conscious decision to fight these negative digital suggestions from trolls, news media, and other paid opinion generators. These are the people who have adapted to new situations and will end up investing their time and resources in ever-newer opportunities. The **losers** will be people who fall into the rabbit hole, akin to an Alice in Wonderland story created by digital mind shapers. If you are passionately for or against your political leader, then you probably already are a victim of this digital manipulation. If you are neutral to them and their contributions, then you may still have hope before 2015. Whoever heard of people not speaking to their siblings, parents, teachers, friends, or someone on the street due to their political differences. Right now, it is routinely converting into life-threatening situations across the planet.

Technology has reduced the size, distance and the reach of global citizens, while at the same time enhanced direct communication capabilities of humanity. Seeking and working on the positive applications of this new world means that you provide the best solution to a tiny group of people who then fall in love with the result and each sees this as social outlet from the predicament they are currently in. This is how digital startups grow and the team becomes millionaires and the founder's trillionaires.

Opportunities: The secret sauce in the future is to find a successful physical concept that can be digitized. Uber, Lyft, Ola are examples of digital replacement of the traditional taxis. Lesson here is just because someone has done it before, i.e., Uber, does not mean the opportunity is gone. You should plan to do it better, i.e., Lyft. In this example you can ride the wave that uber had created. On the other hand, one could digitize a totally new area of operation, like construction. A company called Katerra was launched in the US by leaders from the semi-conductor industry. They saw the opportunity to take chip manufacturing techniques and apply it in the construction industry.

When looked at from above a modern city looks very much like a computer chip. They raised over $1 billion but along the path made a few mistakes. The first being they diversified too much and too fast, due to availability of very large investments, and thus available cash on hand. They started thinking along the US trap of synergy, while they should have stuck to German − do one thing and be the best in the world. They tried to build better air-conditioning, home fittings, etc. thus not only alienating the traditional leaders in these niche spaces, but also spending critical safety funds on non-essential business units. Their biggest mistake was that when finance became tight, they diverted their strategy from being a digital construction startup — something that attracted me to them in the first place, to then hiring traditional construction management and thus trying to become yet another traditional construction company. This fast forward to the past, just did not work. In the end it was Covid-19 that stopped them in their tracks globally as construction cannot be done remotely and the world was in a physical lockdown. In the end by June of 2021 the company declared bankruptcy. From start to finish they burned through $3 billion, making it the best funded US startup to go bankrupt. So, if you get funded and grow, please think German. Focus only on your core product or services and never deviate from that, ever. Even if you have 700 million dollars sitting on the side. Use the funds to become better in your prime play.

It's an easy path to success if you can find you win-win application where your customer wins by filling a Quality + Cost gap, and you win because your customers are your best PR service. For example, my Uber trip from home to the airport is more predictable, the car is cleaner and better quality, its faster from book to arrive, and cheaper than what I used to pay when I booked a traditional taxi service, a day in advance. In addition, if you grow fast and steady there is always the possibility of an exit, or being acquired. Many startups today are designed for exit. Sometimes if the concept is great, it might benefit from a larger organization to globalize, capitalize, and penetrate the market faster. The other option is that you could become a successful local, national, or a multinational organization as a small company simply on the factor of what I call German focus. Germany has the maximum number of successful Mittel stand, or Medium sized, global companies in the world.

We can say that digital transformation has provided a global foundation for everyone to become a winner at an individual level, while instantly giving them the platforms that can take them into the multinational enterprises. While, at the same time the large brick and mortar companies are $*#! scared of these new startups that could disrupt their business seemingly overnight. Guess who their direct threat is *You* and *I* — the people on the street with a new idea and a new way to do things. How many apps do you have on your phone with some of them being your favorite, each was developed by an individual or two.

These digital disruptions are not a gradual process but a violent change, like the industry of taxi companies vs. Uber. The examples are in front of us, sitting in your hand and your smartphones. We're unfortunately focusing on failures, but are more surrounded by individual success stories. The growth of Uber and the crash of traditional taxi services happened almost overnight. Hence, the only way for enterprises to grow is to inculcate or incubate that change from within. Identify an idea and then expand rapidly to become the disrupter before somebody disrupts you.

The two key factors need to become a great disrupter is better user interface – i.e., easier to use, and digitized, i.e., available on your digital devices, Quality and Cost by now are a given. Get these attributes together and you have an idea you can launch. It could also be an existing application that you have the potential to improve.

There are numerous opportunities at a political and national level because as of this year politicians in almost every nation want to drag their citizens minds back by 50 to 2,000 years. We all are familiar with what Russia and China are doing, the communists banned the emperors and now XI aspires to become an elected emperors by order. The only difference is that they don't call themselves kingdoms or emperors as that is politically incorrect. However, if you look at the system of North Korea, there is no difference between the emperor and its president. He has the same power and applies the same rules. In north Kore Jong-un has declared himself as the Supreme Leader since 2011. He democratically gets 99% of the votes.

It is crucial to control the reign of digital disruptions because it is not subject to national or political boundaries. I can quite easily hire a Russian, Chinese, or US trolling company to amplify whatever point-of-view opinions I need to push to a very targeted or specific audience. The CA influence on British public before the vote for Brexit, or the US presidential elections in 2016 or 2020 is well documented. First and foremost, it is important to realize that if we fall prey to the politicians and their digital algorithms, they have the potential to brainwash our minds with their targeted opinions and communications. In the US some of them, political parties, political leaders and elected officials, already believe it is their birthright to do so for their own betterment. The rich will use the traditional digital assets for influencing your mind. The very rich, and autocrats, will buy, or coerce, every avenue that surround you, and encompass you with their alternative reality feeds. Think China and Russia were every digital and media feed is controlled by the system. Now come closer, to the US where media companies, that are paid billions for touting only one political party's version of the truth — to their tribal followers. For example, announcing on a daily national feed that the Biden election is stolen, and then swearing that Trump is the only true and legitimate president. Then getting fake electors to swear they have proof of the true results because the election machines were tampered with all fake paid opinions. This is then followed by the owners of that media telling anchors what fake news to pimp. Then when caught pay almost a billion dollars with not a blink of an eye. Another example, distributing a negative Black message to a White tribe, and a reverse White message to another Black tribe, is all now part of a game, often in the same city and area. The reason is to make tribes fight each other and create the foundation of fear. That's how power corrupts.

We have seen how power corrupts and destroys nations. We are also aware on the power of opinions, especially if made by our selected news media channels. Such opinions give birth to the creation of highly polarized citizens, down to family members within a single house. It is common to witness polarized arguments between brothers and sisters, husbands and wives, and parents and children, as they each have different political opinions, and are getting amplified feeds from their personal digital alliances. Now, after a person is brainwashed, they are encouraged to lose their ability to accept or understand the other person's opinions, and that's how the two people get into a fight, sometimes with guns and weapons, and sometime resulting in deaths.

As we get more and more surrounded by these free and convenient smart applications it is important to slow down, or stop, for a moment and ask who these devices are smart for — the supplier, the Governments, or some hidden buyer. The answer to these questions is complex and elusive; their only purpose is to disrupt and brainwash their masses into total subjugation, when their psychologists, trolls, and algorithms can identify each of us on a 1:1 basis today.

Take Aways

Individual: The greatest shift out there is coming in the area of AI/ML, and now ChatGPT. ChatGPT is a beauty and a beast. Beauty in its simplicity and a beast on how easy if make it to use it for good and bad. Companies and nations that own this will launch the new digital colonization by around 2030. The foundation of this new AI/ML platform will be quantum computing. So what should you do as an individual: [1] Identify the things you cannot change, (a) digital disruptions, (b) trolls in your social networks, Russians, Chinese, or US algorithms trying to figure out your mind, (c) your bank or company going bankrupt, etc. [2] Then focus on what you can control, (a) become the digital disrupter with a digital alternative however small or great, (b) Figure out a way to identify when a troll is trying to trap you. Better still, if you find markers, then build an app that people can buy to identify troll messages. (c) Move your money to larger banks and get out of small banks, especially after the SVB and RB crash in March 2023. [3] Build your own method of 'fact checking' to audit highly explosive texts you might receive via your social networks. [4] Predict the future 10-20 years and see where you would like to fit in that future. If you think regular taxis will die, then don't invest in them. If you think Quantum + ML/AI will be the next winners, then invest in that segment.

Companies: *Unlike individuals, or maybe just like individuals, there are companies that will adapt fast, while there are others that will be most comfortable in their legacy ways. [0] Digitize, Digitize, and use ML/AI. [1] Companies need to upgrade their platforms as the old legacy solutions cannot adapt to new digital expectations. [2] When deploying these new solutions, make it experience-inclusive. In my last 3 projects, IT delivered, and the companies spent years and large funds to make the system do what they expected it to accomplish in the first install. Which most of them did not deliver fully. [3] Make a digital scorecard and put all your partners and assets, and then every quarter, predict if there is a risk with any. Today is March 10th; a Silicon Valley Bank just crashed with $200 billion in play, while the First Republic is getting pulled by the rip currents. SVB was the world's largest Start-up bank, and I got a call just today from 3 colleagues that their payroll would be disrupted as their prime bank was SVB. So, the bank should be one of your attributes of risk. Another could be your cash cow. I know of a US enterprise where its margins are dominated by manufacturing in China. Now, if the world and the US sanction China for any reason, their margins could get impacted in a big way; this could be your next risk attribute. In a fast-moving world, we need to conduct our risk score more frequently than before. Automate your risk scores using digital assets and customize them to your attributes for daily status updates.*

Nations: *For nations and national leaders, there are, again, things they can change and a lot that we just cannot. What you cannot change,*

[0] The baseline is that every government is using and misusing digital assets,

[1] Identify your government type: (a) Communist Russia, (b) Communist China, (c) Democratic India, (d) Democratic UK, or (e) Democratic USA.

[2] Then, focus on what you can change (a & b). In Russia and China, the leader is an emperor, and citizens are only pawns with no rights. (c) In India, there is a good balance of leadership responsibility and citizen rights. Currently, India has the highest digital enablement on the planet. (d) The UK is a great democracy but polarized by the White Brits and their king and the rest of the citizens. They're unraveling from one mess to the next, (e) USA is a total mess. The two parties see each other as enemies along with their members; they are doing more harm to the nation than their enemies could have managed. Unless they extricate themselves from

this digitally enhanced death lock, they could take the country back a few decades or centuries. The US must not become a Titanic or Titan with the belief that 'We are unsinkable' arrogance trap.

The 'P' factor: *Once we get past the shock the rest is very easily doable. Step 1 is to start thinking positively by identifying the opportunities and the good we can distribute across the planet. Build apps that demonstrate the physical detractors that can be scored. Along with that, build mental indicators that score the degree of addiction each individual can check. Then follow through with opportunities and threats for individuals, their companies, their state, their nation and finally their planet. The only living planet that we know of in the entire universe is Earth — let's not work to destroy it. Then realize the amount of assets humans spend in destroying each other in the name of intelligence, and find ways we can decrease conflict and hurting other forms of life.*

Chapter 16 — Navigating the Digital World

Leveraging the Digital Shocks

There is an old saying in investing, which goes something like this "When everyone is selling, that is the time to buy." Its similar in the digital disruption. When everyone is in a state of bot enabled acceptance and in a state of shock that is the time for the brave to go and think on their own.

Digital hooks, baits, and opportunities: Your Social and News media is getting more and more efficient in giving you more and more feeds that may be factually inaccurate and that you have been primed to personally agree to. In the US, Fox News has been caught red-handed faking news for the republican party. It may end up with Trump having singlehandedly corroded the republican party from within and exposed their PR arm in the shape of Fox opinions. Media is one of the primary methods of manipulating citizens to the mental positions that their political customers today demand.

We have reached a point of time where the media giants feel they have a right to thrust their lies down your throat. We all have noticed that over time, we get very few articles, posts, or news feeds that contradict or challenge our tribal beliefs if we stick to one single media or social channel.

The channel then has an addict that they will milk, and milk, till you cannot wake up from their media and digitally induced trance, they pimp as news today. This is because we are slowly converting into the individuals they want us to become. However, that is the start, and then they work slowly to alter the very axons of our minds. It is time we face each of these media giants and other feeds and demand the truth and only the truth from any organization that carries the name News in their header. By now, you don't need to second guess how this happens.

They use [1] **Shared Hatred** — regardless of sex, color, or race, shared hatred is exploding across the planet. One of the best tribal addictions comes from having shared enemies. Since the beginning of time, our brains have been wired to get attracted to fear and evil — an instinct for survival. Social Media, News channels, and religious houses play with this by creating their *Us vs. Them*, by feeding people to their personalized primitive thirsts, mixed with emotional hooks and shared by tribal likes and dislikes, often amplifying them a thousand times. Sharing repetitive negative opinions creates a strange bond of its own by reinforcing shared dislikes. A clear example is when Tucker Carlson at Fox News anchor received 40,000 hrs. of feed from the Jan 6th Capitol TVs. He is trying to prove the

Republican narrative that these fanatic people who visited the capitol were peaceful people and not violent fanatics that we all saw. All by selective editing to distort the overall truth with a politically paid end in mind.

[2] **Filter Bubbles** — wherein the opinion generators selectively filter our feeds toward a planned endgame. Or, the channel you watch conducts *algorithmic focusing*, which slowly attempts to make people into close-minded extremists with zero tolerance for opposing views. The optics are purposely biased toward their highest-paying customers by building alternative bubbles of reality that they constantly feed their tribal members as the truth.

[3] **Polarization** — creates a polarized echo chamber where you hear more and more of less and less. This includes voices that echo with your tribal thoughts first till they move you slowly into their desired amplified echo chambers. This leads to normal social humans preferring ideas propagated by tribal ideological feeds and adapting to the extreme edge of ideologies. Once again, to create loyal, paranoid individuals. Whites will find more whiteness, and racists will have more color biases, while Republicans or Democrats will find more of their internally biased political messages. Creating tribal fanatics.

[4] **Recall Amplification** — The number of times we hear the same message is proportional to how often our mind will recall a particular truth. If your tribal social message daily tells you how 'they' are killing 'Us,' it triggers a survival response of 'We need to do something NOW!' The current domestic terrorism in the US and most parts of the world is fed by this survival bias. You may be surprised to note that the news shown to one person can be significantly different from what's reported to another person in another state. That is why our 'Them' have never heard the news you get on your feeds on a daily basis.

[5] **Confirmation Bias** — When some Republican on the TV says they don't trust Biden and that Trump is still the real president/or that he actually never lost, it reverberates with all Republican viewers who, when asked a similar question, will often respond with the exact same response. When the Responses are identical across different locations, it is proof of recall verification and confirmation bias.

While the digital shock is a far bigger opportunity, it also remains a huge threat. At the center of this digital revolution is the human, who is the individual in societies, companies, enterprises, states, nations and the planet. In the last century they would have been compared to an atom, however today we are now comparing an individual to a quantum particle, a particle that is not necessarily intelligent just having the ability to think. Just like the quantum particle can have twins, individuals can have multiplicity of neural twins, which can resonate and harmonize on identical thoughts across the planet instantly. My intention in this book has been to dive deep into how we can prevent the traps of our intelligence to switch our individual intelligence on or off based on identifying friendly, or criminal targeting bots. The bot is the smallest molecule in this revolution which is today a slave to their handlers. A bot is like a gene, with the handlers who have unlimited wishes to tune their gene.

By now we have delved deep into the very micro, yet very important, segment of digital disruption. Another way to look at this is that — what we are reading here is the pointed tip of a needle in a very large digital haystack.

It is most important to realize that at the center of all this optimistic *Digital Opportunity*, the threat of *Digital Anarchy*, and the pessimistic *Digital Threats* is Data. More and more smart devices and smart things are producing some very intimate data constantly. There are two attributes behind this data — the first is your mind and how you react to each stimulus. The second is data that we produce is a reflection of what's in our mind, because the digital world is driven by data and the digital dust that we create. This digital dust is full of very personal details, and extremely valuable data. Unlike the traditional wealth of the past like land, gold, oil, petroleum, diamonds, technology devices, the new wealth is data. Companies like Amazon, Facebook, Google, Apple, and every successful digital company out there create their wealth from this simple data.

I hope, by now, we agree on how our free applications, smartphones and smart TVs are busy collecting data and making billions of dollars for the data harvesters, and little or nothing for the true data generators. We continue to produce our digital dust every second of every day. Check how much time you spent today on different smart screens and you'll probably get a shock of your life. My current estimate is no less than 6 hours per day.

We start our days on smartphones that distract us till we sleep. We drive in cars with sensors, interact on smart applications. Then return home and watch TV programs on very low-cost smart TVs. We end the day with our smartphones and social networks as our constant addictions. So once again, rethink the word *smart* — it is smartly designed for you as an individual or for your data harvesters who provide all the discounts to these devices and application manufacturers for access to your digital dust.

At the center of this new world is our new AI driven Analytics and Informatics. Analytics is nothing but reporting with more user aggregated visuals. Informatics is getting exactly the visuals you need right at that point — no more and no less. Where there are numbers there is an absolute need filtering and extracting information from these petabytes of data. So, the new center of the universe is Data and Analytics, something I've been building for the last decade and a half for major US companies. Everyone has a lot of access to data, in fact people, companies and nations have too much data. However, most of them don't know how to use their data as an asset. Too many companies are destroying historical data as useless while the smart ones are harvesting their own data as the new goldmine. According to Gartner 100% of companies launch Business Intelligence and big-data projects, yet fewer than 30% of these projects meet business, or user, expectations. Main reason is the lack of business ownership, accountability and/or participation all through the project. Most of the times as demanded by their triad partners on unsuspecting CIO's and program managers. This is like investing $1,000 hoping for a return of $300.

The reason for this is that most of these projects become IT projects, whereas the successful ones are driven by business needs and expectations, with ownership and accountability to match. IT tends to focus on technology, budgets and time, believing that technology can answer every business question – and therein lies the biggest failure of IT-only run projects. Whereas very often business stakeholders who need to own and become accountable are presented with technical complexities that make little to no sense to them. They often flake-out and thereafter let IT take all the decisions. Some IT leaders actually present solutions for this very end — more control and fewer business distractions. The sweet-spot for leveraging data as an asset is to simplify technology complexities into business visuals so business can take intelligent decisions based on what they expect, or need. The success of the modern projects is driven by the end experience of the users of that technology.

I published a book called 'BI Valuenomics – the story of meeting business expectations in BI' back in 2011, and it is as relevant today as it was when it was published. Meaning that despite my speaking at global conferences since 2012, having all the IT and business stakeholders nodding in agreement, the moment they go back to their office they simply return to their old habits. I have conducted many 'Lessons Learned' sessions, and three years later find the company making the exact same mistakes we had agreed they need to avoid. So, the answer is in the deployment of a constant quality and success assurance owner. One of the main reasons is that most IT leaders, and I've heard this from a majority of them, like to write the story and not listen to someone else's needs. As a rule, business inclusive projects focus on meeting, and exceeding, business expectations including a focus on User Experience in caps, which represent the <30% or so, projects that actually meet business expectations. The design focus should be experience.

We have all evolved from Reports, to Analytics and now to Informatics. A report contains a lot of data in Excel like columns and rows, it is mainly used as operational lists. Analytics is a mix of graphs, which we call visuals, with the option of an ability to drill down into the underlying data, and sometimes into the original document itself, like a single Purchase Order or sales order. This further evolved into Dashboards and led to the evolution of applications like Microsoft, Tableau, Qlik, Google Looker, SAC (SAP Analytics Cloud), Power BI and many more. While Reports were mainly designed, delivered, and maintained by IT, the analytics layer came up with the concept of self-service, where business users do not have to go back to IT for every change in the flavor of their information. They could now do a lot of this by themselves.

The new paradigm for this is called *Self-Service*. The architecture for self-service is different from that of an IT driven architecture. However, in order to be able to build this architecture and data flow, business stakeholders have to become part of the design process as only they know what self-service definitions are important to them. The dashboards were an evolutionary mix of graphs and data for management, leading to decision-based analytics. Since then, along with the launch of the hyper-digital-disrupters we witnessed the launch of Informatics. The best example of a great informatic is the Uber app. It has everything you need to catch a ride from point A to point B. Nothing more and yet nothing less.

The current definition of perfection is decided by the user's experience themselves, in their UI (User Interface), and Ux (User experience) definitions. Uber never needed to advertise their application, service or offering. Its perfection was the textbook solution that simply spread like a sticky viral application. Sticky meaning *once you use it, the convenience sticks in your mind, i.e., difficult to go back to the old ways.* Next impact of stickiness is that you share it with your friends, and you thereafter never want to go back. That is how it simply gets viral. It had all the hallmarks of digital excellence that was then quoted, referenced and copied as all great things are. Uber evokes Trust — you know exactly when your ride will arrive. In addition, it is of a higher quality, is more convenient and your ride costs less than it used to in the past. The bonus is that the cars are in better shape and the drivers more pleasant too.

As we look to the future there is a very common playbook that applies to all. This can be for a person, a Company, an institution whether that be a political party, or a religion, and finally a nation state and all that comes with it. In each of these categories most people think that by dipping their toes into some digital technology they have accomplished the task of digital transformation. Most successful digital transformations are not a gradual addition of digital assets, but a brand-new digital transformation mostly built from the ground up. Successful digital transformations are a paradigm shift, quite hard to predict using traditional legacy processes as it requires entirely new habits, systems, applications, new outlooks, alliances, tribes, partners and existential models. It's like going back to the 1960's and trying to predict the future opportunities of computers, by buying a calculator. Once again think Uber, or Amazon, they could not become what they did by buying a taxi company and then digitizing it one-step-at-a-time — the only way is a paradigm shift. For individuals and people in companies the secret is to use your experience and then to become the industry disrupter.

Digital disruption is not a gradual evolution it is a giant leap-frog and a new paradigm of design-strategy and operational excellence. Both of which are critical processes in disruption and undertaken in the form of workshops. Not IT workshops but end user, in persona, workshops. For any company this is not a workshop with your existing IT, it is not a workshop with your existing business stakeholders, but a paradigm shift workshop with design thinkers, ideation generators and in the form of a design strategy workshops that includes, end user current experiences, environment tagging, persona mindset analysis, out-of-the-box design and checks, random issues harmonized into key issues, design, testing and redesign to excellence. All driven directly with your final end users/customers.

Hopefully, by now, it has been demonstrated, to some degree, that most of these digital changes are extremely dynamic and happen in real-time. The digital driven changes are constantly evolving on their own, further influenced by algorithmic and machine-learning logic, teaming in their collective behaviors and finally proving dear grandma's statements that a group, or a family, is stronger together than any individual. Recently, i.e. on November 30th, 2021 is was announced that an AI has already built the first <u>living robot</u> that can reproduce.

AI built Starship: Now let your imagination take these digital capabilities to a beyond every point that we can predict today. Lose yourself and think of a spaceship in the year 2182 that is going to meet the first intelligent life that is outside our solar system. This spaceship is undoubtedly built by AI robots only. These robots act as architects, electric engineers, and StarCraft engineers, and they did an excellent job. They in turn control the mini bots that are kind of humanoid robots and do our day-to-day operations like look after the passengers and the visible ship. Under them is an army of highly specialized micro bots that are super-specialized for highly specific jobs. In each case the bots can change it expertise with an instant software update in microseconds and transform from electronics engineer to a neuro surgeon because the situation requires that. Finally, under them are the thousands if not millions of nano bots which are extremely small auto tuning bots a few nanometers across and all over the spaceship. No human knows how this spaceship works and the spaceship will never breakdown. It has seven layers of force fields that can be repositioned based on externally facing quantum driven AI viewers that can spot any particle that is a danger to the starship with one, two or all nine layers of protection to ward off just one rogue intruder at nano second speeds. If a big meteor comes toward the ship if can place 90% of its shields to protect the ship from that one threat, and every other simultaneous one. As the meteor breaks it can track each fragment in a millionth of a micro second and continue the same protection till every threat is neutralized. This whole tribe of bots are the starship Bot System. Each time a micro-bot is damaged they reproduce two more. The passengers never see them yet they travel across the whole spaceship with their safety assured and their needs met constantly. By this time the designs of the spaceships will be so complex that it will be beyond any single, or even a thousand humans to accomplish what these self-replicating hyper specialized, Quantum Robotic Platforms, will be able to accomplish in a few minutes or seconds. I'm planning to write a fictional book on this topic next after I finish this one.

Between 2030 and 2035, humans in developed nations will be living in a time when our bodies will be maintained at perfection, and all of us will remain in a state of *most healthy* yet *permanently ill*. Humans will get more and more sensors that will track specific ailments a human may have. Some people will become partial androids with many intrusive and non-intrusive sensors that constantly track our physical and mental health in real-time, 24x7. So, these humans will be in the best health since recorded history began. Yet, due to the sensors we will keep getting nano doses of drugs that will keep us in that state of perfection. Each time a sensor tracks a slight deviation, like your heart beat rising due to stress, a nano-dose of a corrective drug which will be automatically given because we were micro-sick at that moment. Because you were given a nano-dose we will call it nano-sick, but because the sensors controlled it you will remain hyper healthy. Thus, humans will be permanently sick but also the healthiest all the time.

By this time, or very soon thereafter, the only edge a human will have over robots is their ability to die, as explained in Isaac Asimov's book iRobot. Our only advantage, which in fact is a colossal disadvantage, is that once we are born, we must die. Birth is our beginning, death the destination, life is the journey between these two points. There are multiple billion-dollar businesses that sell, and continue to sell hope for times before the start and way after the end. Humans as a biological form of intelligence have to restart from scratch each time a baby human is born. For humans the ability to learn, is dependent on their genetic capabilities, and adaptability, a process that is still extremely slow when compared to ML and AI, and soon to come QI (Quantum Intelligence). Humans spend half their lives simply catching up to their last generation, because they are born very prematurely in order for the parent and the child to survive childbirth. They will spend their first half of productive life just to come to par, and then the next half mostly trying to remember what they learnt as it slowly slips away. Some will discover or invent new ideas, but most will simply work on what is known. A computer on the other hand comes into existence with all the knowledge of its predecessor the instant it is born, all ready to perform every task its predecessor could perform from day-0. The next generation of computer's will be ready to evolve from where their predecessor stopped and move forward from that point onward, someday on their own. This will never work for an individual, at any regional or national level, like it happens with Technology. There is yet nothing available for a global ramp up paradigm for human intelligence.

Geolocation, country, religion, finance, and opportunities differs for each human born. For example, two children, let's say a girl and a boy — born to a stone age tribe in the Andaman Islands, two born into a Black family in Africa, another two in Iran, two Shite British born to a UK banker family in UK. Two untouchable babies born to an untouchable in India and another two born to a billionaire in India. Two born to a Uighur and another two to a family member of Xi's inner circle in China, two born to the royal family, and another two to a Bedouin in Saudi Arabia, two born to the Muslim ruler in Saudi Arabia, and two to a Christian family in Iran, A child born to a caste segregated Black family in Georgia vs. two born to a White governor in Florida or Texas, and so on.... None of these children will ever have any semblance of equal knowledge, opportunities or assurances, or to even to their right of life. None of them will be identical in knowledge or their opportunities to learn, thus to their opportunities itself. By age 6 they will have very large cultural and religious differences. By the age of 13 the mega differences will commence. By the age of 16 each boy child no matter what country they grow up in will become the owner of male dominance rights. For girls the delta varies extremely depending on their geolocation. At the same time a laptop built in China, Iran, Saudi Arabia, India or the US will all perform quite identically as all others.

We have used human intelligence through the ages to suppress and dominate women no matter what country, religion or culture we belong to. Hopefully, we will collectively become the change we want our future to have.

In our quest for *Saving Humanity*, the critical fact to remember is that "What got you here will not get you there?" Were currently surrounded by artificial boundaries, fake histories and very fake media, and a global populace that demands peace, but is forced to go to war. This is because we are led by leaders who desire command-and-control dominance, with some who mandate, and some with the ultimate control over the life and death of every human in their nation, and maybe the entire planet.

So, the big question is what can we, that is you and I, do as individuals, and collectively to create a successful future for ourselves on the PEH scorecard, i.e., Planet first, environment second and Humanity third. Everything else is secondary and anti PEH trolls feed us as addicts for some ulterior 'Command and Control' dynamics of artificial boundaries. This must not be the way we use our intelligence.

The easiest to a collective, safe future is to walk on the path of the positive status quo, or the minimalistic path, at the individual level and start using the social network positively by actively participating in global and national non-zero-sum-game activities. Use the smartness of social networks and smart devices for positive results and eradicate the bad players with as harsh a penalty as humanity allows. At the enterprise level, migrate the existing Sales, Logistics, storage, and marketing segments of the enterprise on modern digital platforms. According to Gartner, most enterprise projects for BI, Big Data, ML, and AI do not meet business expectations over 70% of the time. Though these companies will continue to spend billions of dollars, their return is under 30%, including principal and interest – all of which can often turn around the concept of investing $1 million to get back $300k; the rest evaporates into *smoke and mirrors*. It's like paying for 100 tons of something and feeling lucky if you got 30 tons of it. This is standard return across the board and is common practice today due to the make profit first and last concepts followed by the partners who promise to deliver these solutions in the art of flashing lights.

It is important to realize that the true cost of a deliverable is often much higher than the 70% loss cited. When a project goes south after go-live the costs to fix that are far higher than what could have been fixed before go-live. Read my article — The (f)Actual Costs of a failed project.[ix] This is a true-life case of a US company. A lot of companies continue to go ahead with a failing project hoping things will improve but they never do. Fix the project when the lights at the end of the tunnel start to dim, and not after you go live. The digital world is changing very rapidly. In the predicting the future chapter I elaborate some predictions and one that stands out clearly is that the next colonization will be by the Data Consolidators with world class quantum Computing, assisting their AI and ML applications. For the US, and every nation across the planet, we need to accelerate the development to build some digital transparency applications while applying our lessons learned from our legacy, analog rules and not simply by copying our existing defects on ever faster engines. We need to reformat, and re-architecture, a full digital redesign and a conceptual shift from the ground up. Realistically it is not the digital core that is the primary driver of each of these segments but a total change-in-mindset with which we have to ensure that we will truly make a difference in our three prime attributes, being convenience, time, and cost. Quality here is a given, as that is when the digital metamorphosis becomes truly transformational. When you reshape the very core of your mind and reshape the way one approaches a need with a digital solution, not as a personal need, not as my state need, not as my national need, but as a global 'For the benefit of Humanity' one. The enterprise that will dominate the future will be the collective tribal win-win solution designers, which cannot be accomplished without an underlying PEH mindset.

A successful digital transformation cannot just be planned by IT and funded by stakeholders. The design needs a strong Xo focus that looks at the experience of the final end user as their success scorecard. Not an individual focus on what you don't have then go and fight with the world, but with a new digital focus on what we want to be by next week, a month and a year from today and become a born again digitalist.

Remember that life is like a mirror, "Smile into it, and it will smile back, growl into it, and it will simply growl right back to your face, smash the mirror, and the only one that now has no optics is you." The key is to create the perpetually smiling mirror that smiles in all directions, and only then the next generation will make you their viral solution.

Digital transformation is never a gradual process, it has to be transformative. For example, let's build an app, something like Dr. Uber that allows volunteer doctors to be auto-prompted of any nearby emergency where they can volunteer to assist someone in need that could save a life. So many times, do-good doctors want to help, here is an opportunity to help a medical situation as a fixed price volunteer. Its aim is not to make a millionaire, but to volunteer good will. Initially forget the monetization aspect let the idea evolve then let each provider place their charge based on some standard floor and ceiling rates that are pre-established. Digital transformation is no less than a caterpillar turning into a butterfly. It has to be thought from the ground up as transformational, and as a disruption to everything that has been done so far. Once again, the core of your design has to be faster, more accurate, and cheaper, heading toward the *highest quality at the optimal cost*, which has to be below the cost of the legacy solution paradigm. Also, optimal as it is not the lowest cost, but the best cost to maintain the highest quality and continuous improvements required.

True digital transformation may not happen as a gradual step-by-step approach but by becoming a next version of itself literally overnight. As humans we will need new ways of thinking, disruptive modelling based on our PEH criteria's that check for impact on the planet and other people not just about how much profit your idea can make for one person. Doing good is going to become an important criterion for future generations. This will be essential from a design, application, and consumption point of view.

When we think we have grasped the potential and the meaning of the Digital opportunities, when we know all there is to know about the digital potentials out there, its capabilities, its pitfalls, peaks and troughs. Even, at that point when we feel we know everything about digitization and digital awareness, <u>we shall be but beginning.</u>

In the recent example think of how confident we were of Digitization, ML and AI technology and here comes ChatGPT version 3 to 4. Where in 4 months, between Dec '22 and April 23, it has totally changed the landscape of ML and AI in what it can do. As an example, Google announced on April 20th, 2023 that their Bard AI, a competitor to ChatGPT, went and learnt Bengali, an Indian language, without any training, human interface, or ML, i.e., all on its own. It's AI that has successfully bypassed ML and can today translate Bengali for all people in Bangladesh and West Bengal in India into its known languages. In another few years we shall have simultaneous translation available for most common languages. At that point if will be the beginning of a universal translator for planet earth. I'd say give it 5 years and language barriers will become a thing of the past.

Roadmap for an individual: I once had a manager David Mulley who taught me that when a new selling partner comes and tells one about their products focus not on their stories of success but ask about their points of failures. Ask them to elaborate how they handled a situation by sharing their example of when their product or service failed. If the partner says they have never failed then you're facing an arrogant liar. It goes back to an old saying, "Show me a man who has never made a mistake and I'll show both a fool and a liar." You will learn a lot more from failures than by listening to their tales of how they caught the world's largest fish. Apply this rule to your social feeds, your politicians, and your influencers in life. So, let's start with a few warnings and then move on to the Do's.

In general, don't just hope for a better future but dedicate your whole life to creating that better future first for other people and then you will automatically be part of the winning team. The digital future is within the grasp of every reader of this book and every educated person on the planet. Even as we stand on this razors edge where on one side is the possibility of future peace and wellbeing, and on the other total chaos and a very autocratic future managed by oppressive digital surveillance being used by tyrannical leaders like Putin and Xi. As humans our biggest threat to our PEH goals, is from other humans and from them we have to take care. Very deep care indeed, especially if we are the ones that elect them as spokespersons and give them the reigns of our buying habits and manufacturing options, nation, our state, our institutions and thus ourselves to them. Think Hitler first and don't become a bystander victim to the future you might inadvertently create.

1. **Beware of the trending 'For-Life Leader' model:** This is a sign of power corrupting the holders of power, who are mostly the rich. [1] **Politicians:** The Xi and Putin President-4-life success has already created two very arrogant emperors who now do everything to ensure that no one can oppose, replace, or defeat them. **Xi** now publicly oppresses internal cabinet members as he did with the former Chinese president, Hu Jintao, on Oct 21, with a casual *"Not feeling well"* statement into oblivion. His oppression of Muslims and Tibetans has skyrocketed, as have his excursions into neighboring nations. Taiwan remains a jewel missing in his crown, and he would love to do the Hong Kong on it. The lack of any global response to Russia invading Ukraine could embolden Xi to actually walk into Taiwan and take over. **Putin,** on the Russian side, has been a dismal autocrat. His goal is to become the political and economic leader for life. He systematically imprisons leaders of commodities and large businesses so they come under his shadow companies of ownership. His assassination attempt goes unchallenged, as does his imprisonment of Navalny, his political challenger. He buys US and European assets like all Volvo businesses, McDonald's, and Starbucks for cents to the dollar. Now, finally, his unprovoked attack on Ukraine because of a personal desire is an example of where our international organizations have totally failed. The UN and ICJ both sit in structured paralysis because Russia is a UN veto member and thus assumes it is immune to action. This is despite there being a clause in the UN charter that allows nations to punish even veto members for breaking the law, but no nation has the will to activate that clause. In one of his moments of public expression, Trump did mention, in a momentary lapse of joy at the very thought of it, on TV, upon hearing that Xi had won his vote for president for life with a sinister, "...I'd like to give that a shot someday." (Source: https://youtu.be/sfvQYeAgYj8). Just stop and think of that for a while, a long while, and let this sink in. This will be the death of democracy as we know it. So, it is not hard to imagine that if DeSantis or Trump became president, they would try to get their party to help make that decision. The Yes-Men: It has two benefits: President-4-life includes party-4-life. That will mean speaker-4-life, George Santos-4-Life, MGT, and others-4-life. SCOTUS judges with a polarized vie-4-life and all judges with one political opinion for life, too. A president for life will need every 'Yes' man or woman they can put on every seat. Why would they vote? All you have to do is look at Russia and China, and the answer is right there. Only the yes-men survive in these two nations; all others are

summarily and publicly removed by *He's not feeling well* or by getting locked up on fake charges that every judge and supreme court approves. I say yes-men as this is a very misogynic play of hand. Can't imagine Angela Merkel playing a game like this. <u>It can't happen in the US:</u> If the Republicans can win House and Senate, by hook or crook, this can be done. Constitutional changes are allowed if approved by the Congress and ratified by the Senate. There are no clear limits to its powers. Also, beware of the confidence we carry in the pillars of democracy. Jan 6th is a pure example where the unimaginable happened. Americans violently attacked the citadel of US democracy. Not only that, 40% of Americans, i.e., on July 2023, still believe that these attackers were innocent tourists and must be all released, including our two Republican presidential aspirants, who had stated that they would pardon all Jan 6th people charged by their presidential powers. Think very hard about this global trend entering your nation before you vote your future away. [2] **Enterprise:** Seeing all this centralization of control at the presidential level, where we are witnessing a clear desire of the rich, autocratic, powerful leaders force this president-4-life vote. It is not surprising that the CEOs of mega-corporations want a piece of that power, too. There are now CEOs who want to play the CEO-for-life card and possibly graft their voting components toward the same. According to Barron's report on June 23, 2023, Motorola CEO Greg Brown is already pitching this scenario across conservative and Republican media. Now imagine President for Life filling all business leadership positions with his 'Yes-men-for-life.'

2.   **Beware of intelligence traps:** Traditional Intelligence traps are a dilemma of highly intelligent people. This dilemma is born when an intelligent person who has been right in their particular field for most of the time, start to believe in their omnipotent knowledge, and believe they are right in everything they say. They succumb to the intelligence trap. The trap snaps open when they start to believe they cannot be wrong in anything. This has led to autocratic president-4-life, and is now morphing into CEO-4-life. All they need is convince their quorum and then the public. Now with digital manipulation these tasks become easier unless we place clear guardrails on such tactics. In the digital world the algorithm's take the ownership of becoming the truth-sayers in the garb of an invisible friend. Once the algorithm has established trust in your mind, they can then entrap the reader into a modern intelligence trap scenario. Amplifying your desires activates your trust bias based on your perceived personal intelligence traps. Once trust has been established, the bots can start pushing false statements and quoting inaccurate data, thereby getting a tribal buy-in that can be repeated, word-by-word, when anyone asks a tribal member an anticipated question. Now, because you have been verified to be intelligent, even if by a bot, you soon start believing all you hear as the intellectual truth. For a Republican it

could be a What should you say if anyone asks "Is Biden your President" or What to answer if anyone asks "Do you think Trump won the 2020 election?" Once the question and reply are cast in stone, along with their answers, with pure repetition. The answer, then becomes a mental reflex answer when anyone pops that question to the target. Then again, the more times you repeat that answer the more you believe that it is your, and yours alone.

3.   **Beware of Freebees:** Human brain is driven by a strong desire for things that are free. That is why you will see a lot of apps for downloads saying it is free, but when activating it will tell you it is free for the first 7 days then then $x' every month or year. This is the power of free marketing. It gets you into the door, but on a false pretext. For me, I lose trust at the moment I get in. A lot of our decisions are driven by data, and data is more and more being manipulated by local and foreign trolls. But it is most important to remember that nothing is for free, including Google, Facebook or WhatsApp. What we need to do is to build a network of your-physically known-very-trusted core group of Fact Checkers within your friends. This cannot be your religious or political tribe as they are loaded with nonexistent trolls and bots, nor can it be people that you personally do not know. Then expand this to known networks city, state, country and finally, if possible, at a global level of highly verified humans on the other end. Remember biases will always be there, but human biases are different from bots designed to create a predefined bias in their audience. Groups of trusted local-checkers who can locally verify, fact check and flag that when Russian social trolls report, with photos and video, that a 3-year-old child has been put to a fiery stake by the Ukrainian police for reading a Russian book, in some kremlin square that does not even exist. This, for sympathy for Russia and animosity against Ukraine, on a total lie, based on a report of death by fire that never happened. We firstly need to realize this is an emotional sticky statement, and secondly need to have someone in Ukraine verify the truth rather easily. Such statements are made periodically to arouse the readers emotions and get a reaction or even a like or a share, the trolls hope something that explodes into a viral forward. Here is another fake story I have been told. The story was streamed on WhatsApp post that builds a story for sympathy for Putin. It talks of how Putin when he was an infant faced genocide in his village. Men and women were killed and their bodies were piled on top of each other. Afterward, when the villagers came to find the bodies, they found Putin's dead mother in a pile of bodies. However, lo and behold under the dead mother they heard a muffled cry, and little boy Putin was found. The infant was barely alive under the protection of his mother and the weight of all the dead bodies. Without going further, this is an example of total fabrication for an empathetic warmness for whatever Putin was doing in Ukraine. A form of total disconnected empathy and distraction, because some

bots found this garnished sympathy from readers. Build your local and global capacity to sieve such apparent fake from truth. Fact-check everything that arouses your empathy via your emotional doors. Open a blacklist digital proofreader that warns that a particular article has a 0% or 100% probability of it being true.

4.   **Pseudo Popularity:** Popularity is a heavy emotional high for most humans. Politicians, actors, musicians thrive on this as do each of us. A lot of musicians ride on the waves of their popularity as the fuel they need to continue writing their songs, or writing their music. This happens when they sing songs to large audiences, when the audience response becomes their elixir to creating more music. In modern social networks the popularity score is measured by how many followers you have on Twitter. For example, In Sept 2022 the top account is with Barak Obama, followed by Justin Bieber, Katy Perry, Elon Musk, footballer Christiano and Prime Minister Modi of India. The trend of popularity is equal to social score points, i.e., becoming friends with digital and robotic falsies designed to influence your tribal members with a feeling of popularity and with predictable mass hypnosis. When one tweet says a 3-year-old child was crucified it is very different from ten thousand re-tweets claiming how horrible because someone forwarded the post, because it grabbed the emotions of an influential participant when seeing a seemingly in real-life and attached a fake photograph to prove the point. Even though the whole message is an empathetic trap from the start.

5.   **Don't befriend every invite that comes your way** that praises you with things that you want to hear, especially when it comes from someone you have never known. This is a digital hook. Avoid biting into the digital hook sent by strangers or loners, especially ones you do not know and who occasionally feed you *sweet nothings* as predictable psychological bait. Instead, withdraw from your social addiction only to realize that there are thousands of real humans right around you — unless they are addicts too. Most of all avoid speaking to non-human bots that sound more like humans than many humans, meaning anyone you do not know.

6.   **Polarized and binary:** One of the prime threats of the digital and social revolution is the global formation of highly polarized and non-binary tribes. Just as an example in the United States, no matter what Trump does or says there is a polarized reaction to it. One section continues to not only believe that he is right in everything he does but also continues to support his every call to action. At the same time there is an equally opposite polarized population that thinks everything he does is absolutely wrong. The fact is that no one can be absolutely right or absolutely wrong in everything they do. This polarized divide has transgressed nations, states, zip codes, media outlets, social and family groups as well as individuals. Each one of us

has recently been faced with a version of 'if you are a supporter of 'X' then you are a true friend/ or lifelong enemy of mine'? This is not because individuals have changed as a whole, but because there is a social and digital polarizing undercurrent propagated by companies like Facebook and Twitter to influence your minds into making collective directions, as defined by companies and troll. They both profit from such neuro alterations on large populations, by paying large funds to these social networks, and media companies, the trolls by receiving funds from political parties and other rich and powerful.

In a way, digital bots are designed to trap us in their psychological grips. It works to tell us what is right and what is wrong based on their unique programming and our psychological weaknesses. It works to control our minds, then our lives, and finally our lifestyle options. On the bright side, digitization has simultaneously given us untold freedom to break free from this vicious cycle of our lives being controlled by others. It opens new doors to generating a positive influence far greater than its potential of ill use. Once again it depends on who leverages this power, the funds they have, the influence and their purpose. Digitization has challenged all norms and defied most odds, bringing us together as a society that relies on changes. We should embrace positive change with open arms and become more transparent, and more openminded than we have ever been before. We believe in impermanence and that has been our biggest asset because we challenge the world and its rules with more ease now.

Auditing the New Digital IER&N (dIEN)

Humanity will need to pass through the audit of destructive digital farsightedness to survive the current misuse of these assets, and only then arrive into our future digital economy. These laws need to become universal and apply to every **IERN** — Individual, Enterprise, Religion, and Nation, as a form of legal PEH controls. This time with an aim to curtail, and criminalize, the negative trends of digital communications. Every negative digital scraper, algorithm, or troll must be constantly audited by some fundamental laws that protect the planet, humanity and all **dIEN**'s (digital-Individual, Enterprise and Nation), with strict digital code of conduct. These checks and balances must be legally defined, and mandatorily inserted deep in every chip, semiconductor, and smart device with an intent to digitally track the core algorithms and pass a new global digital law like the ISO 9001 that mandates strict controls and penalties. Only this time this is all about every semi-conductor, sensors, digital devices, ML, AI, LLMs like Chat GPT, and Quantum Computing. It should form the foundation laws of future programming. So, tomorrow if China manufactures a billion chips with a hidden instruction in them, they must not be allowed to distribute them, and they must be penalized for that illegal insertion. If that chip is in a billion laptops all billion must be recalled and either scrapped or cleansed to conform to guidelines. However, till we don't have the rules in place 'every fool will define their own rules' and we will keep heading into chaos.

In a world where the human life is the only one protected, and hyper growth is the predictable path, then 'we the people' need to demand secure controls and measures, while making sure corrupt leaders are disallowed to use them to circumvent privacy, or provide a foundation for political espionage, both global and national. We need to prevent bad actors from misusing these assets for their own good and to subvert the good of the citizens. Define, make mandatory, track, and then bury these laws deep into every chip, device and application of the future. Mandate that only the IERN's certified devices, and components can be used, for funding, developing AI, and create algorithms that audit random components, so the law applies to the creator. Penalties should be stiff enough that no company tries to circumvent them.

A **PIA** is any Program, Interface or AI Application.

Law #1: No PIA can, through any action or inaction, digitally or physically be allowed to do anything that physically or mentally harms, or has the potential to harm, any human. Or creates an 'us vs. them' communication, or design communications that has an inherent potential to cause conflict with any other person, or persons, with any intention to directly or indirectly harm any individual, a people, enterprise, religion, or nation. It must thus protect every IERN (Individual, Enterprise, Religion, and Nation)

Law #2: Our God of Sustainability: Every PIA must adhere to the overall **HOPE**, (where HOPE will become the House of Planet Earth), guidelines except where such action conflict with the first law, and not harm any IERN. In the **HOPE** concept the ultimate loyalty of every PEH and IERN philosophy is the strategic benefit of the planet. Where Earth becomes our mother and God while we live on this planet, and the term changes to HOPM (House of Planet Mars) for a future community that Elon Musk builds on Mars. We begin to collect funds to protect Earth as it is the only thing that gives us everything, we need to stay alive, and with zero expectations for any returns. Our planet is our true mother and god bar none. All funds collected will be reimmersed back into out planet.

Law #3: Every PIA must protect its own existence so long as such protection does not conflict with the first or the second law.

Take Aways

Individual: Time has arrived when we need to take ownership and accountability of our own actions. We have the tools, we are supposed to have the intelligence so now we need to use if for the benefit of the planet first, and then the human environment. Build your own scorecard that identify hatred messages, Filter Bubbles, Polarization or Recall Amplifiers. Use your head and don't fall into the trap of confirmation biases that are not yours.

Companies: *Companies have the maximum to gain. All the digital tools are being created to make better and faster products and services. Digital cores, Streaming data, Single global databases, Machine Learning, Artificial intelligence and now ChatGPT. Next will be Quantum Computing, quantum Security and Quantum threading. The Bots are highly intelligent workers, but only if used properly, so as companies we need to target all efforts for the good of humanity, and globally boycott those that cause harm. Humans have a semblance of what is right or wrong to some degree. A bot and current regenerative AI together have no such foundation. They will mostly do what they have been programmed to do and then what they collect from human interfaces and documents. We the humans are responsible for their actions and programming them. We need to become responsible for globally auditing and policing them too.*

Nations: *National leaders and all elected officials have a duty to assist in catching all the bad players and reward the good ones. We can quite easily let our nation go the way of China, Russia, or Iran, where the citizen is the property of the leader. We cannot follow the path of the US, where political parties are busy fracturing their own nation and echoing the playbooks of their enemies. Leadership has disappeared as each leader plays a mind-grab game with their citizens. It is now in the hands of 'We the People' in democratic nations to stop the bleeding that is draining the energy of the best of nations.*

The 'P' factor: *The P factor is how we can use digital assets to stop humans from turning this planet into an I-me-My autocratic society and my food factory. Up until the day when some humans continue to believe all of god's creatures are placed on this planet for them to eat, we cannot be on the right path. Plan one of these days to start by* **No-Mam,** *i.e., give up eating all mammals as step one of our repentances on how we treat non-human life on Earth. Remember that 96% of mammals on this planet are represented by Humans and the mammals we grow as our food. Of the 45, around 80% are birds. Of this, around 72% represent the birds we grow for our food. Think of the animals — we eat their children, their babies, and them, while at the same time, we term ourselves as intelligent.*

Chapter 17 — Some Taboo Topics

"The illiterate of this century will not be the ones who cannot read or write; they will be the ones who are unable to adapt. Those who cannot learn, unlearn, and then relearn". Alvin Toffler

Thought for the year: I have often heard, from one source to another, that *The Planet is Dying.* Let me tell you our planet is not dying; it is simply adapting. Our planet is too big and we too miniscule to kill it. Our planet will ultimately revert to its base rich ecosystem. In this process, it will remove all toxins and forms that are working against this goal of a healthy planet. So, the earth ecosystem will adapt, only that if we continue to fight against that harmony, we will no longer have a place in that future earth. So, the planet is not dying. We are.

It takes very little courage to accept the status quo, to support ignorance, or to follow every guideline because we have been told so, i.e., believe in something because we have believed in that for the last 'X' years, or because someone tells you this is how it is. However, it takes a lot more courage to fight existing biases and social beliefs because you believe the dogmas to be untrue. Only when we rise above, the doctrine of our politics, community, nationality and our religions shall we take our first steps toward a peaceful future. Just like, in 1610, when Galileo stated that the sun was the center, and not earth, the Church did what the rich and powerful have done, and continue to do, since known history. They acted like Putin, or Xi would today, they locked up Galileo, and censured his facts, all in the name of their God. Only when truth and facts become the foundations of our beliefs, can we venture a semblance of PEH or a 'Saving Humanity' mindset. We need to delve deep into our personal unknowns, and challenge the traditional dragons that were created to make enemies out of fellow humans. That made a German soldier kill a French soldier in the 2022 movie — *All Quiet on the Western Front.* It mirrors what happens in every war, and battle, on artificial boundaries, where the rich, powerful, and old leaders first entrap the youth, and then send them to needless slaughter and to death. All while they sit at home making brave speeches, eating good food, wearing the best clothes, and drinking the best wine. While protecting their children from this task. Power, that allows a leader like Putin to lock Navalny into prison and no on in a nation of 145 million, can ask even a single question, or when Putin goes to war with Ukraine not one person in a population of 8 billion can stop this barbaric act. Even our basic elementary education teaches us that WHEN we don't know something, our first step is to acknowledge our ignorance, and only then to proceed with the process of learning. Then comes the stage where we think we might know something, and then suddenly we are faced with many new points of views. Some, we will agree with, while others will challenge our every existing belief. The question then becomes whether at that point we are brave enough to look at the real data and new information against our societal understanding of culturally instilled assumptions. This is the most difficult leap and the place our mind dreads looking into. This is because maintaining a status-quo is far simpler than to challenge it, thus a majority of minds find it more convenient to stay with the current situation. In 1610, if Galileo had asked someone why they did not believe his hypothesis, the reply probably would have been "...because the Church says so, and as the Church represents God and God cannot be wrong." Gandhi faced it when he challenged the British with their own laws in India, and proved to them that they were doing wrong.

Martin Luther King faced it when he challenged the White supremist in the US. George Floyd became a publicly executed murder victim of the hidden caste system in the US where Blacks are treated as bad, if not worse, than the Dalits are in India. Do we dare to look at the native Indians and the genocide and ethnocide that happened in the US and Canada that is only now being exposed. When Vancouver and California want to stop the caste classification in South Asian communities the question is whether this is truly a class equality exercise or yet another political endeavor. Have these proponents yet considered the total deprivation, by an American caste classification, of our native Indians right here corralled into American human camps with a total denial of their rights to vote or be part of the US society. Have they taken time to questions the culturicide committed by the Roman Catholic Church, across Canada and the US, along with the ethnocide committed on the native Indian children in Catholic convent schools designed to wipe out the Indian from children and hand back a good Christian.

Something we are today blaming China of doing to the Uigurs. Have we considered the caste classifications, of the Blacks and Muslims here in our US of A in that same breath? If we have not then we need to find the political influencers that are behind all these demands, so that we can fix matters at home before we go picking up our AK-47's and shooting anyone we don't like. When our hate overflows this shooting can include children in our own schools, in the US, followed by radio show hosts that want to broadcast that the whole 'Sandy Hook' massacre was fake-news.

Finally, politicians are creating, amplifying and leveraging all these polarized views, and it is right now fracturing the very fabric of the world's greatest democracy that had made the US a great nation. But now this nation is being torn, fractured, and ripped purely on party political agendas. On Jan 6th in 2021 it erupted into a coup that is still being unraveled with one party calling it a coup, while the other is trying to prove it was a peaceful demonstration by tourists. The interesting fact is that despite all the live coverage and evidence the republican party continue to digitally call it a peaceful protest, and the democratically elected president as an election stolen. The US democracy is today standing on the edge of a sandstone cliff and the politicians are doing little to save her, but more interested in simply becoming an echo chamber of their party politics, with their media engines to match.

In each case the majority mostly opts for the easier mindless option, i.e., the status quo as that needs little thought and has the masses behind ones back. It's easier to support Putin in Russia that oppose him, like it's easier to support Trump and GOP if one is in a republican state, or to support Biden in a democratic state. Supporting the status quo, is flowing with the river with mindless obedience. One does not even have to think the river simply takes you for the journey. The same goes for India, France, Germany and other democracies. The outlier voice has the most difficult time if they do not echo the mindset of the masses.

Fear of the new, is like the fear of the unknown. It is the greatest fear of all and it has been challenged, imprisoned, crucified and pushed aside for centuries by those who wish to control the masses. This fear has been consistently filled with fake stories, biases, and beliefs. Truth is never influenced by race, sex or color as it is based on pure facts and what the data tells us. But truth and facts are cold and do not evoke high emotions, or passions as fake fears do. So, instead of praying for miracles, or hoping that your covid will disappear because you have the blood of Christ, as told by a Florida pastor, one should respect facts, wear a mask, and hopefully live a little longer.

We are almost at the end of our book. Therefore, it is important to take shallow dives on what we consider taboo topics.

Taboo topics can be defined as topics that people do not like to discuss as these are topics that many people are very sensitive to. An example that comes to mind is that in the UK there are men's clubs where any discussion on politics, religion or women is prohibited. A famous author then asked, "Why should he join a club where no interesting discussions could be had?" So we go way beyond politics, religion and women.

Emotions are the area of future growth. As polarizations are amplified, as populations grow, we can easily predict more conflicts. If you put five people in a room from different nations and religions, you will probably get five different interpretations of the taboo they are not supposed to discuss.

If we want to build a perfect world tomorrow. We will need answers to many questions like who will fight for the digital transformation, the digital controls, the digital balancing acts, and the digital security to make certain acts of digital manipulation a criminal act — a global, local, and in the US federally. We need an answer to who will make rules and regulations that will be fair and applicable to everybody on the planet so that no one would be above the law.

It seems difficult to get answers to such questions. There appears to be a special law for the rich, and powerful in every nation. Thus, we, the golden brick, the individual, must fight to bring digital transformation. It is taboo to discuss digital transformation because we cannot depend anymore on our political leaders.

If you look at the government of the US, you will realize that the last ex-president did two things. Firstly, he tried to cancel everything the Black president had done before him. He did not consider whether the thing was good or bad; he just canceled it. Secondly, he strongly denied climate change.

For a long time, if you look at history, the Republican party has been trying to fight climate change since the mid-1960s. We can see a consistent pattern of Republicans going into global meetings, like the Paris Accord, and the Kyoto protocols with a singular intention to eliminate the questions. All funded by the US Oil and Gas industry. But Donald Trump was the first Republican who had the guts to publicly announce, "let's kill it." That's why the US was no longer a part of the Paris accord up until Biden brought it back. So, the question is whether these people will fight for digital Liberty and the basic requirements if they get paid by some trolling companies? Will they fight for what will be good for our future, nation, our planet, and for our children's children?

The second thing happening globally is that some countries like China, Russia, Iran, and Saudi Arabia mask or delete essential data. China deleted the Tiananmen square data so people in China do not know it ever happened. In Russia no opponent to Putin has any data or media support. This was elaborated in the movie Navalny that won an Oscar, but nothing else happened in the global political arena, nothing from leaders, nor from the UN that was drafted to prevent such happenings. A lot of governmental-full-control propaganda is blasted through the media channels within China, North Korea, Russia, Iran, and Saudi Arabia. Now we know this also happens in Fox news in the US, and every nation out there. Fox news is no less a propaganda machine than some other media outlet.

That's not the world we want to create. In the US, right after he withdrew the USA from the Paris Accord, Trump stopped the EPA from sharing data on local weather and climate with the sole intention that if universities do not have data, they cannot make any predictions or warnings. It is not right because it is more like forcing blinders in front of educators and scientists so they cannot see what is happening, all for a political agenda. Similarly, if you are a person who has sight like a university, then the government propaganda and their policies will turn you blind, so you can't say a word about climate change.

In conclusion of both points, it can be said that the digital transformation and revolution will have to start in the US. Within the US it needs to start in California where most great things have started.

Thus, we need to demand the US federal offices define global laws and controls for digital access, then simply roll them down to every state without any option to not deploy the prime directives. This must then percolate down to individual manufacturers of components and parts used in the digital landscape to deploy as a foundation mandate. The put into place tests, check and measures that these controls have been installed without any back-door loopholes. We also need to make sure that companies in the country are not allowed to make a profit from private data. All private data, besides, should be masked as it happens in Europe. European laws are very strict and strong, we now need to make it stronger.

Saving Humanity: As a human race we're literally screwing our planet for short term gains, while placing the future on the altar of politics, religion and profits. We are made to believe that humanity is the most intelligent life form. In 1961, Dr. Frank Drake and Carl Sagan introduced the Drake equation and began a search for intelligence across our relatively small Milky Way galaxy. This resulted in SETI, or Search for Extra Terrestrial Intelligence, and a very large radio telescope in Palo Alto that points to the heavens, searching for intelligence. 59 years later, in 2021, I published 'Beyond Drake-No Earth 2.0', available on Amazon Kindle, asking critical philosophical questions, that no tenured astrophysicist, could either associate with, nor ever dare to add a brief comment in the article. I carry that as a badge of honor in thinking outside the fence mindset. Read it and let me know. My article, versus the formula presented by Frank Drake are both based not on pure engineering, or astrophysical facts, or available data, but on assumptions. I wrote my unqualified paper 59 years later with more information of the heavens, than Frank had in 1961, and added some philosophical question into the mix, that I thought were relevant in auditing intelligence. As the foundation of Frank Drake and my hypothesis was intelligence, I started by questioning our interpretation of intelligence itself. In my article I take the POV of an alien intelligent lifeform, with 100% of my personal human biases of course, minus my assumption of what a truly intelligent lifeform would be looking for. To this I added the only known quorum of life and intelligence, both the center of the universe in our articles. I, then added new paradigms to question if our interpretation of intelligence was demonstrative or hallucinatory. Questioning the rigidity that we currently apply in the scientific community where authors are disallowed to present or discuss their nonscientific philosophy assumptions. Despite the fact that Frank Drake's whole equation was based on assumptions to start with.

Geopolitics: As a baseline we need to establish some founding assumptions. In the next 7-10 years, i.e., around 2030-33 the two technologies that will become the new competitive differentiators, is AI and Quantum Computing. Just like the big powers went to war to control energy the new battle will be in these two areas. The competitors will be USA and China. US that has all the knowledge and discoveries both in AI and Quantum computing, and China that is being fed all these technologies for free by the build-it-cheaper that the world is currently suffering from, in their quest to make higher margins, and profits. This IP drain, even if it means handing over their brains and IP to China to manufacture it a little cheaper with communist labor. The new colonists of the nations will be USA and China. Russia is out of the game because their internal distraction of Putin trying to amass Russian wealth has taken Russia down a path of geopolitical tunnel-vision dominated by the whims of Putin.

The US: is the dominant world leader; however, it is currently in a strange mode of political self-cannibalism. The two political parties are busy trying to destroy each other with little or no consideration for the common citizen or for the good of the nation. Historically, civilizations, leaders, and nations rise and fall. The question is as to whether this is the start of the decline of the US due to infighting. The query we need to ask ourselves is whether the US is doing a Nokia to itself. This is in reference to when, in 2007, Nokia controlled around 49.4% of the global mobile phone market. They got lax because they thought the global mobile phone market was theirs to keep. Their rise was swift, but their decline was a straight fall off a cliff. Apple, with its iPhone, disrupted its market, and by 2013, Nokia sold its mobile phone business to Microsoft. While the US is busy with internal conflicts, in one instance coming close to a civil war, their internal animosity is still on the rise. During this total internal distraction, both China and Russia are working hard to disrupt the Petro-dollar framework by providing oil on non-dollar negotiations. This could weaken the US. This infighting is also damaging US optics and its global image. The final nail is the unhealthy US dependence on low-cost manufacturing out of China and letting China become their manufacturing hub for all US products. The USA is today made in China. There are a few products we buy that are not made there. China has both the US and Europe in a firm grip as their production hub. They do this by first becoming their lowest-cost producer, then their lower-cost competitor by selling their Chinese-manufactured clones. It was around the year 2000 that China became the US, then the world's Factory, by promising ultra-low manufacturing costs and then keeping to their promises. Since then, the US and Europe have become addicted to manufacturing at very low costs while keeping their prices the same. This resulted in higher margins and profits, which is like a drug addiction to most capitalist businesses. The global consumption nations still continue handing over to China their priceless family jewels, i.e., IPs and manufacturing technologies, with just one goal — profits. They still have no consideration that they are actually feeding Goliath with a constant stream of funds, IPs, Technologies, and global markets. China, on the other side, took all these designs, stole more from Europe and the US, and started manufacturing their own competitive products. Now, China is sending their ultra-low-cost products to the US and the world via Chinese apps because they do everything at a mega scale, while shipping is close to Zero due to them sending all kinds of goods to all nations. China successfully became a global factory. Today, in 2023, the US is manufactured in China, and China sees the US as its enemy. This is a very difficult situation from which the US will need to extricate itself. Currently, the US and Europe are actually funding all the arrogance and atrocities of China by making them the mega factory of their nation. The US is the biggest spender on defense. Now, with the help of global funds for manufacturing, China is today the second largest spender on defense, all made possible with US and EMEA donations. When countries make bigger and better arms, there is only one direction in their mind — domination. So, while the US is making a lot of noise about the uncomfortable military growth and threats of China, they themselves are encouraging their companies to manufacture in China and thus directly fund their defense spending. The US is currently at the forefront of AI, Quantum Computing, and robotics, followed by China. We must use the UN to mandate global rules on AI and Quantum computing while we are at it, especially the recent regenerative AI options for safeguarding dysfunctional developments by any player.

China: On one side China has played its cards well and has established itself as the manufacturing hub of the planet. Whether you buy in USA, Italy or Germany you end up buying products made in China. The global dependence for Chinese manufacturing is very addictive, similar to Europe's dependence on Russian energy. On the other side Chinese leadership has managed to get emperor status, followed by Putin in Russia. Funded by the continuous inflow of manufacturing funds China has been on a global spending spree. China is today providing debt financing to strategic nations so that they can own ports in foreign nations. Their next goal is to control the global supply chain, so no nation can deliver without their supply dependence. They are financing projects inside and outside its boundaries and committing human atrocities all around its borders. While, we don't like all these genocides, and ethnocides being committed by China, our dependence on their cheaper products is far stronger, than our willingness to stop China from their atrocities. They are spending big funds on Weapons, AI and Quantum computers.

They are the world leaders on automation and catching up on robotics very rapidly.

Europe: Had great potential but they never got past their artificial boundaries. Today, their culture, language and differences are more important to them than their choice of joining hands and making Europe a better place. So, in the game of competitive advantage of nations they are missing in action.

India: This puts the third major player in an awkward position. India has traditionally been a Russian ally. But, if Russia is going nowhere in the next phase of dominance should India align with them for cheaper oil, or align with China for cheaper manufacturing, or should they align with the US. The second problem with the Indo-Russian alliance is that I dread to think of a situation, when in the event of an Indo-Chinese conflict, all critical Russian weapons coincidentally stop working at perilous times. If Russia has its own Munroe doctrine, then they may have a digital option to disrupt the efficiencies of their weapons when they so desire. This would then test whether the China-Russia alliance is stronger than the Indo-Russian one. India needs to make strategic alliances with 2030-2040 in their horizon. India has yet another strategic requirement. So far, since the mid-1990s India and Indian companies are the developers-brawn in many technology developments across the world. While China invested in becoming the world's manufacturing hub, Pakistan on becoming the world leader in terrorism, India invested in becoming the world leader in IT developers and people. Unfortunately, the Indian manufacturing is still fighting for 'Make in India' as their goal with the technology, and royalties still owned by the developed nations. Around 62% of weapons used to come from Russia and today around 45% still do. In this game, India needs to play the China strategy; it needs to move from *Make-In-India* to *Made-By-India*. India needs to start manufacturing their own weapons that, no external technology-owners can subvert. It also needs then to start its own defense industry and start to export 100% Indian weapons to nations that cannot afford US, Russian or Chinese weapons. Secondly, India needs to capitalize on their world-class *Systems Integration* companies to soon start to work for Indian developments at international rates to become the backbone of *Made-by-India*. India is not a main player but a big influencer of democratic future.

An Alien-Assessment of Humanity: In the next few pages I ask you to once again imagine you are an alien lifeform visiting planet earth for possible contact with the Galactic Living Guild. You, an alien, have your checklist and as the alien lifeform you are invisibly auditing humans on your intelligence scorecard. Your results will determine if this life form is intelligent, if it needs education, membership, or exclusion from the galactic intelligence community that you the alien belong to. Some of my high-level alien audit on humans of earth are based on facts. **The score is based on a 1 to 5 scale with 5 being exceptional and 0 being catastrophic.**

[1] <u>Self-respect as a species:</u> Humans officially spend around 2.2% of their global GDP, i.e., $2.113 trillion, on building tools for killing each other (we call it defense) each year. The unofficial version of this is probably double this amount. Our passion for killing each other leads to building bigger weapons than our neighbors. Does this make the human race cannibalistic, as we gloat in our ability to kill each other even if we don't eat each other? Bigger weapons translate into political arrogance. In some leaders, this arms superiority results in activating their desire to dominate their neighbors. China took over a successful Hong Kong in July 1997 in their desire to consolidate a 50-year promise that they never kept. Now, their eyes are on Taiwan as a strategic diamond in their backyard. The Chinese interest in Taiwan is based on two facts: AI and Quantum Computing are the next global areas of dominance. Taiwan is today the global leader in semiconductor manufacturing, and that is a skill that Xi wants for China. On February 24, 2022, Putin attacked an innocent and peaceful nation, Ukraine, while the other intelligent nations watched it on their TVs. As of March 2023, Putin started recruiting criminals and forcing his citizens to fight for his imaginary cause. China just announced that Taiwan will soon become theirs, so here is us learning from the future but unable to do anything other than listen. Many humans are aware of this Faultline and spend substantial time and funds developing good things like safety apps, smart cars, Quantum Security algorithms, rockets to the Moon, digital assets, medicines, AI-driven medical devices, safeguard inventions, and everything we love to use on a daily basis. But the overall effect is that while the common citizens are busy making good things, the rich and powerful are busy amplifying global hate, any form of divisionary *Us vs. Them*, rather than a global collaboration. Humanity is not a cohesive or globally collaborative species. **My score is 1 out of 5 = 1.** Total 1/5.

[2] **Planetary respect:** The leaders of Earth's most developed nation, the US, denied climate warming and used their collective political influence to defund any discussions on this topic. Since the 1960's their Republican representatives have attended the Climate warming talks with one agenda — shut them down. This was done because the US oil and gas industry-funded them for their political positioning. All this is when the planet is evidently tethering toward a cliff of planetary weather-related disruptions. These global climate changes are already impacting weather in the form of hurricanes, floods, and droughts on a global scale. California has started its flooding-to-fire extremes, while Europe has stopped many river cruises due to a lack of water. We are at, or near, the razor's edge of no return. Thousands of fish died in the river Darling in Australia, which is just the beginning of global warming and its sudden impact on life

itself and our food supplies. The unpredictable weather will soon impact the earth's food belts by either drowning them or drying them in predictable floods and droughts. Humans continue to burn the planet's lungs, pollute the water sources, deplete the fish with industrialized fishing, and are instigating the greatest extinction of non-human life forms that belong to the planet if Earth is indeed our mother and God. Humans do little for the planet nor respect what the planet is giving to them for free. This includes the planetary systems and every life form that lives therein. **My score is 1 out of 5.** Total 2/10

[3] <u>**Sustainability of Energy resources:**</u> Soil, after water, is the most crucial foundation for life on this planet. Humans are depleting their planetary soil at a 50 times faster rate than the planet can replenish it. For example, this is being done by unscientific farming practices that remove the natural clumping of grassy soils and replace it with human farming that promotes the creation of particles of otherwise naturally clumpy soil. Thus, the healthy soil is turned into fine particles that can be easily eroded and blown away by water, wind, and rain. What is left is the eroded non-farmable soil. This encourages the spread of deserts. In addition, humans are destroying the very lungs of the planet by setting fire to natural forests that the planet has built for millions of years as its living air conversion stabilizers. Plants and trees are the lungs of the planet. Soil is the basic skin and the foundation on which all earthly life forces grow; plants are what keep the soil intact. Humans are depleting the valuable soil at a rate that is not sustainable for their baseline growth. The combined effect of climate and soil erosion will rapidly drop food production to an unpredictable point of no return. **My score = 1 of 5=** 1. Total 3/15

[3.b] **Potable Water**: Fresh Water is the primary critical source of life and survival. Humans are not only polluting the global oceanic water supply with lethal chemicals but eroding the replenishment capacity of the planet to produce fresh water. Global climate, along with uncontrolled population expansions, and a very limited supply of fresh water is already taking away the planets ability to replenish fresh water naturally. How precarious is this — most humans don't realize that if we take all of the fresh water on Earth and make one single drop it will just about fit northern California. That is all the fresh water the planet has. That water is critical for all life to survive. Take away this drop and no more crops, no more animals and certainly no more humans. Then add to that no more snow, no more rivers, and soon no more oceans. As we deplete or pollute water it has a direct impact on every other natural resource our planet provides. This includes creatures earth gave us, trees, shrubs and grasslands which are the lungs of the planet, and crops that we plant to sustain the growing human populations. Humanity is slowly raping this planet in the name of keeping you safe. **My score = 0.2 out of 5;** Total

3.2/20.

[4] **Leadership Scores:** Leaders can be of two types. The peaceful unifiers and the dangerous separators. Humans have had leaders like Buddha, Dalai Lama, Gandhi, and Martin Luther King, who worked on dismantling artificial boundaries for the benefit of humanity. We shot two out of the four in this example. However, in the last two millennia, leadership has gone from bad to worse. If we take the top four countries and score them on a scale of peaceful unifiers on one side and dangerous dividers, we get this 2021 scorecard.

	Disastrous				Beneficial	
	1	2	3	4	5	
US	1					Trump
Russia	1					Putin
China	1					Xi
India		2				Modi
Bhutan				4		J Wangchuck
Tibet					5	Dalai Lama

I see Russia sliding south the fastest on my alien scorecard, followed by the US and China. The US used to be a global lighthouse nation, but it has now fractured itself on politically polarized motives. In the US, it seems reasoning has fled, and the nation is busy now with internal terrorism and may lose its leadership position by personally handing it over. **The top 3 nations score 1 out of 5.** Total 4.2/25

4.2 out of a score of 25 is an utter failed score. This is 16.8% marks and no one in the universe will ever let this student in any university or their galactic guild.

The top 3 powerhouse nations are performing like rich spoilt children with their 'Mine is bigger' arms arrogance, and secondly by misusing their weapons-for-profit, and their new found digital toys now in the hands of blatantly irresponsible leaders. Unfortunately, they are also the veto empowered, founding members of the United Nation, which due to their self-interests is getting more and more impotent. In the middle is India, a large democracy straddling between the two extremes. On the positive side we have Bhutan that keeps our hopes alive, and then Tibet, a non-nation, where China continues to commit genocide and ethnocide while its leader, the Dalai Lama, continues to preach peace from Dharamsala in India. These last two nations are the balancers on an otherwise disastrous world. So out of these 6 nations from a total potential of 35 points we end up with 14 points. If we consider only the developed nations then the score is 5 out of 15. If we take the bottom two nations then their score is 9 out of 10. So collectively, Bhutan and Tibet score an *A* grade, and the top 3 nations score a dismal *F*. You the reader has the full right to disagree as this is my alien audit score but then you need to try a wear a neutral alien hat and give it an honest score and post the number on this global score of only these 6 countries. Or you can create your own group and give it your alien score. Alien means 'zero' bias on color, nationality, sex, religion or a dominant species considering itself as the only intelligent species on this planet.

With this baseline now let's move to the topic on hand — our topic for this chapter is Taboo. Taboo topics are the ones that will be very sensitive to most readers, while having the potential of also being exhilarating to others. These are topics that have high emotional, and belief content. A pure example of a taboo topic is religion. When an Indian novelist, Salman Rushdie, wrote a book, Satanic Verses, that an Iranian Ayatollah Khomeini considered as offensive to his religion, he went and issued a fatwa, or a religious decree, in February 1989 for the author to be killed by any good Muslim. This was followed by many of the global religious members. The book was banned but, due to the ban, it sold more than 746,949 copies placing him in the league of Tom Clancy and such authors. Most of that popularity was because of the publicity the fatwa created. The Fatwa was issued in 1989 which resulted in many stab wounds on the author, by a believer in New York in 2022. Two of my editors even recommended I do not push these religious buttons. However, I strongly believe that as more and more people start becoming global citizens in the next two decades, we must join hands in order to play a big role in redefining these artificial boundaries if we expect any changes to the status quo. We need to make major changes in order to alter the very fabric of our beliefs. Religion is the most powerful taboo item on this list of topics, and many people are most sensitive on this. In many countries religious non-conformity is the largest reason for deaths and executions. Going back to modern Iran we now have a situation where pretty young girls are being locked up on trumped, no pun intended, charges by powerful religious police. There are now reports that these girls are put into jails, raped and then conveniently die or disappear behind closed doors. There is right now a revolution ongoing in Iran and we need to choose to help save the young women in Iran, or not.

Unfortunately, almost every religion preaches that only their god and religion is the greatest. Practically every religion place woman as second-class citizen, with fewer rights. Some even preach that their god will give them a golden step to heaven if they protect their religion, convert someone to their religion, or even kill someone who is not willing to convert. We need to decide whether we need to support this one God theory or a become democratic about religions and our support for the rights of women.

Let's review whether religious beliefs are artificial constructs or absolute facts. Different religions have dominated humanity for many thousands of years. Beliefs focus on convincing our survival-instincts and fears, by taking them to a realm beyond life itself. Every religion builds their realm beyond death. They all promise wealth, comfort, and some even 72 virgins in that realm, for performing required tasks in this life. Because no one can return from death the fable continues in every religion, bar none. All these beliefs take birth in our non-logical emotional side, which has been repeatedly proven to be more powerful than what our logical side brain tells us. Our left side of the brain is the logical side of our decision system, whereas the right is responsible for creativity, imagination and beliefs. As a Hindu if someone tells me that 22% of India is Muslim, it is a left side mental construct, thus it does not evoke any passionate, or emotional response. Now, if someone reports, or sends a WhatsApp text, that in a certain region people are forcibly converting 16year old Hindu girls at the rate of 5 every day into Muslims, now suddenly that becomes a very different game, it's a right brain trigger. It the emotional trigger that has been activated.

Similarly, if I am told to murder a certain author because they said something wrong it is one thing. But if I say that your god has ordered you to go kill that man and that he has a special place for you in heaven when you do, where 72 virgins await your arrival, and then add to that — while forgiving every mortal sin you have committed before this great deed. We the authors, of such constructs, have suddenly changed the very rules of the game. In this scenario some people might get very influenced with the promises, and commit crimes against humanity, and even want death to be awarded to them to reap the harvest, that awaits them on the other side. In the digital future we need to establish rules for the criminals and their accomplices. The person who picks the knife or the person who publicly instigates them to pick that knife.

For example, if the algorithms, knowing me as a wheat colored, Indian origin US national, starts sending me and my friends' fake texts with photographs — that people with my skin tone are being shot all across the US in daily feeds, and that the media actually wants to keep it politically quiet. Then I will be a fool not to start to get worried. When I share it with my friends, they say they have heard the same thing on their WhatsApp group feeds, and some even have other fake photographs, and videos, from people they don't know proving this is happening. By now my mind perceives that there is a risk to life, i.e., my life. My right sided brain will now get into survival mode. This mode will shut down my mind and body from thinking about anything that is not a threat to my keeping alive. In this example imagine this is all a troll needs to feed a targeted person by some group from across the planet targeting Indian Americans. 70% of US Indians, on social networks, have been targeted in the US and most of their Facebook contacts in India are blasted the exact same news but as if from another Indian source. This is called news validation. Then one day it tells us that some of the actual perpetrators are gathering in some city street to kill some Indians, going peacefully to a temple celebration. Now, some of the brave ones will pick up what they can to go on a protest march. Some may carry hidden weapons. Now, at this exact same time there is an opposite message being fed to the White pro-nazi Christians that the Indians are getting together to kill some Christians, under the masquerade of a peaceful Indian temple gathering. Based on these opposing feeds both parties will come together in anger and with the least amount of emotional spark this could become the start of a trigger that takes a decade to wipe out. This is exactly what happened in Charlottesville, when 'good people' on one side came in contact with a colored group from the opposite side. All built, and programmed artificially, by an algorithmic bot, in order to bring two polarized tribes into a state of psychological aggression. People died on that day. Most were pre-programmed with hatred, and reactions, that were not entirely their own.

Belief, depends on imaginary concepts that sometimes can be invalidated logically, but most often not because they are driven by emotions. Most humans have over twenty thousand years of experience with emotional decisions, around six thousand years of belief decisions, while only around one thousand or so of brain decisions based on pure data. Just to give an example – for over 5 thousand years the people of Europe were part of a group, that fought the invaders from the south. They formed an emotional color-biased trust between people of their skin color. Around 4 thousand years ago a multi-god, multi-sex, religion of Hinduism started, then around 2 thousand years ago the concept of a single male god above all else was started by Rome. Rome started unifying people under the name of Christ. In Europe this became the prime religious bond between people. Around 15 hundred years ago Islam was started in Saudi Arabia. While today in the US we all believe that our constitution says that all humans are equal and no one is above the law, and that Black or White or any shade in between, all deserve equal opportunities. This is the brain. However, the heart tells us that 50% of the US lives in the gray areas of this conservative constitutional law. In our mind and constitution, we're all equal, but our heart has been converted into believing that was quite different. If a religious fanatic stands to gain by amplifying the concept of *Us vs. Them*, they will play with our hearts for their gains. It is important to remember that the most powerful and emotional belief is religion. We have over 2,000 years of history that has engrained religious beliefs into all humans as the most powerful 'Us vs. Them' foundation. If you think your god is the only true one, or better than mine, then you are already an addict.

Each of these taboo areas are constantly maintained in a state of very emotionally charged, and highly addictive, and they create very strong tribal loyalties, and boundaries. They are the original launchpads of creating this 'Us v/s Them' environment followed by an "either or" mandate. The *Us vs. Them* has resulted in Muslims rewarding their followers with *Jannat*, the Islamic heaven, if they participate in the holy war or Jihad, i.e., the act of killing non-believers that do not want to convert to follow their one God. Christians have the Spanish inquisition on their head, along with the various Pope's during some not so proud moments like the one against the Templars for the benefit of the king of France, and most recently, the pope directly supported the pro-life, anti-abortion movement in the US thereby denying women a right to self-determination. The church supported enslaving the womb of a woman as property of State and the Church.

So as a reader your digital opportunities could be in building the next generation solutions that have not yet been built, but what attracts developers with passion is the social good factor. Every company that focused on building anti-oppressor solutions has consistently attracted better workers, more dedicated passion, less fear of getting caught — something that will sustain across space and time.

Politics in the Digital Future

Politics is an area that, along with religion is most archaic about the potentials of digital transformation. For thousands of years, politics used to be a fairly distant and political subject starting with tribal chiefs, then kings, and emperors. Recently it evolved to presidents and prime ministers. This was status-quo up until the mid of the covid-19. The Covid-19 global lockdown placed most humans in a rare position of getting exposed to digital neuro massaging at a level of 1:1 total focused attention, with the advent of companies like Cambridge Analytics, people like Steve Bannon and its clone IT departments popping and developing their own algorithms in the game of politics as a business of money and power. They have become the new digital priests who could control the minds of the masses and become the new kingmakers across the planet for a few coppers in their pockets. During this period, it was also realized that politicians in developed and developing nations, after they get elected, could allocate millions, if not billions, and now trillions of dollars to politically sponsored programs. It is quite possible that the allure of power, and unfiltered funds, is what probably attracts many of them to become a politician. This can now become not only a big incentive, but also a big attraction to diverting publicly allocated funds and generating slush funds for getting internal political party support. Trump changed politics in the US, as did Modi in India, along with Putin in Russia, and Xi in China. Politics now has become a digital art form to an old desire to accumulate more power and wealth. Outlier nations like Saudi Arabia, North Korea and Iran are not part of this discussion as these are states of political oppression and not about the politics we're discussing here.

During the Egyptian reign the priests were a very powerful influence in approving or rejecting the Pharaoh. Similarly in 325 BC in India, Chandra Gupta was declared a king by a brahmin, an Indian priest, who was known to be a kingmaker, i.e., this brahmins name was Chanakya, and he still reverberates in Indian politics 2,300 years after his reign. What is often forgotten to be mentioned is that he was also known to be a king breaker. In Western political authorship, what is quoted most often is Machiavelli and his book *The Prince*. However, after reading the volumes of books written by Chanakya, the reader will soon realize that Machiavelli was but a kindergartner compared to Chanakya. Chanakya was a religious power-broker who kept his kings in power by deceit and cunning. Every king, lord, leader, and politician has had a softer or harder version of a Chanakya who helps them keep competition out of the way. This is personified very well in the British TV series *Yes Minister*.

Some of the most powerful politicians who traditionally used their political staff hit-men, today are using digital-tools to do the same far more effectively. At the far right, of this misuse are nations like North Korea, Iran, China, Russia, and Saudi Arabia where political opponents are locked up and murdered without any public, or international explanation. Closer to home on August 17, 2016, with 88 days until the 2016 presidential election, Bannon was appointed chief executive of Donald Trump's presidential campaign. Bannon left Breitbart, a hyper-rightwing news outlet, as well as the Government Accountability Institute, and Cambridge Analytica, to take this job. After his victory, Trump handed him a key to the White House. Soon he had to leave the job in the White House, reportedly after a power struggle with Jared Kushner, Trump's son in law. After that, Trump had a revolving door of 'Yes-Men' who, seemingly were let go the moment they challenged any of Trump's wishes.

A huge part of politics is studying and learning from History. The problem with history, is that it is a version of the truth written only by the victors. The point that worries me is that the word History contains the very words that defeat the quest for truth in history. It is a version of His Story — His being the mostly male leaders and, the story being the interpretation of truth as defined by the winning males. Let's start with India as an example — in the 1960s when one read the history of India it was mostly a version of history left behind by the British colonial writers. Our initial leaders were British educated and inculcated humans. They replaced their Suits with local clothes – local emotions, but their brains remained quite British. They simply inherited and continued the same policies but under a new class of people. Our Indian history is now slowly beginning to change but the British linguistic roots run deep in the English-speaking leaders in positions of power. The same thing was happening across the planet in the US where history was being written by the White writers. In fact, the problem was so severe and the divide so total that in the late 1930's when US wanted to author a book on the negro problem in the US they could not find a single person, White or Black, to write this unbiased book. What this resulted in is the committee finally found a Swedish economist with low to no US color biases, to travel to the US to author the book. The book itself was so painstakingly detailed and exceptional that one of his researchers Ralph Bunche, later received a Nobel prize. The book 'An America Dilemma' sold over 100,000 copies but never became a must read at schools and universities where it could have done so much good. If we now travel to modern US, i.e., July to Sept 2022 we still see states trying to mask history by not allowing any discussions on the Tulsa massacre of 1921 when the Black wall street of rich Black business people was burnt to the ground in a sheer racist operation.

Many successful, and rich Black people were killed by the Whites who could not bear to see their wealth. No one was ever brought to justice for this genocide. In fact, this massacre was totally erased from history books. In addition, states all across the US now do not want to teach the truth about the White man's tyranny by slavery, or the true Black events in history, and their systemic genocide, so we go and write the White version of history of the US. The reason given by most is that they do not want their White children to feel guilty of crimes done by their ancestors. This is the exact same reasons Tiananmen square does not exist in the history books in China.

It is termed as political re-writing of history. The governor of Florida has by 2023 taken it to the next step of whitewashing history. He has canceled all Black AP classes and has banned a whole bunch of books that deal with slavery and racial inequality. All this reminds one of the burning-down of the library of Alexandria, sometime attributed to the Mongols, and at other times to Caeser. However, a similar event is happening in 2023 in America the supposedly greatest democracy on the planet. The extraction of books by politicians because it does not confirm to their political agenda. This is exactly why history is important for all to learn, sometimes to question, and sometimes to get them to write the absolute truth, and nothing but the truth.

The proliferation of blatant lies: In the digital world of the future there will need to be some fundamental criminalization, when elected leaders tell blatant lies. This includes people, parties, media and all other people of influence, based on clear strike-3 rules. In the US, and across the world telling lies has become synonymous with politics. The more outrageous the more optics it receives. Along with lies is the explosion of paid media anchors who will replicate the lies as media checked opinions, that they broadcast to their viewers and listeners.

As a start no politician, or political party, or any member thereof, should be allowed to make a statement, in jest or otherwise, that is proven to be a lie Absolute, and undeniable, lies have become a fashionable extension of politics, at least in the US. Elected politicians, and their media counterparts make outlandish statements with zero to no repercussions under the protection of freedom of speech. The republican party and Trump have been continuing on the message that it was Trump who actually won the 2020 election despite all evidence proving the exact opposite. But unfortunately, these lies continue with some voted republican representatives, and senators, where the system allows them to continuously broadcast these fabrications, thus becoming an accomplice to such deceits. We the people need to place lies as a crime and a clear strike-3 rule with lies that must be supported at a federal level. 1st lie is a warning; 2nd lie is blacklist from any political post for 3 months. 3rd lie is banned from a political post for the rest of their lives. A lie is when a court proves it beyond doubt that the person knowingly stated a lie. Think of George Santos, our Republican, as proof of why this is needed. When someone lies to this extent, they can be coerced to sign any document that the controllers want them to sign simply to maintain their current position — that cannot be good for any nation.

The Xi, Putin, Bernie Effect

Our greatest danger in the future is the ever-lasting desire of great kings, emperors and Leaders to, firstly become emperors for life, and then after that to spread their philosophy across the planet, i.e., world domination. In this list rulers stand out. It starts with Genghis Khan, then Alexander the Great, on to Julius Caeser, then Napoleon Bonaparte, onward to Hitler, then Joseph Stalin, and most recently starting with Putin and moving to Xi. Trump was seduced by this idea briefly in one interview. The need to rule the world has been a desire of many great people, and all of it always went in one direction — defeat and abandonment. Despite this each subsequent leader thinks they are the one that can accomplish this goal.

The Stuff of Nightmares

b.1 Intelligence & Omnipotence traps: The intelligence trap has been explained briefly before, and it is the result of highly intelligent and powerful people falling into the trap of "If I am right most of the time in my small area, I then become right all the time, in every area." It leads to leaders, CEOs, and people putting their own beliefs and rules above all logic, data, and opinions. These people do not listen to data, experts, or other opinions and barge through with their own edicts and orders, often of a logic cliff for their company and themselves. The crash of Blockbuster and Nokia can be attributed to intelligence traps. The recommendation of the Republicans to not take the COVID-19 vaccine and that of the ex-president to inject Clorox as a COVID-19 cure is not the greatest example of an intelligence trap. **Omnipotence Traps** are at another level altogether. It starts with intelligence trap and proceeds to an omnipotence trap. Putin in Russia, Xi in China, and the Ayatollah in Iran are examples of their own omnipotence traps. When rich and powerful people are trapped in this dilemma, they start to believe their omnipotent control over laws, justice, and often the life and death of their citizens is their intelligence. In most cases, the common citizen sometimes allows and often encourages them to undertake the unthinkable. Like Putin (a) first trying to assassinate Navalny, his political challenger in the new democratic Russia, then locking him in prison believing there can be no repercussions. (b) Secondly, he attacked a peaceful nation, Ukraine, on his omnipotent beliefs, followed by sending criminals and then sending the 21,000 Wagner Mercenaries, which is an unofficial army within Russia. Wagner forces are used for Russia to do things that Russia can deny in operations abroad. In Ukraine they deployed 40,000 prisoner recruits. In addition, these mercenaries received a barbaric bonus to steal to their heart's desire and rape whoever they wanted to. In Syria, they worked for Bashar and his requirements. When the US attacked Syria, it resulted in hundreds of Wagner forces being killed. In 2019, they entered Libya to help with an assault on Tripoli. In Central Africa, the Wagner mercenaries helped quell civil war for authoritarian government leaders. In Mali, the government pays around $10.8 million a month to help fight an insurgency by killing hundreds of civilians. The stories continue in Sudan, where they help suppress democracy for their payment of Sudanese gold. These are the dogs of war deployed by Russia

*My Digital recommendation: We the citizens of Planet earth: Need to build a global scorecard and track examples of **Intelligence Traps**, and **Omnipotence traps**. Share photos, and videos, as the final mode of truth. Share data and numbers. Identify fake reports and data. The scorecard can slowly become self-scoring by the number of people who agree to someone's accusation of either of these categories. Each authorized scorer must be validated and authenticated by a secure global process, to avoid it being run-over by trolls of all types.*

b.2 Caste, Racism, and Colorcide: I see caste and race very closely linked. If we search for either, we get a highly inaccurate and biased historical perspective, i.e., depending on which country publishes the stats. For **Racist** if we look at the US position India is the most racist country according to Washington post & Business Tech, and US is not even in the list. If we look at the UK stats from the evening standard then Iran is the most racist, with US at #19 and UK at #21. So, with racism one thing becomes clear it is a story of the country reporting it. Now, if we look for **Caste** it is reported that India is at the top of the list followed by Bangladesh, Bhutan, the Maldives, Nepal, Pakistan and Sri Lanka. According to this report there is zero caste systems in Europe, China, Japan or the US. Historically when we think of caste atrocities our minds have been trained, by the systemic history written by the developed nations, to take us instantly to India where there has been a formal caste system for centuries. However, the very same historical bias also distracts us from the US caste system, as elaborated by Isabel Wilkerson, in her NY best-seller book *Caste*, in 2020, and as proposed by Dr. Martin Luther King Jr, that the US caste system is very strong, alive and still operational today. The US caste system disproportionally thwarts the rights of their Black citizens and sits in the middle of national color-code. Yet the leaders, the system and governor's do not want to accept their White caste system thriving in the US. The first case in this is George Floyd, where justice was served based only on undeniable digital proof. The next is of the

'Central Park 5' who were charged and jailed for 13 years based on false Charges. At that time, Donald Trump released a White man's view in a full-page ad in the New York Daily, asking the accused be sentenced to death — a KKK-like edict. On the Right is a Twitter plea from Yusef Salaam requesting that justice must be served. *Yusef Salaam, one of the exonerated 5 in the NY rape case, is a perfect example. The case was referred to as **Central Park 5**. The 5 were accused of raping a White woman jogger in 1989. Trump's ad in that year exposes the caste system in the US and the White man's omnipotent rights over the lives of innocent Black people. Reminiscent of the Jim Crowe and KKK justice call for lynching Blacks. They were all acquitted in 2002, after 13 years in prison and false charges brought on them. In 2023 Yusef Islam won elections in Harlem, and he plans to climb the difficult hill of fairness and equality as written in our constitution.* It's now a miniseries on Netflix, "*When They See Us.*" Just like sex is a big divider, where being a woman gets the person all the wrong ends of every deal, and every nation bar none. Color is a universal cultural taboo that we need to extricate ourselves from. According the Spencer Mills we all come from Black ancestry, so let's join hands are eliminate this criminal bias on caste and color as our moonshot for 2030. Finally, heading back to **Colorcide,** we find that, Color and caste is a global cultural, and racist, issue. Most broadcast about India, where it is traditional and centuries old, and the US when in 1619 20-30 enslaved Africans were brought to Hampton, Virginia. It soon became the worst color-based class differentiator on the planet, with remnants of this still prevenient today, as demonstrated in current George Floyd and the Central Park 5 cases. The next case is the Tulsa color-code, and racist massacre. When between May 31st and June 1st, 1921, on Memorial Day, where White supremist burned down more than 35 blocks in Tulsa or a flourishing Black community. This is called the *Black Wall Street Massacre.*

What is most surprising as a mix of caste, race, and color code is that they never recorded:

1. How many people died. Reported estimates are 150-200 Blacks dead, with more than 800 injured.

2. Around 10,000 successful Blacks were left homeless.

3. It took all of 75 years, in 1996, when a commission was established to study the riot.

The shocking fact is that the US collectively erased this massacre from all its history books, because history is written by the victors. Another incident is the Jallianwala Bagh Massacre that took place on April 13, 1919, in Amritsar, in India. This is when Brig. General R, Dyer surrounded civilians, Women and Children in a garden and shot around 1,200-1,500 and seriously injured another 1,500 unarmed Indian people.

My Digital recommendation: We need to work collectively and one age group at a time and eliminate this CRC, curse of caste, race, and color from our mindset. In the US even though the constitution wiped out the two, the prevalence of the bias never departed. In fact, we see very clear Republican US politics now creating color and caste barriers in their electoral process for the coming elections in 2023 and 2024. We need to launch 'Dignity of Humans' that aims to eradicate the CRC, as actively as we did to eradicate polio and malaria.

b.3 Human Organ Trafficking: Also known as the Red Trafficking. We've all heard of human trafficking, and till I was writing this book, I assumed it was only about sexual trafficking young girls to brothels and the global sex trade. We shall treat human sex trafficking as a separate section here. However, in the process of writing this book, I've discovered another worst trade, i.e., the Human Organ trade. Worldwide, this is a $1.4-$3 billion industry. According to AcamsToday, the current average cost for organs in USD is Corneas — $30k, Lungs — $150k, Heart $130k, Liver $90k, and Kidneys, $62k. What brought this to my attention was the arrest of Ike Ekweremadu, 60, Beatrice Ekweremadu, 56, and Dr. Obinna Obeta, 51, in the UK for illegal Organ trafficking. This happened in March 2023. Ekweremadu is a senator in Nigeria suffering from the omnipotent trap. The report says that his daughter, Sonia Ekweremadu, was diagnosed with deteriorating kidneys and needed regular dialysis. Dr. Obeta was tasked with finding the perfect kidneys for her. Using the Nigerian digital medical records, they found a perfect match for a kidney donor in Nigeria. They found a boy in Lagos, and with a promise to get him a job in London, he was flown to the UK. As a bonus, he was told to tell customs, doctors, and others in the UK that he was a cousin of Sonia. This was necessary to justify the victim's temporary visit visa and his free kidney transplant. It was in the hospital that the victim realized that he was in grave danger. He immediately went to a local police station and reported the incident. The Old Bailey court found all guilty and awarded all three to serve imprisonment in the UK. The Nigerian govt tried to interfere and ask for all three to be tried in Nigeria, but the UK refused as the crime was committed and reported in London. So, the big questions are (a) Who are the big suppliers in this nightmare trade, plus who are the buyers. (b) who all, the rich and powerful, in your country, could be undertaking such organ trafficking. (c) I am certain no one believes that this only happens in Africa, so what about the rest. On the **Supplier side, #1 China** is reported to be the largest supplier of illegal human organs. They have industrialized the organ transplant business. It was transplanting 13,000 livers as far back as 2004. In fact, China provides offshore medical services for transplants with a guarantee of the perfect organ on arrival. In my mind, totally unverified, I imagine China having access to thousands of captive prisoners, plus the medical data of thousands of potential people across China, the Uighurs, Tibetans, and people under watch. When an organ is needed, the person simply disappears, and a fresh organ is thus available next to the bedside of the customer. **#2 India** is a leader in illegal organ supplies. According to Al Jazeera, India is the top exporter of Kidneys, with around 2,000 kidneys sold per year. India is a confirmed illegal transplant hub, though because Qatar is a friend of China, there is not a single word of China in their article. **#3 Iran** is the only country that allows the stealing of organs. They allow the legal buying and selling of human organs. Maybe this is the reason pretty girls are charged, imprisoned, and die in prisons in Iran. **#4 is Pakistan,** which has a low population but proportionally a high number of illegal organs harvesting. The general pattern is nations with high population or density, poor or under control, and people who can be bribed, coerced, or tricked into organ transplants. Other seller nations include Angola, Brazil, Canada, Colombia, Costa Rica, Eastern Europe, Ecuador, Georgia, Haiti, Israel, Kosovo, Libya, Mexico, North Macedonia, Peru, Philippines, Russia, South Africa, United Kingdom, and United States. On the **Buyer's** side are the US, Canada, the UK, EMEA, and basically all the rich developed nations. The buyers, logically, have to be the countries with money to pay. The US and Canada, as of 2018, did not include organ trafficking as a form of human trafficking. In my field, we have a saying, *"If you don't track it, then you cannot report, control,* *or improve it."* So, is this by intent, error, or protecting their rich voters?

The transplant business or tourism is thriving. This form of tourism currently bypasses global laws or lack of local laws. General Supply: Demand trends are from Southern hemisphere to North, Female to Men, and colored victims to Whites.

My Digital recommendation: I worked with a Biotech company that successfully put a QR code on every bottle of their very expensive, or life-critical medicines. This was done to prevent fake medicines, and selling of expensive drugs, under their brand, that were not what the customer was buying. Just as an example during the Covid-19 peak, families were willing to pay 10, 20 and even 50 times the price for Remdesivir, a drug that could prevent death from covid. I can say with certainty, that millions of people paid more than the MSRP for that medicine, and thousands of the bottles were fake drugs. BTW the company in question, i.e., building the QR solution, was not Gilead, this is just an example. The QR company made it mandatory for all global doctors to scan the QR code, on the bottle, that auto-filled a form listing critical things like batch number, SKU, — autofill, hospital, country, patient, etc. Now if we can do this with a million bottles, then with a little political will, we can certainly, and easily do this for every organ that moves through the global system, from any point A to B. Even if it is from Shanghai to Hong-Kong, or Bangalore to Delhi. Xi and Modi will need to sponsor this, as will other national leader in the list of suppliers, and buyer nations. Global airlines, WTO, Customs, and Transport organizations, must not allow any organ to move from one point to another without full blockchain and QR code verification. The metadata must contain Patient name, Age, amount paid, contact number, Photo, extraction doctors name, customer, amount paid, etc. If required I can have a team build this solution in blockchain + QR rather rapidly. But there has to be a global will that needs to start somewhere. For example, the US can deploy this solution and not allow any organs to come in or go out without this verification. Fact — every organ for transplant has to travel from point A to point B. we need to place checkpoints at each of these points just like we do for drugs, guns and gold. It's a simple process, and all it needs is global commitment. If an organ reaches any transit checkpoint whatsoever without these documents, the supplier and buyer contacts must be immediately placed on a lock-down status till the matter is resolved. They can also mandate that if any US citizen goes abroad, they fill out the same tracker.

b.4 Exotic Animal Trafficking: unfortunately, a lot of this is done for food, i.e., when the rich want to eat things no one else can get. The next segment is the rich who like to collect exotic animals in private Zoos, as their personal hobby. This is an illegal trade of exotic animals, and parts that is worth from $15 to $20 billion. Animals include Exotic colorful birds, Elephant Tusk's, Tiger parts, Rhino Horns. Exotic turtles, bears, etc. Illegal animal Trade statistics report that Birds represent 42.03%, Mammals 25.83%, Amphibians 10.92% and reptiles 21.22% of global illegal trade. The **Sellers** of exotic animals are the nations with exotic animals, **A shallow deep dive:** The world's largest illegal trade is in Pangolin. Their species are the Chinese Sunda, Indian, Philippines, African, and Malay, where they are now reaching extinction levels. One of the culprits of the Covid-19 outbreak was the consumption of Pangolins as a delicacy in China. Their meat is an expensive delicacy and their scales are considered as Chinese medicine. Pangolins are shy, harmless animals that roll up into a ball when approached. They eat ants. In 2019 alone, Chinese customs reportedly confiscated 123 tons of Pangolin scales. An average pangolin weighs around 4.5 to 7 lbs., and one pangolin produces around 500 grams of scales. There are 1 million grams in 1 ton. This catch of 123 tons represents the killing of 246,000 pangolins, catering to a unique Chinese delicacy in just 1 year. Pangolins are thrown into boiling water alive, just like lobsters are in the developed nations. Despite being banned in 2016 the Chinese meat market was full of pangolins when Covid-19 struck. In our list of exotic animals trade it is our good ole China that sits at the top as the #1 global buyer. As we all donate more and more money to China their rich want to eat more and more exotic animals. Chinese will eat anything, and everything that moves, reportedly even cockroaches as medicine. They are today the largest buyers of illegal animal trade. **The Buyers:** While US, Europe and Japan have placed strict bans on trade, buying and selling of ivory, animal skins, and any trade in illegal and exotic animals. It is China that seems to be the global harvester, and leader, in this trade. They eat every exotic animal and are the reason for most of the illegal animal trade worldwide.

*My Digital recommendation: I would like to give two quotes from Gandhi here, **"The greatness of a nation, and its moral progress, can be judged by the way its animals are treated."** In this measure, China is at the bottom of the swamp. He also said, **"I hold that the more helpless a creature, the more entitled it is to protection by humans from the cruelty of man."** Plus, a quote from Enzo Maiorca, **"Until man learns to respect and speak to the animal world, he can never know his true role on Earth."** At the top of global trade, we must place much stronger controls on the trade of exotic and helpless animals. We need to have stronger penalties; I would propose stronger than drugs, as this kills innocent animals. So, if the punishment for drugs in Singapore is death, so should it be for exotic animal trade. This is an area where "We the People" can contribute directly to the global initiative on **DORA (Disease of Research Animals)**. I can see kids loving this from its day of launch, as will my daughter, who loved Dora the Explorer. We can launch, contribute, grow, and report directly to this global database. Each time you see an exotic bird, animal, or food being mistreated anywhere; you can report it into this database. Switzerland has already provided a living case for this. In 1842, the Swiss enacted the first constitutional law against cruelty to animals. It lay* *dormant till 1978 when someone called for its activation. In 1978, Switzerland became the first country to recognize animals, with a* **constitutional provision** *warranting the protection of "the dignity of the creature" that was approved by 80% of their people. They call it the* **Animal Welfare Act.** *In 1992, Switzerland became the first and only country to re-consider the 'Dignity of Animals' enshrined in its constitution. This basically stated,* **"No one shall unjustifiably expose animals to pain, suffering, physical injury, or fear."** *This law was forgotten till someone made a case for lobsters and won. On March 1, 2007,* **when it comes to cooking lobsters, the Swiss collectively stated that "we feel your pain."** *Based on this, a Swiss Judge banned:*

a)   Transporting lobsters in fresh-water ice as Lobsters were not freshwater or icy-cold creatures. Instead, the transport had to be in salt water, not below 4 deg C.

b)   Lobsters could not be displayed in fish tanks with their claws in rubber bands.

c)   Lobsters must be rendered unconscious, with strict guidelines, before being thrown into boiling water, and never thrown live into boiling water as it is a very painful death.

Taking it as a joke, an omnipotent French Chef in Switzerland arrogantly said, "The Swiss can make rules, but if you want to eat French perfection lobsters, visit his Swiss restaurant to select and eat the freshest lobsters." Within a week, the Judge stated that any restaurant that disobeyed his rules would stand to:

a)   Be shuttered and pay a penalty, and

b)   The owner and the chef jailed for up to 3 years in jail.

By the next morning, the French restaurant and chef followed the Swiss law and toed the line. It's as simple as that. So, we already have a proven roadmap. Now, it's just about putting it in place and then slowly expanding our desire to give animals their due right to dignity. **I say (A)** *let's launch our US version of Dora in California and then export it from here to every state across the US and then the world. Let's start with Lobsters as our first step.* **(B)** *I would expect the only vegetarian nation to adopt this rule in a jiffy with very wide support from a majority of Indians.* **(C)** *Let's get our US to lead a global initiative to make exotic animal trade a criminal activity globally. If we can use Interpol for drugs and weapons, it's now time to take care of exotic animals with the same passion.*

b.5 Human Sex Trafficking: is the recruitment, transportation, transfer, harboring or receipt of people through force, fraud or deception, with the aim of exploiting them for profit. This comes in the form of sex and labor trafficking. On the sex side this is normally on children under 18. Unfortunately, sex trafficking is initiated, by someone the child knows and trusts. Forced underage marriage is a form of illegal sex trafficking too. So, on April 24th, 2023, when a US judge, Mike Moon a republican from Missouri, USA, wants to give rights to a 12-year-old girl to marry, he is probably seeking approval for illegal sex trafficking and child marriage, in his republican state. Surprise, surprise, just learnt also that Senator Moon is a deacon in the Southern Baptist Convention, which is being investigated for a massive sex abuse scandal, on handicapped children, and their coverup? In the US today, this is classified as pedophile sex, and is very illegal. Now more recently This is creating an Afghanistan or Pakistan in USA. Another example that reeks of a religion is when 13-year-old US pregnant girl So Tyree, was forced to marry her 32-year-old rapist and child molester, as the US version of saving family honor. I thought this only happened in Pakistan, but looks like were importing family values from our allies in Pakistan. In Italy, in April 2023 when a 66-year-old janitor put his hands inside the pants of a young 17-year-old school girl, the judges in Rome concluded that less than 10 seconds did not constitute sexual abuse even if the man's hands were inside her pants and briefs, touching her buttocks. What kind of a judge, or a group of them thinks less than 10 seconds of touching a girl does not constitute sexual harassment. How come the pope is quiet about the city that is the shell of his country. Italian men always had a bad reputation but this is the absolute limit of indecency. These are not a one-off report, many states, including California have no minimum age for girls to be married. A 2019 report stated that 92% of sexually exploited children were under the age of 13. This is probably the market the US judge was pandering for. Sweat shops and child labor also falls under this category.

My Digital recommendation: This is another area where we can contribute. I know of a set of very smart coders who are giving their time and loving it to catch underage sex trade in the US. As an experiment they started in LA and using ML and AI, and traced a house in LA with no police record. When the police raided the location, they found 32 underage girls that were being primed for sexual exploitation. So, build your own application, group or company. Contact the police and work in partnership with law. Collect data on a topic of concern, often from arrested people and scans of their cellphones to identify innocent words with dual meanings. Then use MLAI, and now ChatGPT LLMs to identify nefarious transactions and activities, all in real-time, and report them to the police.

This is just the tip of the iceberg.

If you have ideas, share them on the Digital Shock blog so we can all develop new PEH solutions. Or simply send me an email at harig.auth@gmail.com with a subject line starting with Digital shock:

Sex in the Digital Future

Sex and sexual preferences are extremely powerful directives of procreation in all forms of life, including human desires. Nations, Presidents, and humans throughout history have committed atrocities of various kinds based on this one desire. Some go as far as one of the most powerful emotions after fear of death. Due to its powerful influence on humans, we find that religions, cultures, and politics have for millennia used this as a tool both for control, and oppression by defining male domination rules and regulations that allow them to control their tribal members, with an undue regulatory-control-emphasis on women. Somehow, no matter which religion or nation, most of the control and domination guidelines are toward women, and most of the benefactors of rights, and lack of definitions are for men. Unfortunately, most religious, constitutional and legal books have been authored almost entirely by men. Men who want to rule the sexual game, by suppressing the rights of women. This is a constant across humanity. The tribal controllers could be a nation, but the reigns are in the hands of their religious, and political leaders. Throughout history, kings and their royal houses, and unfortunately catholic priests and including their bishops, could resort to allowable debauchery, participating in rapes, sex with boys and other things banned to the common citizens. The Church and politicians created two major divides. The **first** divide being between men and women with men writing all the holy books, its laws therein and their god given role in dominating women throughout the centuries and even today. The **second** being the creation of a fanatical 'Us vs. Them' environment in their religious sermons, and political discourses that allowed both religion, and politicians to wage wars indiscriminately across time. Today in a state of declining attendance at Churches, and a decline in citizens participating in voting, these two institutions are now fighting for survival. In their fight for their survival, they both are trying to dominate you. One to dominate your soul, while the other using the holy + political powers to dominate your mind, and now in case of women their womb.

There is a **third,** still a persuasive invisible party, which is media. Media over the centuries has become the extremely rich, and powerful paid mouthpieces of customers with deep pockets. In every country, they are protected by their democratic constitutional rights in the form of free-speech, or freedom to report the truth. However, with the corruption of institutions, this in the US has now morphed into become freedom to push false opinions for payment, mainly by political parties. Media today are paid-mouthpieces for one political party or the other — bar none. Their richest customer is politics, and the second richest being religion. Even today no media house can go against the will of their political or religious patrons. In 2023 Fox news is now finally being exposed, due to digital proof, of supporting political lies for payments and at the expense of the health of the nation.

Sponsored exploitation of women and sexual rights: The domination of women can be taken into two forms the first being physical and the second mental. Then there is the mix of both. **1. Physical Abuse:** At a high level I'd like to share two very extreme methods of sexual control. The **first** is in China and Japan. It is the practice of **Foot Binding** or Lotus feet.

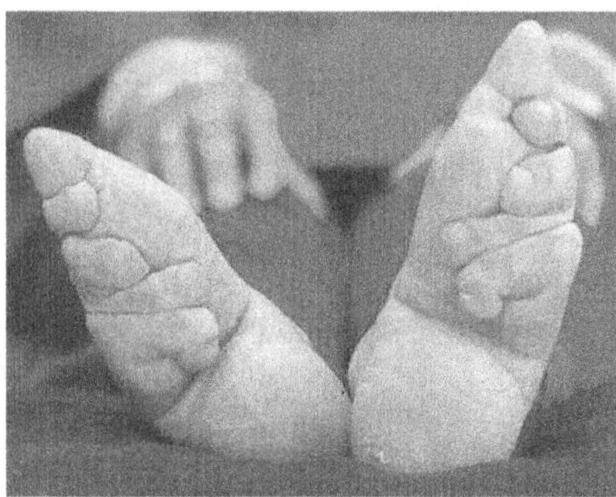

In this practice when a girl attained puberty the bones of her feet were broken and bound tightly subjecting the woman to total discomfort in a process called Lotus Feet. I had heard about this, but it is only when I saw the image above, of what it does to a girl, that the shock hit me. While both China and Japan compare this to putting a Corset on a woman, I however, believe this is very different. I have confidence that this was a form of female control undertaken by the parents, and/or future husbands of ensuring their girls, or women did not go out of the house, or see other men. It is a physical cage they built by making the act of walking itself painful. This was banned in China in 1912, but was eliminated only in 1949. In Japan it was banned outright in 1915. The **second** is FGM, or **female genital mutilation** as practiced mainly by Islam. Though it is also practiced by some Christians mainly in Africa, and one Jewish community in Ethiopia. So, this seems an Islamic-Afro practice. This is prevalent even today in many Islamic nations and is a form of sexual physical control. This involves removal, in part or the whole Clitoral glans and hood. The believers state that this is akin to the male circumcision but it is not. This mutilation removes the pleasure glans of the woman as placed there by the creator. It is possibly a dominating way of stating that pleasure in sex only belongs to men.

2. Mental Abuse: There is probably not a single nation on this planet where women have not been abused in one way or another. Societies, religions, cultures, and man all have systematically conspired to subjugate women to their lowest possible status. **Systemic Brainwashing:** I can say with confidence that in 95% of the nation's women are taught from birth that they are different from buys. In many nations women are taught that they need to educate themselves but only so they can marry. Some, so called home science, colleges for women teach subjects designed for a woman, and not professional subjects like they have in men's colleges. Religious houses worldwide teach women to remain in their place, while most religions do not allow priesthood for women, i.e., they cannot be the layer between God and humans. **Women's Rights:** In my research I cannot find a single religious book where women have written any part of any holy book. Nor, any country that is fair and gives equal rights to women. Interestingly, over 20 countries have enacted a 'Marry your Rapist' law as a way out for women. I recently found this prevalent in the US too. In 9 countries rapists go free if they marry their victims, so in these countries if you want to marry a pretty girl who is refusing your advances, you simply go and rape her. Now she is denied any other option except to marry her rapist. In many countries forcing a wife for sex is not considered as rape, but an ownership right of the man. In Iran, Afghanistan, Pakistan and Jordan a man can legally kill his wife, sister, or daughter if he catches her with another man under their 'Honor Killing' laws. But no country allows the reverse of this.

Sexual & Birthing Rights: The recent *Roe v. Wade* case in the US is a severe decision reminiscent of the days of the papal-sponsored and Spain propagated Spanish Inquisition. During this period, in the Middle Ages, Spain conducted barbaric cruelties in Latin America on non-Christians, and a competitive covering of their women inside Spain as a form of holier-than-though optics to their Arabic counterparts. Getting back to abortion rights and the US, this abortion decision, supported by the pope himself, was undertaken by the highest court, by the country on top of the hill that is 100% related to state-based-sexual exploitation, created on the guidelines of religion, politics, and state. Unfortunately, it is a voted selection as the leaders of these states have been democratically elected to their positions and thus represent the will of the majority of people across that state. So, technically no woman of that state has any right to complain if they hit an abortion barrier — other than to say next time, I will vote for someone else.

Krux's Law — In case of rape: Here is a recommendation for women. Something that needs to be taught to every girl and woman on the planet. It is impossible to realistically deliver the goal the Texas Governor proposed, i.e., that he would eradicate all rapes in the state of Texas. So, the reality is that there will be that possibility of rape, and along with that possibility comes the absolute denial of the rapist. Here is what the law recommends. In case of a rape, (a) save all underpants. Dry them and place them in an airtight Ziplock for posterity. Never throw these away or wash them, as most women do in disgust. This underwear contains DNA evidence that is better than digital proof. (b) Scratch the person with your nails. Once again, remember where you scratched the person, collect their skin from under your nails, dry it, and place them in a separate bag with the dry underwear. This will one day let you prove beyond reasonable doubt that the sexual abuse happened.

With the advent of the new digital order, and the two-year covid lockdown from open discourses we are now facing a world where sexuality beliefs, and optics, have been totally altered. Meeting people has gone from slow physical introductions to instant gratification of swiping left or right. This includes casual, for fun, or serious sexual encounters. Sex has been a religious + Political enforced controlling mechanism for centuries. In our new digital world, we must realize that sex is a powerful, and God given urge for procreation to all mammalian life, and many other forms of life too. The Catholic church, and other fanatic religions are squeezing their fists ever tighter by placing one draconian rule after the other associated with sex, and their guidelines on dominating women. The Church possibly assumes, in their omnipotent belief, that by making abortion related decrees, even if there is no mention in their bible, they will attract more true believers, and fill their Churches across the planet once again. While, the republicans assume that by enslaving the womb of every woman, in America, and misrepresenting sexual preferences they will ensure the birth of more voters, and garner more religious votes. Then right now I can predict that both are on a very wrong and slippery path of holier-than-thou righteousness.

The New Anomaly — Digital Military Warfare

Just like space was the *'domain of nations and governments that a private person could not dream of,'* rule was dashed to the ground by Elon Musk. Similarly, there is a momentous disruption happening in the field of weapons and warfare. Unfortunately, conflicts, battles, and wars seem to have been woven into the genetic code of our 'intelligent' human genes. Humanity has been in a state of peace for only 8% of known history, and most of that peaceful time has been in the now. The 2008 Stanford research, titled 'The Reasons for War — An Undated Survey,' is a great expose on why humans go to war. Almost all are artificial and created for dominance. If we look at the history of war, we see two patterns every few centuries the motive for war changes, but the underlying reason is always power, and dominance of one tribal definition over the other.

Since the beginning of history, battles, and wars have been a political solution to resolve some territorial, economic, religious, internal issues, revenge, nationalistic, civil/revolutionary, and/or a defensive difference as conceived by the political leadership. Whatever the reason a war, until recently, was fought on a two-dimensional ground, with people and weapons that evolved over time. The next chart proves beyond reasonable doubt that wars are waged on the personal priorities of the leaders of a people. Most of the time these leaders have been men, so we can term this as testosterone-rattling. Like the recent US vs. North Korea with public statements of *'Mine is bigger than yours'*. The reasons change but the killing of young, healthy, and brave men has continued since recorded history. As a human race the unnecessary culling of the bravest and strongest makes no

sense, from the outside looking in.

Coincidence 1: It almost seems that old gray-haired men, consistently sit back and send our healthiest, strongest, fittest, young men at the peak of their youth into wars of one form or another. *Coincidence 2:* The rich and powerful mostly never send their own children to war, it is always someone else's child preferably of a lower class or strata. *Coincidence 3:* The reasons keep changing, as demonstrated above. The only thing constant is 'Let's go to War' by finding yet another reason. *Coincidence 4:* Unfortunately, conflict unites citizens of all nations. Too many leaders use conflict as a means to fasten their seat-belts on the political throne.

In the last century the main reason for war was about forging political ideologies on others which resulted in the two unnecessary World Wars. In this century were predicting Digital and Political polarizations that will be the main cause of both external and internal conflicts. External conflicts like we are seeing with Russia and Ukraine, with each side trying to digitally influence world opinions and the physical landscape. Internal conflicts like what's happening inside the US and their Jan 6th coup with each side digging their heels deeper. Also, on what's happening in Iran with the religious leadership using their position to rape girls in jails, while banning their athletes and even executing some for actions against the state. These internal conflicts can be seen in every continent and will only escalate unless controlled.

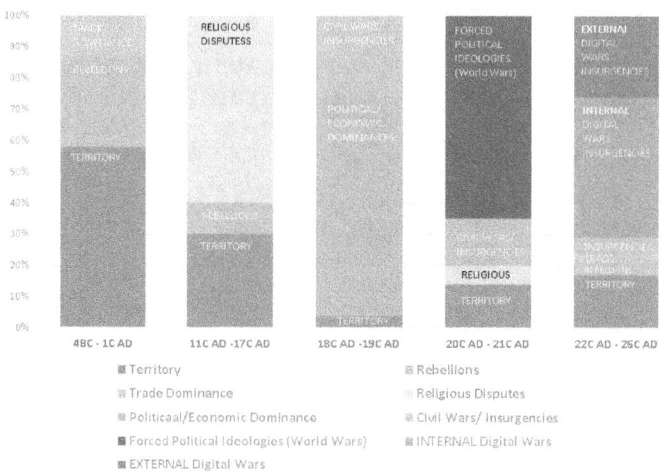

One big change we are witnessing in the final bar on the right representing the year 2022 to 2030, In 2023-2024 we will see the emergence of two new type of wars. The first is an internal war being waged by political leaders and parties by pitting their own citizens against each other. Post covid one can almost believe that something has altered the human neurons to become intolerant and care less about their own people, including themselves, and even their own lives. My belief, is that this is the result of the global Covid lockdown that created a worldwide addiction to smart digital devices, that were then misused by internal and external players, to retune not only the conscious, but also the unconscious minds of their target audience. If we look at Russia, China, North Korea, Iran, Saudi Arabia we know that their leaders and politicians care less about their citizens and more about staying in power. But when this new digital opportunity was harnessed by religious and political leaders of nations like China, Russia and USA, we are now reaching critical mass, of digital controls.

A reason for all this conflict is the hyper growth of humans on planet earth, with minds that can now be controlled remotely and in mass.

The room theory: Some sociologists believe that as the human population increases, it will result in basic resources becoming scarce, and proportionally increases the frequency of conflict, battles, and wars. This includes increase in disease, and wars as a direct consequence of population increase. It works on the principle that if you put one person in a large room the probability of conflict for a fixed time stands at 0%. Add just one more person and the probability of conflict increases by 2-fold, or 50%. Place 4 people and the probability of conflict increases by 16-fold to 100%. As of today, our global population has grown to around 8 billion, living in 195 or so nations. This is a minimum of 195 artificial geo-based landscapes, with nation-based traditions, customs, religions, boundaries, flags, uniforms, and nationalistic pride that bubble forth to create a thousand macro-reasons that a politician can use, to start one form or another, of conflict. Within each nation, let's take the US for example, we have thousands of artificial reasons for conflicts — these could be by political party, my White color vs. your Black skin, my Republican View vs. your Democratic View, Biden is a fake president vs. Trump never lost the 2020 elections, the media I listen to like Fox News or CNN, religious, education including what books a school must have or can no longer keep, sex including enslaving the womb of women, color like Christian White or Iranian White or Saudi White vs. all other colors, or something as simple as my point of view not matching with yours. Each is a point for disagreement. It is important to remember, that an assassination is the ultimate form of disagreement, as is every physical conflict.

For our new digital future, we need to plan, and strive to bring the probability of conflict on a strategic goal of zero. The United Nations was formed for this exact goal, but it has failed miserably as the winning nations that formed this organization have now become the biggest reason for unhindered conflicts.

The last two decades have totally changed the art of warfare. War has evolved from being only 195 polarized national beliefs, to 195 fanatics plus thousands of profiteers, in the decimation of digital polarizing solutions. Once again these are companies like Cambridge Analytica and its hydra clones that operate legally, and illegally from across the planet. The buyers of these services are some of the most powerful political parties and individuals on the planet.

As a planet, humanity is currently living in a world of a continuous, and constant digital war, being fought, both internally against its own people, and externally, i.e., against any nation where someone is able to pay for their needs.

Almost no nation is ready for this, there are no rules and regulations, and it's the export of a digital wild west out there, taking the world-order back to the days of the US cowboys and native Indians. The only difference is that this time, it is the digital cowboys, who are controlling all rules, regulations, communications and empathy, and once again with the poor common citizens becoming the native Indians, and getting massacred. This time the profiteers have no way to distinguish between who the Indians are versus the cowboys. They are all potential targets based on who pays them to do what.

In 2022, with the attack by Russia on Ukraine, we entered the era of broadband warfare. Here, all that we are talking about in this book came to a head. Warfare changed from a governmental arms race to micro-segmented algorithmic devices run by smart individuals. In September 2022, Google's CEO, Eric Schmidt, had a meeting with senior Ukrainian officials. His 36-hour-long visit was exploring the role of technology in future warfare. The art of War was witnessing a monumental shift from every war so far. Schmidt visited Ukraine not as a CEO of Google but as a private billion-dollar investor in military technology startups. His trip is a reminder of how individuals in tech startups invent micro-segmented ideas, from fast AI and ML-driven algorithms run by very smart individuals, versus the slow-moving governmental armies, which are all coming to a head with technology at the new border of conflict. An all-pervasive, individually defined border that does not believe in national, religious, or cultural borders but in just doing what they think is right, like defending their nation from unjust aggressors

When you think the breakthrough digital thinking, think platform. If you plan to launch a drone company, think way beyond drones. Think satellites, Artificial intelligence, Machine learning, facial recognition, remote identification, geo-mapping, and pulling all this together with a smart drone in the sky. This smart drone will get into alert mode as it crosses the geo-markers of the border, it can self-recognize enemy vehicles versus friendly ones. It can even auto-recognize enemy personnel by their uniforms versus the friendly ones in battle situations. It can further recognize a senior officer from its vast library of faces and ranks of the enemy generals and officers. This drone can travel alone or in a swarm of tens or hundreds. It can travel at night with infrared or in the day time with stealth paints and designs. The possibilities are as endless as the imagination of an individual designer. Why I elaborated this is because, today an individual digital technology person in Ukraine, is now being forced to build smart digital weapons in his personal fight against the might of Russia. The new domain is the personalization of digital war assets. These new weapons are not made by the for-profit, weapon factories of the G6 nations, but by skilled individuals that have the knowledge and feel the pain that drives them to develop highly intelligent drones in Ukraine.

This is similar to some individual in San Francisco designing the first Airbnb app, or another developing the uber app. It is similar to Netflix overthrowing traditional brick-and-mortar Blockbuster to bankruptcy. These app based new technologies have been so successful that even NATO has launched a $1 billion startup fund for new innovative ideas in fighting traditional weaponry. This is only the beginning of the new world of individualized weaponry. Senior Ukrainian ministers have requested Elon Musk, a civilian, to re-route his star link satellites to provide broadband internet across secure bands to the Ukrainian digital experts. Elon Musk, a civilian living in the US is a hero in Ukraine today, as stated by Schmidt.

We all know that the US has provided their HIMAR (High Mobility Artillery Rocket System), the precision rocket launcher made by Lockheed Martin, but it is critical to remind people that this is not a weapon or a system, it is a platform for warfare. HIMAR is part of an attack platform with rockets as only one component of the overall solution. The rockets by themselves undermine their true capacity for strike accuracy. The HIMAR targeting system requires real-time information and its inherent intelligence feeds from satellites in the sky. This is enabled by a very complex and structured total platform, with communication, networks, sensors, control modules, guidance systems, and eyes-in-the sky technologies, as a harmonized solution with capabilities to be able to operate optimally. The HIMARS and the supersonic missiles will not win the digital wars of tomorrow, it will be determined by who can process more information at a faster rate, more and more of it in Realtime. Once again it is not the technologies but the micro segmented expertise of technologies and passionate personnel, often common civilians, that will amplify the operational competence of these weapons. Somewhere in the future it will be AI and ML that will drive critical decisions with humans as simple collaterals on both sides.

Just like micro-businesses disrupted brick and mortar giants in the retail business, we will not instantly start to see the micro focused technologies take over the weaponizing of nations and armies. But it will happen, slowly at first, and then as it reaches the 'tipping-point' will totally change the art of warfare. These micro companies are today fighting unfair oppression not because some country demands that from you in a command-and-control matrix, but because someone has faced unfair pain, personal deaths, and who has the expertise, the passion, and the brains to disrupt the operational capabilities of a battlefield across the planet. This new digital intelligence will come with a single person's aim to save their nation from an unfair oppressor to the way people want to live their future lives to be.

Satellites used to be national properties of very rich governments, when smaller nations could not dream of having one of their own. Today a very large number of them are owned by private individuals and companies. These micro satellites, some as big as a shoe box, can be rented or donated by the providers to do things like track and understand Russian troop movements, get daily snapshots of Chinese incursions on the Indo-Chinese borders, identify a rogue individual who shoots a civilian on a Ukraine highway and track his conversation with his wife, when he chats with her in Russian about his shooting that civilian. All this will help the world track war crimes through massive open-source, or close-loop, data collection, with undeniable digital and often visual proof.

Remember, in the world of IoT and IoE the center of each universe is data. This data has information on most data-generating-things on the planet. Now with regenerative AI we can assimilate all this data in seconds and give logical data driven answers very rapidly. Soon law books of nations will get digitized and AI will assist law with unbiased recommendations based on the digital interpretation of law. Imagine a future where American and European football will be refereed by AI. When International law will be refereed by AI, and when the United Nations charter will be applied by AI driven logic and UN laws will be applied on nations without consideration for Developed or undeveloped, Veto power or a developing nation. Laws and charter will become standard. Today most international organizations are run by the rich and powerful for their gains and their ability to punish and oppress lesser nations. Every year we hear of great organizations like FIFA, Siemens, Panama Papers, Russian money laundering, etc., where the rich and powerful routinely make tons of money and then have to hide them in some corrupt havens that specialize with this offering. Humans are corruptible by nature. AI in the near future will ensure rules are applied without bias or graft.

Until recently, artillery fire was directed by helicopter pilots who would look at where the shell fell, where the enemy was, and thereby guide the next shot toward the intended target. I know this as my brother was an AOP, Air Operation Post, pilot and one time his co-pilot got shot right in the seat next to him. These pilots were targets for the enemy as their helicopters were big, made a lot of noise, and their movements were slow. This made the mission of AOP a very hazardous task. Now, this whole AOP helicopter is being rapidly replaced not by one but by many micro drones that accurately track the target via a geo-positioning platform, that will then automatically not only send very accurate artillery instructions, but essentially guide the shells themselves. Already simple artillery is being computed, corrected and precisely aimed by the intelligence buried in these drones, while future ones will be precision guided bombs and missile that will hit the precise target each time, every time. They will instantly follow through with self-correcting, battle damage assessments. Along with this are the *loitering munitions*, which are armed drones that can simply loiter in the air for hours and then instantly drop on their target. There's a bunch of micro companies like AeroVironment that make switchblades, Blackhorse providing cyber solutions, SpaceX with their micro satellites, Bayraktar from Turkey making the TB2 drones, Palantir conducting very complex data informatics as the Barakat data analytics experts, Clearview AI that specializes in facial-recognition, etc. Just on the alternative micro-focused attack systems — a group of Ukrainian hackers infiltrated the electric-vehicle charging stations in Moscow. They used their digital skills to hack, and thus disrupt physical objects in space and time. For a brief period of time no electric car could get charged in Moscow. For whatever its worth, Ukraine is a half generation ahead of the Russian war machine. Do you want to be the technology company that future generals will call for assistance or a platform to provide that the next oppressed nation calls for help when a new oppressor is on their horizon or in their territory.

So, for readers you have two choices if you plan to head in this direction, the first is the US way, and the second is German. In the US way you collect a lot of money become successful with one idea, then collect more funds and diversify and try to do all the things inhouse. I have helped US companies grow to $1 billion and then $2 billion unicorns and then just go bankrupt – simply because they wanted more and more wallet share of their customers and extended themselves too thin while the sun was shining. They took these extreme synergistic US decisions as their strategy. The other is the German way, a country with the highest number of Small & Medium enterprises, called SME's, a segment in which Germans are world class. Germany followed by Italy have the largest number of world Class SME companies on the planet. I recommend a German view where you become the world's best in one single thing and then work on making it better each and every day, with a clear understanding that no company can become the best in everything — which always leads to the company that ends up being the best in nothing.

Religion in the New Digital World

Historically we humans have had around 40,000 years of tribal nomadic upbringing. Followed by around 4,000 years of religious and geographic tribes. Around 3,000 years of stronger geo political influences like states and countries, and around only 200 years of liberal, equality based, democratic and communist influences. The last 60 or so years of technical innovations, and around 15 years of digital disruptions. Digital disruption is today the newest power game of global competitiveness.

In each cycle the pace of change becomes faster, and our options to adaptability become more influence-stringent. This means that more and more of these changes got thrusted upon humans as time passed, and now we have discovered a method where we can fine-tune human minds via their smart listening devices. At this rate, the gap between the smart-device users, and the ones who don't, will become ever wider. In this case, the smart device users becoming bigger tribal captives of the leaders, and the non-smart-device people being somehow forced to either use the smart devices, or remain free from tribal suggestions, and thereby representing true freedom of thought and action. The true democratic citizens. The non-smart-device people will thus remain free from this influence till they are somehow integrated, it could be something as simple as a smartphone and WhatsApp given to them by a parent, friend or their work.

In the last two thousand years, we have learned that feelings are always more powerful than data. Emotions can convince people far more effectively than can pure scientific data. Not only that, we have also learned that feelings are based on chemicals, and electromagnetic interactions inside our brain. Then there is the human mind itself. For example, two people faced with the exact same situation, will predictably behave differently based on their personal likes and dislikes. Their reactions will, determined both by nature and nurture, depend on their personalized brain-chemistry adaptations. When two people come face to face with any situation their response can be, sometimes similar or poles apart. By conditioning we can safely assume that if two Americans pass by an American flag they might be inclined to smile and even salute the flag. However, if two Iranians, in Iran, are passing that same flag they might be inclined to pull it down and burn it. By political, media, and religious, stimulations societies can change the instinctive response of each person within their tribe. Over the last two decades, this response alignment has been weaponized with highly personalized digital stimulations. Each of these responses being a mixture of culture, tribal manipulations, leading to chemically changed neural web that can now be programmed. Each chemical is a neurological response to genetic programs. Each neurological response can be programmed by social events and/or situations. Most of these events can today be triggered by social networks remotely for their tribal followers, in a lot of cases, these social events themselves are false and created artificially on a 1:1 basis with a specific person of a tribe as their focus. All collectively toward a common goal but fine-tuned at an individual level.

For example, a friend of mine took 19 years to realize that he liked men more than women; from one point of view that is 12-14 years of wasted time, and pure anguish due to religious, political, and social pressures. In the wrong society he could have been chemically castrated, like the good ole Britishers did to their gifted and now famous Alan Turin even as he helped them win the second world war. All this happened despite his being a genius, and the father of cypher, and binary computing by building the first physical code breaking machine that helped Britain decode the German enigma machine codes, and helped win the war. Today there is an annual prize given in his name, but during his life it was the religious definitions of sexuality that marked him as an outcast. Even today people can be locked up or stoned to death as a response to cultural, religious and/or social justice in many countries, now including the US.

Another friend of mine joined a social audio-meeting social networking group after June 24th reversal of 'Roe vs. Wade' episode, and slowly joined groups and talking sessions with likeminded people. He invited us many times but we declined. Today he is a MAGA follower who is fighting for his church in ensuring that America prohibit the option for an abortion across the country. He states that if they win, they would like to take this message across the planet. In one of these group meetings that I participated, when a woman said that she respected his groups rights to their Christian beliefs, but they must respect the rights of her beliefs that they were wrong and could not force their opinions on her. The reaction from the group, all males by the way, was three-fold, the first being that *then your religion is wrong*, the second was *then you should convert to the greatest religion on the planet*, and the third was *if you don't agree with us then she should leave the country.* I was rather shocked that my friend kind of agreed with their logic and even participated in calling her names. The room, I realized, was a tribal amplifier. Quite shocking for me, but this was a live conversation that I decided to silently listen to, just as an observer, see how others were thinking on this topic, and then silently depart.

All of the free will, the conservative fanaticism, and autocratic masses can be found in Developed and developing nations. The intensity increases in communist and autocratic states as it does in religious nations. All these decisions are fundamentally based on social cues, and electro-chemical induced responses. Most often, these responses are not something that humans are born with but a mix of nature v/s nurture. There are the genetic foundations of nature, but with humans its mostly the nurture, or external influence, of parents, environment, culture, religion and politics. The study of identical twins separated at birth demonstrates a myriad of nature commonalities, and it also makes us realize the impact of nurture on a human being, in making of a myriad personality as to who they become.

These experiments are most illuminative when done with identical twins separated at birth and then studied as they grow up. One statement that sticks in my mind is that in the case of identical twins, *"The closer they stand, the more dissimilarities one can perceive, and the further apart they are, the more similarities become obvious."* There is an example of two identical twins separated at birth with one taken to Pakistan and the other staying on in India. The twin in Pakistan was brought up in a Muslim school called a Madrassa, where the only subject he mostly studied was all contained in a single book. Education was about memorizing just one book — the Koran.

While in India his twin went to a catholic school followed by a masters in engineering from a prestigious institute called IIT. His MTech was in electronic engineering and then joining a well-paying company, and is today participating in a major startup in India planning to shake the world. In a comparison of nature vs. nurture the two brothers are 100% identical in nature, but by nurture almost 100% different. His brother in Pakistan is a fundamentalist, who spends his life helping the Al-Qaida in Afghanistan glorifying the rule of Islam. They both, by the way have an affinity to blue shorts, to humor and have the same light brown eyes. As people they are poles apart.

In both these cases it is important to realize that their natural genetic programs were systematically hacked by their environment and the opportunities it created for each of them. While this example, of the twins, is one of societal hacking by physical influence, our digital trolls have today taken this to new levels and can have as great an impact on the minds of intelligent people at a global level.

Religion is the world's oldest, and most powerful tribal operation on the planet. Followed only by nationalism and collective beliefs of all types. Religions started with animalistic, then emotional multi-god-religions, and finally the monothetic religion that basically ended with an imaginary single male who directly represents a guardian and punisher for all followers, i.e., God. Very early in the evolution of religions, priests became the untouchable intermediary layer between the unknown and every tribal follower on the planet.

While religion has been the greatest unifier, for humans to collect toward a common goal, it simultaneously also became the power center of the minds of the common people, on whom every nation stood. Over time religion too fragmented into yet one more religion, so that another aspiring leader could then form a separate tribe of religion, nation or both and become their leader. Often a group of nations were brought together by a more powerful entity like the pope, who in the name of their common god became the king maker of nations, i.e., Europe. Then again there were nations like UK, who when the pope did not allow Henry the 8th to marry, he took the whole nation into becoming protestant with a church of their own. That is why Christianity, Islam and Hinduism have their sects and independent beliefs. The Sunnis are very different from their Shia brothers. The catholic is quite different from their Protestant counterparts, Orthodox as are the Mormons. They will rarely go to the others Church. Sometime these sects will go to war based on their religious differences.

Most religions work on the fundamental principles of prohibiting anything that a human must do in their life. They then proceed to deem that as evil. Most religions also have books that were written only by men and somehow tend to oppress women in the name of their god. It took thousands of years, and in each century, religious scholars honed their skills further. In Christianity, each new version of the holy book could have small interpretations and changes to suit the author and their handlers. By the time the new religions evolved, each of them fought for their positions, first within their tribe and then globally. They took existing stories' ethical beliefs, honed them a little more, and then made them their own. Some included allowing enslavement and often killing of people belonging to other tribes. Others proceeded by ensuring their own members did not enslave their own kind or kill members of their own tribe. In almost every religion, there was always the path to climbing the 'golden stairway to heaven,' and each subsequent religion came up with more and more stringent rules and regulations on how to reach heaven. Some went as far as their gods demanding all people convert to their perfect religion, and if they did not, then it allowed tribal members to kill them as yet another golden step to heaven. This has been propagated by Christianity, Islam, and a host of other religions across time. Most religions discouraged one from looking at your neighbor's wife or daughter, while others stoned their daughters, mothers, and wives for talking to another man. Some discouraged its members from eating any animal whatsoever, while others conveniently wrote that their god had actually placed animals on earth for humans to consume and herbs so they could flavor the animals they killed. But mostly all religions prohibit their member from cheating their fellow believers, from desiring someone's wives or daughters, where even the thought is a sin from telling lies and other such things that a human invariably must do at some time of the other. These turned into directives, and each religion has its own version of truth.

Normally, it is considered OK to speak about your own religion, so I will start with my religion, Hinduism, a type 1 religion. Hinduism is a very ancient religion, probably the most ancient major religion surviving today. Luckily, so far, the Hindus do not issue fatwas, or a Christian bull, that fanatics then pursue as their sole purpose in life. They have no stairway to heaven or a religious guarantee to 72 celestial virgins. For example, in 1498, Pope Sixtus issued a papal bull, very much like an Islamic Fatwa, authorizing the Catholic Monarchs to name inquisitors to enforce religious uniformity in order to expel the Jews from Spain. Based on this bull, Spain sought to use the Pope-approved Inquisition to increase their absolute power over their global colonial regimes. This launched the Spanish Inquisition in Latin America, which we are all aware of. Ferdinand and Isabella used extreme torture and dismemberment of the non-Christian citizens if they did not convert to the catholic religion. This was their interpretation of a papal order as they realized that fear was a strong motivation to maintain absolute power. Not yet, but don't be surprised, as humans have the intelligence to corrupt even the best laid-out ideas and philosophies with one singularity — 'Us vs. Them' polarization for control and profits. It is important to realize that the Pope actually authorized the Spanish Inquisition. Spain just happened to use the bull to their advantage. It's like we must not forget that the US laws authorize the legal purchase of semi-automatic weapons by an 18-year-old, the shooter of kinder Garten children in the Sandy Hook massacre, just happened to expand the law to fulfill their madness. If the guns were never available, the children, all of them, would still be alive today.

Hinduism can be categorized as a type 1 religion. It is ancient and started with the quest to explain nature, the elements, and the sky as first representing strong forces, and then designed gods who represented these elements and human emotions. They believed every living thing on earth was created by God and had its own right to live. Hinduism is a religion between type 1 and type 2, i.e., a polytheistic religion. Hinduism started as a religion that prayed to elements and then evolved to become a religion with a thousand gods. It is possibly the only religion with goddesses in a world dominated by male gods. She is the only religion surrounded by male gods across the planet. Their goddesses are more powerful than their male gods, which is demonstrated by the fact that whenever any god is in trouble, they invoke the goddesses to solve their dilemmas, which they always do. Like every religion on the planet, Hinduism considers itself a prime religion. We, too, believe our gods are the only true gods, as does everyone else. Within Hinduism are their individual subsects like Buddhism and Jainism, each trying to jostle over who is older and more ancient. Though my Jain wife or a Buddhist may not agree that they are subsects of Hinduism. Also, the way Hinduism absorbed local customs, gods, and deities and amalgamated them into the broader mix that we call Hinduism today.

For most of the time, Hinduism has been a religion of *ahimsa*, or peaceful coexistence, while they often carried out mixed edicts that were mandated by priests and powermongers. Ahimsa is the ultimate 'live and let live' point of view of the Hindu religion. For example, it believes that every animal has not only a soul but also a God-given right to live. It believes every living thing, from plants to ants, from rats to humans, all have a soul. Unlike other religions that relegate the soul only to humans, Hinduism respects all of god's creatures with souls of their own. Due to this, many Hindu sects, religions, and families have not eaten meat for thousands of years, holding on to this compassion for animals and their right to life from one generation to the next.

Hinduism is also the only religion that does not allow conversion. So, no Hindu can take a sword and ask someone to convert to a Hindu. There hopefully will never be a Hindu Inquisition or a crusade for conversions. They believed very early in their religious evolution that if one rewards, or gives the rights, for conversion of other humans, then bad things will be done in the name of their gods. So, as a religion, they simply disallowed conversions. Imagine if the Christians followed this, then there would be no Spanish Inquisition or their religious wars, and Islam would simply stop all fanatics from murdering and getting their virgins in the name of their god. This prevented Hindus from going on any religious wars that most other dominant religions continue to this day.

Almost every religion, and thus the nation, still consider women as subservient to men, a sex that was only fit to stay confined within the house's four walls under the rule of the man of the house. Some more while others less. This outdated thinking exists even today in many societies and cultures. We also see lingering traces of this ideology in every patriarchal society, with matching mindsets as well. Imagine that in June 2022, the US went and emulated the laws of Pakistan and Afghanistan right here in the US. Not just laws but a decree issued by the SCOTUS or the Supreme Court of the US. They went all out with their religious doctrine, interestingly toward enslaving a woman's womb and denying them all birthing rights because their wombs were suddenly made the property of men, the church, and the state. Interesting because the US woman is free in Republican states, but it is her womb and her birthing decision that the political system has enslaved.

In many of these societies, women are still prohibited from doing most of the activities that men can freely participate in. Till very recently, women were not given the right to an education; in Afghanistan, this is still true. In many countries, women cannot inherit the family wealth; it only went to the sons — this is still true in many countries. They cannot go out on their own free will, unless accompanied by a man of the family. If found with a man they could be imprisoned, caned or even stoned to death even today. Sometimes murdered by their own family members like their father and brothers, as an act of saving the family's honor. It's called honor killing and it exists in every nation across the planet in one form or another. Women couldn't work, vote, or even buy or own property. They couldn't write or publish. This used to be a permanent structure of society, which established a woman's place, directing her where she should be, or more where she could not be seen. It all started with men writing almost every religious book, existing today, in the name of their gods and as authors they allocated all of God-given rights to themselves and delegating women as objects for the use of men. In April 2023 a US judge now wants to allow 12-year-old girls to have rights to marry. This is the dreaded specter of child marriage now entering the US Christian male dominated society. Do you think the judge wants to protect the rights of 12-year-old girls or the pedophile obsession of 30+ year old men that want to own 12-year-old girls as their brides. When all the holy books are written by men the advantage is always biased toward the male. When the religious leaders are men, once again the advantage is always male. When a priest can only be a man then the religious bias is further amplified for the advantage of men. Once again all of this is done in the name of their individual God and religion. When the pope supports the anti-abortion politics in the US despite not one word on abortion in the entire bible, or when one of the Pope's says pedophilia' is like rubbing two hands together' it is on behalf of a male god, a male pope and a book written by men. When the leader of Iran, Afghanistan or Pakistan imprisons a young girl, allows rape in their prisons, and encourages fathers and brothers to stone their daughters and sisters to death, it is done in the name of a male god, a male religious leader and based on a book written by men. Let me know if I'm logically interpreting anything wrong here.

Let's deep dive into Iran — in many countries especially in the middle east this system still prevails. In September of 2022 a very pretty girl, Masha Amini, was taken to prison in Iran, all because a few strands of her hair were visible outside her mandatory head covering. She was taken to prison by their religious police, where she somehow conveniently died. This resulted in a massive social unrest that the male dominated country just does not want to happen. So, Ali Khamenei as the ruler reacts violently and uses all the male dominated omnipotent religious forces available to kill the possible evolution of a woman's rights in Iran. It is not unique only to Iran, in the US if a girl gets raped, even if she is only 10 years old, she has no rights on her womb and there is no discussion yet on the rapist — the man. Now they want to legalize the official rape of 12-year-old girls with the excuse of giving her the rights to choose, just like girls have a choice in Afghanistan and Pakistan to choose her husband.

These things obviously changed with time, but the change was painfully slow. In fact, if any woman chose to challenge these boundaries or break free from these *norms*, she was severely punished by the people, the society, and the government. This depended upon the severity of the woman's *crime*. In some countries, even today it could be something as simple as talking to a man, or attending a soccer game in Iran, that can get a woman imprisoned or stoned to death by her own family — in the name of religion and family honor.

Give yourself a ten minutes honest break, get out of your tribal loyalties and just try and review what all men do in the name of their god. There is one belief that since recorded history more people have died in the name of religion than under the banner of war. Just let that sink-in, in your intelligent mind — just because we have been convinced of the religious singularity that our god is the only true god and better than theirs.

Thankfully, in some countries, these attitudes were challenged by the masses, and some of these societies began to change. The sense of permanent male-dominated societies faded, and the idea of impermanence replaced it. However, as soon as such things happen the misogynic side of powerful people want to drag women to their rightful statuses by immediately passing new laws, like the 'Roe vs.

Wade' in the US. These laws are mostly supported by the religious scaffolding that authorizes such demands from the holy leaders. Who would dare object if the pope himself decries abortions? An extremely large contributor to this change was uncompromising women fighting for their fundamental rights. While, at the same time, there is this shocking misogynic support by women on all such matters. For example, according to the 2023 US Gallop report, from a **women** POV, 15% of women want to disallow abortions under all conditions, 40% want it legal under all conditions, and 46% want it legal only under special conditions. In the same poll the **men** reported, 12% of men want to disallow abortions under all conditions, 27% want it legal under all conditions, and 59% want it legal only under special conditions. Just for the record, lest we forget men don't get pregnant, or have babies, but love to pass laws on women that could. In **2023** 55% of women are pro-choice, and only 41% are pro-life, while 48% of men are prolife and 47% are pro-choice.

In the meanwhile, with passing time women could finally get an education and work. They could finally buy property and own land. They could finally cast a vote. Just for the record women got voting rights in the US on August 18th, 1920, this right was given to women 144 years after declaring independence. While in India the women got their rights to vote the day India became free. As time went by, they could study engineering, and fly fighter jets, and slowly do most things that men assumed as their birth right, a form of earned right. Of course, the fight didn't just stop there and is an ongoing conflict even today. There is literally no country where women have the exact same rights as men, though there are semblances of this coming closer in some Scandinavian countries. But, even there dive a little deeper and it is but an illusion. However, this change needs men to be able to stand up alongside with the women and demand equal rights which is today an uphill climb in every society. More importantly, it needs women to band together globally as one sex and demand equality and not fight against the rights of every other woman. There is a saying that "A woman is often another woman's worst enemy," and only a woman can break this dilemma. Then in the US, we started seeing the growth of women CIOs and their positions of power. Within the male-dominated corridors of the powerful and rich, one could almost hear the grinding of misogynic wheels turning, all wanting to bring back the days of male dominance. Then suddenly, the men in the USA started feeling impotent compared to the rights of men in Afghanistan, and they even said so directly. Then in a spate of planned Christian fundamentalist revival, they took a sudden leap backward, supported by catholic intolerance, in retracting a 1973 law allowing women the right to abortions. I have written more on this topic in 'Being Woman,' an article on LinkedIn, where I let data speak for itself.

Despite all the power shifts through society at all levels, two power brokers have remained dominant. Religion and politics. And they will take of their gloves if their position is ever threatened.

When building on your new ideas think like Elon Musk. First, think with passion — something you will bet your life to achieve, then think big, then multiply that by a thousand times. What Elon has repeatedly achieved, nations and global multinationals could not achieve with their billions of dollars. So, all it needs is passion and commitment, and we can all get there.

From the beginning of societies and empires, we lived in environments that have been heavily swayed by religion in temples, churches, and mosques, along with the politics of kings and emperors, and now presidents and prime ministers. In most cultures, religious leaders have remained more powerful than the king. That is the reason, from Egyptian Pharaohs to Saudi Kings, from Russian presidents to the president of the US — all of them have always kept religious leaders on their right side. In most cultures, the power politics between religion and kings went through their peaks and troughs of power retention. Thus, in the US, during election time, one sees politicians start to visit churches and hold the bible in clear display even if it is held upside down. Or, in Saudi Arabia, where the king will call their business leaders for interrogation but never the heads of religion. Now, with our digital freedom and global reach, we have started challenging that as well, like the women in Iran reaching out through social media about the death of a pretty woman in the clutches of their religious police, i.e., the start of the next revolution.

According to weforum.org religious institutions collected over $1.2 trillion officially in the US alone. The first catch is the word officially, as they have held this number since 2018, when I started collecting this data. The second catch, is that a lot of this money is collected in cash, and like all cash transactions it is up to the establishment to report their collections. The third catch, is that it is all tax free so it is not audited by IRS, except for one clause which is that someone from inside the church has to report and invite the IRS in to audit some wrongful activities. All of this is tax free as every political organization allows religious houses to collect money tax free. What politicians get in return is a guarantee of votes for whichever party pays the church the maximum incentive, that too is tax free and remains untraceable. So, even though we know that during covid people donated more to their religion than before, and for that some preachers asked their parish not to take the covid-19 vaccine as "The blood of Christ would save them." quote from Florida. Out of this, <u>$826.5 billion was officially donated to the US churches</u>. The rest is donated to other religions. This turns into more net profit than Google, Facebook, and Amazon combined. The reason for this is all their assets are heavily subsidized, their income is tax free and most of their collection is in cash. So realistically one can easily assume that the actual US collection of funds is around $1.5 to $2.0 trillion of pure profit. One school of thought states that anything a human MUST do religions have made that a sin. So, here is a list of all thing's humans will do, a very religious POV: **You shall worship no god but me**, *instantly making all other religions your enemy.* **You will not make idols, thus** *differentiating themselves from idol worshippers. Though Christians did end up and make idols of Christ that sits in every church.* **You shall honor your parents** — *this is universally good.* **You will not murder.** *Though every holy war, every crusade, every inquisition and every bull or fatwa leads to murder, so this needs salvation. It ended with not murdering their own tribal members, all others were exempt from this edict.* **You will not have sex outside marriage,** i.e., *this bans adultery, and only allows religiously sanctioned sex by marriage.* **You will not steal,** *even if you go to holy war or fight another religion — something not possible.* **You shall not bear false witness against your neighbor.** *However, the US Whites have borne false witnesses against Blacks for centuries and continue to do so. While the Quran allows their believers to bear false witness to save a fellow Muslim.* **You will not covet,** *or desire, or try to be a better person, a richer person, or get elected as a leader. Looking at your neighbor's wife or husband and having unholy thoughts is a sin, having sex — the most powerful urge of youth, without religious sanctioned is a sin.* If every human will do one or many others things in their lives, what does one then do. The official answer is that if you sin you must come to the priest for salvation. The Christians, and Muslims ask for confessions, the Hindu's to repent and make amends. However, in a Catholic church, unfortunately if a child confesses, their priests often become their sexual abuser, and the child their prey.

Here is the kicker question: I need any priest of any religion to answer for me. You can give it a try in your mind too. It is a question my 9-year-old self asks when we see the volume of tax-free money that goes to these religious houses. ***"Out of this official $1.2 trillion collected in the name of their gods, can anyone, Pope, Bishop, Clergy, Mullah, or priest, show me just $0.01 that actually reaches Jesus, Allah, or their respective God?"*** All of this wealth, mostly in cash, is collected in the name of their respective gods, but where does it all go if not to their God. We know for a fact that none of it reaches, or has ever reached, their god physically.

Spoiler 1: all this money collected is also tax free in almost every nation. In countries like Denmark, Austria, Finland, Italy, Sweden, Croatia, Iceland and Switzerland Churches have to pay only a 0.4% to 0.7% tax. **Spoiler 2** — It is not the church that pays this tax but the members of the congregations who have a tax that is intended to support their church financially. How can it get better than that? Why, one may ask, because the politicians who pass these laws need these religious leaders, and these leaders need the politicians. You can now imagine why these conjoined twins work the way they do. Despite what the constitution says they sleep, connive and frolic in the same bed.

This is like donating to get a seat on a rocket to Mars in the first century, for building the rocket. It includes donating some money every week for two thousand years with not a single rocket being built, no launch in sight, or no one accountable to deliver, or build any such rocket to anywhere. And the believers continue to pay $1.2 trillion for that celestial rocket to nowhere each year. I know of friends who do a direct deposit from their salary to their church to the tune of 3% of their salary every month for their place in the grace of their god.

In this new digitally connected world, we all need to have guts to finally firstly face, and secondly to ask questions that we have never dared to ask before. However, questions like this shake the very foundations and our dependency on faith. It also shakes the humongous revenue stream of thousands of years against organizations that have extremely deep pockets and endless reach. But it is our collective duty, to try and throw light on questions that worry us so we can ourselves find the truth.

Guess which is the richest country on the planet with more power and wealth than any other. You guessed it Vatican. Imagine they collect over $800 billion from their religious houses in the US alone. What would you do to own and propagate such a dynastic business.

It is not surprising that the forces of religion will fight very hard to exterminate any idea that could directly impact their tax-free revenue stream. They have each built very high vertical silos of individual beliefs that they echo in their sermons, along with socio-religious algorithms, and more recently digital outreach programs to preach and quash any such ideas before they can possibly germinate. Over the next thirty years, or so, religious institutions will continue to lose public trust mainly due to the direct belief that their followers will do anything they say or ask them to do. Trust in religious institutions will continue to erode as our new generations start to ask some very basic questions. It is because of this reason, and lack of clear answers, that most traditional religions ban their followers from asking any such questions. People must continue to challenge, and raise questions against religious beliefs that are not PEH friendly, and begin to question the credibility of the way things are being done which comes from blind-faith. While, with passing time, and digital trolls we shall witness religious tribes become more fanatic, zealous and dedicated to their old scriptures and older way of life. Our micro segmentation algorithms will create a multiplicity of fanatical believers, and non-believers and the winners will be the ones with more wealth, power, influence and often weapons. Guess who has control over all four, including the power to influence justices in the US Supreme Court itself.

Religion will continue to get more micro segmented. We will witness the evolution of micro cults and followers. Religious intolerance across the planet and examples of *'Neo-Nazi kills two roommates after converting to Islam'x will continue to dominate global Christian optics.*

Religion was, is, and will probably remain male dominated with a single representative male god for the foreseeable future. In such religions women will continue to be treated as an accessory to their male counterpart, because it is the will of their god.

As far as the speed of communications as an impact of coalescing emotions is concerned — in the old days if something happened in one part of a country it was communicated at the speed of a walking man or that of a horse, that too only if the information was critical. Up until recently a rape of a child in a church, or the killing of a Black man by the police, was never a social concern of great importance. Today, with the assistance of digital social networks when a Black man is killed by the police the live video gets virally distributed across the planet in a matter of hours. So, when used positively the digital revolution is the greatest boon that humanity has ever enabled. Unfortunately, while the rape of a child is still hushed by the church, the priests and the parents as their proof of being good believers.

In the near future we shall see parents not complain to the priest when their child is raped by clergy, but it being shared over the digital domain, when the church can no longer hide behind closed doors, or relocate the priest in question, to yet another new location. Our digital strength must bring justice of the people, and law just as it was doled out to the police in the murder of George Floyd, or Tyra Nicholas, and many others like this. A 2015 movie called 'Spotlight' won an Oscar when it exposed child molestation coverup in the local Catholic Archdiocese in Boston. They found thousands of victims and criminal priests but the political-religious nexus ensured most went free to continue what they were doing to their parish. Also, despite these facts that parents continue, to date, to take children to these places of worship and trust their delicate kids in the hands of potential pedophiles.

When we talk of religions there is one religion that kind of stands out. Here, in all fairness I must state that being a Hindu, I am subject to a bias, and readers must read these lines with that in consideration. Hinduism still remains the only religion where goddesses remain more powerful than their male counterparts. However, in the real world it did very little to elevate the status of the common women in the society. India stands as the only nation, with reference to the Hindu religion, that has not executed any non-Hindu for the sake of conversion, or directly in the name of their god. In fact, the world over whenever a religion was persecuted India was the promised land, where they could find protection. Even when the Jews from Poland, when Hitler was persecuting them, and when most fellow Christian countries in Europe denied their ship landing rights lest they face the wrath of Hitler. It was Hinduism that allowed their ship to land in India and practice their religion in peace. Hinduism solved this problem by disallowing conversions, i.e., if you're not born a Hindu you cannot convert, or be converted, into a Hindu. So, there was never any incentive for any Hindu to go fight a battle to convert a Muslim or a Christian to become a Hindu. Nor was there any need to convert a bunch of Polish Jews forcibly to Hinduism in order for them to land and thus live their lives. So, when a great king Ashoka went on a rampage of conquest, he used Buddhism as the religion for his salvation. According to history he returned a very broken man, seeing all the dead people in the battle fields, and became a peaceful Buddhist. A recent, very absurd firefly example, of the creation, misuse and decline of a religious sect can be found with a godman in India by the name of Ram Rahim. He was born Gurmeet Singh and formed a group called Dera Sacha Sauda in 1990s. He used his religious cult, acting, writing, media and even directed a movie with him in a leading role as a god man. Ram Rahim built a strong base of dedicated followers. In 2015 he was ranked as the 96th most influential person in India. His power was his vast followers and his customers were powerful politicians who wanted the votes of his followers. The highest bidder was rewarded with him announcing the party and person all his followers should vote for.

Due to his single-pane-of-voters he became a political magnet. Many prominent politicians visited him for his public blessings. Translate that into the reality where politicians came for him to ask for the votes of his believers with ever higher bids. The highest bidder got his voice and thus the votes, all tax free. In 1998 he was accused of 2 rapes but because of his political clout nothing ever happened. It took until 2017 for him to be convicted of those same 2 rapes. In addition, he was charged with over 300 castrations of his followers, and two possible murders. It was later learnt that many ardent followers donated their young daughters to the service of Ram Rahim, who had an elaborate ritual that orchestrated who he slept with, when he desired so. Very similar to that of Harvey Weinstein in the US.

He was involved in the voluntary castration of his male guards and devotees who he believed might be competition to his sexual desires. Even the current prime minister of India, Modi had once praised him, he reciprocated by supporting the BJP. In 2002 he was reportedly involved in the murder of a journalist writing about his hermitage. Once again, nothing happened. However, there is a limit even for religious atrocities and on August 28th, 2017, during the time of current PM Modi, he was sentenced to 20 years in prison for the two rapes and the murder of the journalist.

Though, this is an Indian micro-religious-cult case it is important to note that such distasteful incidences have been happening in almost every country, and religion. There is not one religion that can today claim that not one of their priests across all the centuries committed a rape on their followers or parishes. In the US alone almost 1,700 priests and clergy have been accused of sexual abuse of children. But nothing happened. The church is protected to a very large degree and the priests are mostly transferred to another church, to continue their vile activities. In 2002 five Roman Catholic priests were convicted of rape and sentenced to prison, due to which The Boston Globe received a Pulitzer Prize in 2003 for their article *'sexual abuse of minors by catholic priests'* into national limelight. It was made into an Oscar winning movie that visualizes that case. In Pennsylvania alone more than 1,000 children were reportedly raped by Roman Catholic priests. However, we all still regularly go to our temples, churches and mosques in total disregard to the crimes being conducted in the name of God, in God's house, by Gods representatives, to Gods believers, and to the most vulnerable citizens of every society — minor children.

And here too parents to date keep donating their children to this institution that has confirmed pedophiles. If anyone, but a priest had committed such an act they would internationally get the most severe prison sentence possible, however most of the priests simply get transferred to yet another church to continue their favorite pastimes. If any media journalist reported any such activity happening in Saudi Arabia, Iran, Pakistan or China the journalist would probably get summarily executed by royal decree or executive orders. Sometimes, even if it meant getting the person, if living abroad, into a national embassy, or consulate, to eradicate them from existence.

Just five hundred years ago, we lived in a society that was heavily swayed by religion of churches, mosques, and temples, along with politics of kings and emperors. Everything that we did, revolved around the pleasure of our religious, or political, controllers who defined good and bad, and doled out punishments that today seem rather barbaric. Now, with our digital reach we have started challenging that as well. In almost every case, some parent probably took their child to the priest, so things are similar across the planet, when people get so blind in their belief system that they are even willing to feed their very own children into the pedophile's mouth.

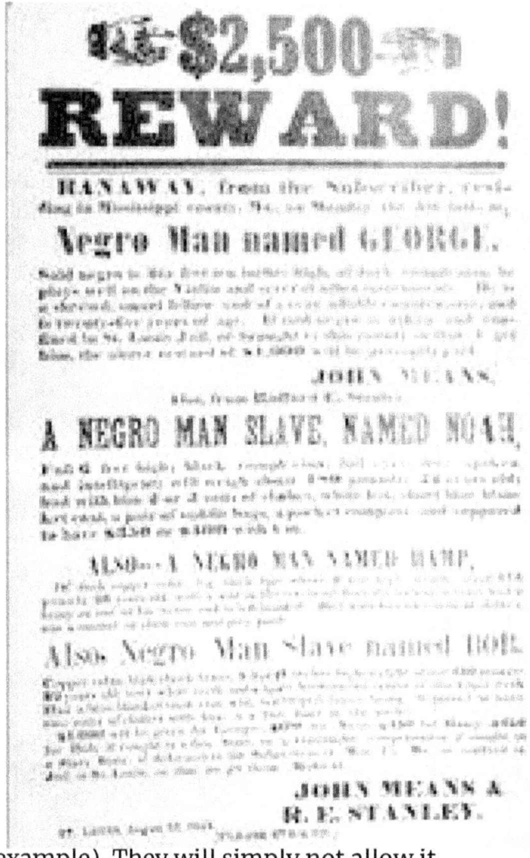

And continue to do that one child after the next. On June 24th, 2022 Texas passed the heartbeat law, which means that after hearing a baby's heartbeat, the woman has zero rights toward getting an abortion. This means it is a crime to get the child aborted. Even if the child came from a rape or incest, and even if the child has congenial defects and must be aborted, and in one case even if the child will not live for childbirth and carrying the child can kill the mother, (true example). They will simply not allow it.

Slavery is when a person, or parts therein become the property of another person. For example, as recently as 1860, any White person could catch and lynch any Black person on charges of their own, with no repercussions. A Black person could be charged on any imaginary charges and lynched with no repercussion on the White oppressors. This was because every Black person was the property of some White man. Property in the US, at that time had no rights, just like today a house has no rights, the lobster caught by a fisherman has no rights. If a slave escaped, there were bounty hunters, whose sole profession was hunting these slaves and bringing them back, dead or alive, in order to collect their bounty.

See the bounty hunter poster from 1852 — There are many Western movies glorifying this bounty hunter. In the year 2022 Texas wants to give the same right to the new bounty hunters who can bring a female child, girl, or woman if she plans to or has undergone an abortion. In 2022 onward, imagine this same poster but with the photograph of a girl, 10, 14 or 20 years old, with a reward of $10,000 when brought in. This is for Christian shaming the girl and her family, and getting the Church and State to pay for their misogynic oppression. By getting pregnant, the girl has then misused the property that is hereinafter owned by the Church and the state i.e., her womb. After that date of getting raped, or whatever, her womb now belongs to the Church and State. So, guess where these bounty hunters will hunt their victims — your social networks. Again, who pays these bounty hunters, your local church and the state. So, a very loud WARNING here if you, or anyone you know, is planning an abortion keep that information very close to your heart, and extremely far from any social network as that in itself can be taken as evidence for a potential blackmail, or even a prosecution. For $10k a person will do a lot of snitching and this is just another legal bounty hunt. Some believe that God will even bless them for this.

The fundamental background for the US anti-abortion law is religion, the Catholic religion, and evangelical Christians. The sanction comes from the Pope himself. The makers of the law quote the Bible. Despite, no one on the planet can find any mention of abortion in the bible. The real problem with this is that, the US Constitutional right-wing fanatics have combined areas that were supposed to be separated. It has opened up floodgates of arguments and controversies that are factually anti the constitution yet undertaken under the banner of the constitution itself. First try to understand the reason why the US courts could have taken this decision in the first place. On the fact side even if a woman is raped, and it results in pregnancy. The question now becomes, as to whether she should have a right to abort the child that was forced on her by a rapist, get the rapist to face severe criminal charges, or if 13-year-old should be forced to marry the rapist to save her family honor. The baby shouldn't be forced upon her if she wishes to terminate the rape or pregnancy. Think, if the girl is a bright 13 years old in school, or a brilliant 17-year-old girl raped in college. Does any religion, or State have the right to not only ruin her life, but also somehow pardon the rapist. No law on the planet, should be allowed to force an unjust outcome, on either of these children, because if they do her life will be forever changed, and never the same. Such a law will be a betrayal to all women, to the nation, their laws, and the religion. The female's age doesn't matter for the male dominated religious slavers. She can be 40 or 10; it still wouldn't matter because, in their minds, she is the property of men. What we need to ensure is that she must have a minimum right over her own body.

But the Pope sitting in the Vatican Catholic Church and your local politician tends to differ. Here is a quote from the Pope just before Roe vs. Wade was rescinded by SCOTUS. When a journalist from the Jesuit publication, America Magazine, asked the pope on Sept. 15 about "a woman's right to choose" and giving Communion to politicians who have supported pro-abortion laws during an in-flight press conference, *Pope Francis responded that "Abortion is more than an issue. Abortion is murder." This statement resonated across every catholic church across the world.*

Factual note: [1] The position of the pope has less to do with Jesus and more to do with Rome. If the Vatican is Fox News, then the Pope is Murdock, and the anchors are the bishops and the priests. For example, (a) Pope John XII became pope at 18. He transformed his residence into a brothel. He even had sex with his sisters. He finally died when a husband caught him in bed with his wife. The husband beat him so badly that he died 3 days later. (b) Pope Boniface VII quoted that pedophilia was no more problematic than "rubbing one hand against the other." (c) Pope Alexander VI became the pope by bribing the cardinals. He was a member of the Borgias family. He was famous for having fathered at least nine children out of wedlock, throwing orgies, and even engaging in incest with his daughter Lucrezia. (d) The <u>list goes on and on</u>, but the worst would be Pope Clement V, who is not even on this list. Pope Clement V arrested the grandmaster of the Templars in 1307. He did this based on a request of the bankrupt French King Philip IV so that they could together extract money the templars kept safe as the world's first bank on the planet.

He did the ultimate crime of using politics and church for personal benefits and nothing to do with saving fellow Christians or the world. Or, even anything written in the bible.

1.   Now the current Pope is once again taking political sides with comments "abortion is more than an issue. Abortion is murder." This is the Popes personal statement, pope the man. It has nothing to do with Jesus or the Bible, as it is not based on anything written in the bible, but totally based on the ancient roman male concept of dominance and enslaving women once again. This has nothing to do with Christ or his teachings.

2.   This whole catholic anti-abortion stance has absolutely nothing to do with life, or pro-life, or the Bible. This is proven as Pope, the bishops and cardinals and the total Catholic paraphernalia broadcasts, a planned edict for lashing women to unfair rules of slavery and submission with only one purpose — total dominance. The Church & state are now funding wild-west bounty hunters across republican states to hunt, and bring to religious justice, young children, girls and women who challenge their instructions. The republican party has decided in upholding the catholic laws with the church, and the Pope's edicts. They, together will echo the same message to their parishioners, and the politicians hope that this will translate into a bounty of votes from their reciprocal agreement.

3. Another reason proving that this is not a 'for life' endeavor is that while the pope and his global anchors prohibit abortion as pro-life, they sit on dining tables with cuts of veal, calf, piglets and lamb meat on their tables. Each of these cuts are all from god's own creatures, that were living that morning or a few days ago, and deserve their true pro-life sympathies. How can any person eat the babies of so many animals and state that they stand for pro-life.

4. All this rush to enslave a woman's womb when there is not a single word on abortion in the bible. The question then becomes as to what does the pope use as the foundation for his quotes if not the data from the bible. History has proven, that each time such atrocities turns out to be a reciprocal agreement between powerful politicians and the pope.

5. Finally, there is the age-old male dominance over women — the lower caste human in Christianity from the beginning of creation, a class system enclosed within the catholic church doctrine.

The degree of religious truth: everything quoted here is from the 'King James Bible.' According to the bible, earth was created first, then in Genesis 16, God created the sun and the moon, followed by the stars in the heavens. He then created the whale and all creatures in the ocean and land. In G-26 he creates man, in his image. Only then God created a woman from a spare rib of Adam, for one single purpose — quote, *"And the Lord God said, it is not good that the man should be alone. I will make him a helper suited to him"* unquote. So, God only created man in his image, which is so strategically convenient for men since. Thus, in Christianity a woman is a helper to man, and created only to help and serve man. God then allocated dominion over sea, land and every life form therein to man. He then declares that every herb is created for man to flavor his meats, so killing animals is approved within the first week of creation. According to Christianity man was not born from a woman but created in the image of God. However, a woman is but a helper to man, created in man's image.

My Hindu mind believes all this has to do with the age-old male desire to *'let's put women in their place yet again.'* Hinduism is possibly the only living religion that is truly pro-life. It is the only country where millions of people respect and take care of all life created by God. They do this by not taking the right-to-live from animals — this is pure Pro-Life. Millions of Indian families have not eaten any kind of meat for thousands of years. The extreme in this pro-life attribute are the Jains from India. Vegetarian in India applies to all of God's creatures, the cow, the pig, the turkey, the chicken and everything we put on our US dinner plates every day. Honest question to you the reader — do all of God's creatures not also have a right to life, or for you to not put their children on your dinner plate on a daily basis? Are they not God's creatures too? Do we have to throw Lobsters and

Pangolins alive into boiling water because we are told they taste better? Is that truly Gods will?

Now, you need to decide if you're 100% correct in your religious opinion, or that I may be just 10% correct in my logic. Which means that you are now 90% correct in the belief that religion is not an artificial boundary or the most profitable business of male dominated ideologies. However, the truth as written by our male authors, mandates rules, as spoken by God. The very difficult step is to challenge your traditional intelligence, put your foot into that slightly ajar door, and have the guts to widen that thought and let your mind take you where it does. If your end result is that your god is the only true god and your pope/ bishop/ and priest is the only truth on the planet. Then be happy and smile in that belief. But also give that right to other religions.

For people living in Republican states in the US, just don't get raped by family, friend, stranger, enemy or get pregnant, and then come running to my side of the discussion. Remember while all courts have been trying to delete abortion rights of every woman across the US, there is not one SCOTUS decree, or court order that mandates finding the man, and punishing him for the greatest crime against humanity, or being the prime reason for the pregnancy. If you do get pregnant the hounds of religion will be following you very closely. It's like the Pope, and his cabal who believe pedophilia, rape and sex is like *"rubbing one hand against the other"* stated by Pope Boniface VII. If this were not true, then begin to wonder why, according to the 2002 Jay report, which states that there were more than 11,000 allegations of sexual abuse, mostly of young children, against 4,392 priests in the US alone. This is an average of 2.5 rapes by each criminal priest, that were ignored by the church. In a 2019 report 1,700 priests and Roman Catholic Church members stood accused of child sexual abuse with little to no repercussions from the church or state. Nothing from the Church because they probably think that this is their God-given right. But, nothing from politicians, or the law can only mean one thing – because they can negatively influence a large percentage of votes if any politician tries to punish them. The big question is as to why is the frequency of distribution, of rapes and modern enslavement of women, so high in the Roman Catholic churches, and republican states, governed directly by the Pope. My unqualified opinion, is that by siding with the winning party, which is most often the party that can pay the maximum, they get assured protection on their political, pedophilia, slavery of women, and the most important is the continuance of tax-free status in almost every country on the planet.

At the same time that the US was busy enslaving their women and banning abortions, India just declared that they will allow no questions asked up to 24 weeks for an abortion. Only if the abortion is a gendercide, then it is prohibited. This means that if the mother is opting for an abortion after finding out that the fetus is a girl then it is illegal. India also just, September 2022, banned marital rape. While in the US no one is talking about the rapist, i.e., the man without who there is no rape. They are still busy putting all kind of religious, and criminal thorny fences around every child/girl/woman who is their mother, sister or daughter.

To be transparent with you, I must confess that I was appalled when the news of this law broke out. It distracted me for over 3-4 months while I researched this topic. The result was my article Woman 'in LinkedIn. I often wonder on how we claim to live in the 21st century, and still act as though we were living in pre-medieval times.

So, when I retweeted my opinions about abortion and how I was against the law, I openly announced to Twitter and Facebook that I was pro-choice. The digital trolls now know what my preference is in the pro-life vs. pro-choice dilemma. Thus, they have the data, and they understand the way my mind works. Now, they can use me and my thinking as a menu for targeting pro-choice feeds or for hate mail from pro-life tribes. If I'm prone to being activated, they will test and then activate me for their tasks.

There is a 95% probability that if you are a Hindu, you believe your gods and religion to be superior to all others. If a Christian the same, If a Muslim still the same, if a Jew still the same. Yet, we need to realize that each of them is both right and wrong at the same time. It is time for us as the human race to grow up from these separatist ideologies, and beliefs if we want to save humanity. Without this we'll continue to have the governor of Florida erasing Black history from books of education, or presidents in US saying it's the blue-eyed blonds from Scandinavia who US wants to welcome. Erasing Black atrocities is not the path forward to future generations of the US, it is teaching truth in history that is most important — before we kill the trust in books of history totally.

Destroying humanity: Before starting this section, I would like to propose that we make a commitment to change our world from the path it is on. If any reader thinks we have taken PEH to a perfection point do send me an email and we can discuss this further. I am calling this discussion 'Saving Humanity' or investing in our future.

Here are some predictions for the leaders. People older than 60 will have below 20% chance of agreeing to anything written in this chapter. This is because although we are all born neutral, with each passing day we become more and more biased and rigid to our social and cultural artificial boundaries. At the beginning of this century, we introduced the concept of social networks, after which humanity and the minds of humans were altered forever. The nation group is between the age of 45 to 80, and they will resist most of all. From what

I write here, I predict 80% rejection and 20% acceptance. In the group between age of 35 to 50, i.e., the enterprise group, the chance of resonating to this chapter increases to around 50%. The main influencers of tomorrow are the readers between 12 to 30 years old, my prime target for this book, and I hope most of you take home something to think about that will change our PEH score globally. I am hoping 80% of you will agree to what is written here as this is your future we together need to protect, and that you need to create from this day forward. It is you that needs to become the change you want to see on this planet.

This section is written to encourage these youngsters to realize that the future of humanity is right now in your hands and some fundamental needs of how you can participate in 'Saving Humanity' goals than your parents or their colleagues have ever done. I know of 70-year-olds with a mind of a 25-year-old, i.e., inquisitive and fact bound, while I also know of 17 years old buried inside centuries of beliefs and enveloped by the artificial boundaries we shall briefly discuss here. No judgement, on who is right or wrong here. Just a hope that we join hands and eradicate all the negative energy that is starting to bubble forth in every nation as it turns into a swamp of power, politics and money.

Saving Humanity: Background and the rules of engagement.

Rules of Engagement: the 6 steps to improving our world in a positive direction and away from the path it is on currently.

On June 1st, 2017 the US president Donald Trump announced that the US would cease all participation in the 2015 Paris agreement on climate change mitigation. Contending that the agreement would 'undermine' the US economy and put 'US at a permanent disadvantage.' US leadership cared more about party politics and its economy than about the future of humanity.

On February 24th, 2022 Russia invaded their peaceful neighbor and have continued to kill innocent civilians, rape young girls, as Ukraine fights for her territories. The reasons, Putin has provided to annex parts of Ukraine are "…Because they once belonged to Russia…" and "…that US was bringing NATO to its neighborhood, by allowing Ukraine into NATO" — as his excuse for his forced occupation. According to OHCHR, as of May 8th, 2023, 23,606 Ukrainian civilian casualties, with 8,791 dead. Putin, as a leader of a UN Veto member has released criminals and drafted them to go into Ukraine on a rampage of raping, and murder of civilians while we the planet can do nothing. Russia is a UN Veto member so no direct action against Russia nor a global sanction, can be initiated. Even though there is a charter item that allows member states to call for an emergency meeting to censure Russia and even take away its veto rights. But who will fight against the big boys. It almost seems that the UN was created to protect the big, rich, and powerful countries, and help them sell their arms, and ideologies across the world. They seemingly have no interest to control, or punish themselves under any condition.

On October 6th, 2022 our esteemed UN called for a meeting to vote on China's abuse against the Uighur Muslims. Something very interesting was exposed on this day. The Uyghur facts can be googled, but at a high level it is about China using modern digital technology to identify, arrest Uighur men and women and then place them into large psychological transformation camps where they teach them how to be better Chinese citizens, a process of ethnocide. Here, Muslim women are forcibly sterilized so they bear no more Muslim children. In the UN, the motions were for debating the abuses in Xinjiang against the Uighurs. Christian nations like USA, Canada and Britain led the motion. The motion was defeated. Now here is the interesting dilemma. The west requested the motion to protect Muslims in China and debate on the truth of women being forcibly sterilized. Yet it was the Muslim nations led by Qatar, Indonesia, the UAE and Pakistan that rejected the motion and any discussions on this topic.

In September, 2022 hurricane Ian tore through Florida and US media was forbidden to report global warming as a possible reason for the hurricane in Florida.

On September13, 2022, a very pretty woman 22 years old woman was arrested by the morality police in Iran for not wearing her hijab correctly. For too long, the bearded men at the top have oppressed women into hopeless submission. It is the new generation that wants change. The UN immediately called for sanctions against Iran, because the big boys can, and Iran is not in the big boys' club.

Now imagine, we have all these global organizations, and international courts like ICJ, or the Internal court of Justice. At an executive level — *the founding principles of the United Nations are to maintain international peace and security, develop friendly relations among nations, promote social progress, improve living standards, and uphold human rights for all.* I think they have failed in mo, in the two recent cases of Russia & Ukraine, and China & Covid-19 policing. As an organization the UN has been compromised by corruption, and global-power-politics.

The fault is not that of these leaders, or the nations. The fault is collectively ours, each member of planet earth living in any democratic nation. We are letting all this happen, we are feeding the monsters. In a democracy we get what we collectively vote for and that seems to be not working entirely well too. *"This is our collective responsibility. If we don't serve for the good of the planet and our people, we don't serve our children their future. If we don't serve our communities, we destroy humanity itself, thereby destroying ourselves."*

Humans have jumped over every hurdle that technology has thrown at us and we can jump this one too. But the question is whether we will cross this one early or late in the game. Form a global digital club for saving humanity with members from across the planet, with no separation of color, sex, religion or nationality and each pursue their life energy toward *saving humanity* on clear PEH rules.

When I was young, our goal was to study hard, get a good education, and then do an MBA that taught us how to make more profits for ourselves and our companies, with not a single subject or discussion on how management decisions impacted the environment or even how it could kill the very planet that is our mother and should be our God. We then joined companies that worked on the principles of open market economies, i.e., manufacture in the lowest cost nation and sell to the highest paying customers. However, in the last decade, fewer and fewer children, at least in my environment, are going for the traditional *"Let's just focus on making dough, even if it destroys the planet"* options, and instead are now heading toward an environmental, social, and be-good work environment. Similarly, more and more children are becoming vegetarians or vegans as they are now resonating with the idea that every living thing has a right to life.

The Christian belief is that God created animals and fish for man to consume and herbs so they may flavor these meats. Women for man to use as help and enslave the earth so man could pillage and profit from what nature gives us. All of this is not flying well with the new generations, so one can very easily predict a decline in attendance and donations in Catholic churches in the future. So, while attendance in their church is crashing across the planet, they are hoping to create a few more Christians by enslaving the wombs of all females. If that works, then every woman in every state across the US might have their womb declared as the property of church and state, even if they are not Christians.

Officially the Pope supported the pro-life and went as far as saying, *"Indeed, is it right to hire a hit man to solve a problem?"* Thereby referring to the doctor that performs the abortion as an assassin, and a hit man. Which is exactly how the catholic church and every US republican state governor took this statement for their political optics, and advantage in enslaving, not just Christian but every woman in their states.

Something to think about is **Religion vs. Science:** Or 'Facts vs. Beliefs. The Roman Catholic church has been rigid from time immemorial against scientific facts that disprove their statements. Here are the 10 example where Church and Facts disagreed.

1. ***Heliocentrism:*** *The scientific discovery that the Earth revolves around the sun contradicted the geocentric model that the Catholic Church*

previously supported. Initially, the Church condemned heliocentrism, and on April 12ᵗʰ, 1663 Galileo Galilei faced persecution for espousing this view. However, it took the church all of 329 years to acknowledged that heliocentrism is consistent with scientific evidence, when in 1992, Pope John Paul II declared that Galileo's condemnation was a mistake.

2. **Evolution:** *The theory of evolution, which indicates that all species have evolved over time, including humans, challenged the literal interpretation of the biblical account of creation that Adam was created 6,000 years ago. The Church initially rejected the idea, but over time, it has adjusted its stance to allow for the possibility of evolution while still emphasizing the concept of divine creation. Republican schools want to not teach evolution in their schools in the US*

3. **Contraception:** *The Catholic Church teaches that contraception is morally wrong, while science has provided evidence that contraception can be highly effective in preventing pregnancies and reducing the transmission of certain diseases. The Church's position on contraception has remained consistent, with a firm opposition to its use, though some individuals within the Church may hold varying views. The church views contraception as a human attempt to decrease the size of their herd.*

4. **Embryology and Personhood:** *Advances in embryology have led to a better understanding of fetal development, challenging the notion that personhood begins at conception. Some scientific findings indicate that while a fertilized egg may have the potential for human life, it takes time for certain essential characteristics to develop. The Church has generally maintained a position affirming the sanctity of life only from conception*

5. **The Age of the Earth:** *Scientific evidence points to an Earth that is approximately 4.5 billion years old, contradicting a literal interpretation of biblical texts that suggest a much younger Earth. According to the Catholic Bible, Adam was created on the Sixth day. If Adam was created 6,000 years ago and God created Adam and every single land animal on the 6ᵗʰ day, then earth was created 6,000 years ago, too. This is found in the Book of Genesis 1:26–31. My ardent Christian friends state that under no way is earth older than 10,000 years*

6. **Divine Intervention:** *Historically, belief in miracles as evidence of divine intervention was more prevalent. However, advancements in scientific understanding have provided natural explanations for phenomena previously attributed to miracles. The Church recognizes the importance of discernment and proper investigation before declaring a reported miracle as officially recognized.*

7. **Soul and Mind:** *The scientific understanding of neuroscience has raised questions about the nature of the human mind and the concept of a soul. Some scientific perspectives see consciousness and mental processes as emergent properties of the brain, which challenges traditional notions of the immaterial soul. The Church maintains that the nature of the soul*

goes beyond mere brain functions and cannot be fully explained by science. It is only a matter of belief as proposed by every religion.

*9. **Gender and Sexual Orientation:** As scientific knowledge about gender and sexual orientation has expanded, it has challenged traditional understandings within certain religious communities, including the Catholic Church. The Church often reaffirms its commitment to teachings that view gender as binary and sexual acts as limited to heterosexual marriage. The Pope and the catholic church have routinely supported Sexual statements that do not exist in the bible, like their comments on abortions.*

*10. **Origin of Life:** Scientific theories on the origin of life, such as abiogenesis (life arising from non-living matter), have raised questions about divine creation as described in the biblical account. The Catholic Church's position generally emphasizes the belief that God is the ultimate source of life, leaving no room for exploration of scientific theories on how life emerged.*

It is crucial to remember that the Roman Catholic Church is a diverse institution, and the perspectives and reactions of individual Catholics and religious authorities may vary within the broader context of these scientific challenges. The Church's response to scientific discoveries continues to challenge each data driven scientific inquiry advance, when it contradicts papal, and biblical statement.

Here is a set of 6 tips that we need to follow as an individual, a company, a university, a religion, a political party, a nation and a planet. The fundamental rules are exactly the same. The checklist exactly the same, and our path is our personal option. Our baseline is PEH, Planet, Environment and Humanity.

The universal checklist. If we don't provide basic growth potential and care to every human, we may never truly harness the power of the only living planet in our universe. Not just a living but our greatest giving planet. Our universal checklist before we undertake anything has to become −

[a] Is it good for the planet;

[b] Is it good for the environment;

[c] Is it good for humans;

[d] Is it good for other life forms on the planet

[e] Bonus: You put your checklist attribute here

Does it make you feel good as a PEH contributor? Stay **PEH** focused — **Self-Scoring examples,**

(A) You need to decide about a job offer from a cigarette

manufacturer. Should you take the job on a PEH checklist. Do Cigarette manufacturers have a PEH focus= No. [a] No; [b] No; [c] No; [d] Average; [e] N/A its individual.

(B) Take your Manufacturing to China to improve profits? Because they can manufacture at .3 the cost by using forced labor; [a] No; [b] No; [c] No, they use communist slave labor; [d] No Chinese eats everything that moves; [e] Yes, 50% more profits. Don't try to make more profits at the cost of slavery, genocide, and ethnocide (Tibet and Uighurs).

(C) Get a new MBA in a US University: Does the university have a PEH focus? Yes/No [a] Check, Few US universities are teaching sustainability over profits, but more of legacy books and lessons. — Its mostly Profits above all else. [b] No, US MBAs have almost no concern for the planet. The environment has become a learning topic only in the recent 5 to 6 years but is not the critical overarching process in future executive decisions as yet. [c] No, Little concern for general humans. It focuses on the very rich humans and teaches how to make more profits for them. [d] No US MBAs has relatable concern for other life forms or extinction of species. [e] No, because most MBAs still train students on remnants of the industrial-era management by adding the ML/AI aspects of the digital disrupters as a profit mechanism by teaching students to make more money for their corporation for less cost (i.e., outsource to a lower cost nation, i.e., China with slave labor) — the industrial era mentality. These are globalized courses and are often not localized for specific nations. The question is how does an individual use the PEH principles when deciding whether to accept a job offer or not. One can use the same principles when selecting Enterprise or national decisions. Here is the job offer and the PEH decision process.

(D) Mercenary job in the Russian army: Does Russia have a PEH-NO. Is being a mercenary aligned to [a] No; [b] No; [c] No; [d] No because this is the worst job on the planet today. Only good if you are a prisoner in Russia, and a rapist by nature; [e] No because you'll probably never get to spend anything you earn. Don't join or hire from this country.

(E) Offer to become a drug mule: Does the drug cartel have a PEH focus? Are the job aligned to [a] No; [b] No; [c] No; [d] probably yes, but then you'll be blackmailed for life; [e] No because you'll have to keep doing this till you get caught, or get shot

(F) Join Palantir as a data analyst. Does Palantir have a PEH focus= Yes. Is the job aligned to [a] Yes; [b] Yes; [c]Yes; [d] Above average; [e] Yes. Palantir builds software that empowers the good guys catch bad actors. They augment human intelligence not replace it. They fund internal research on 'Catch the bad guy's ideation research.'

(G) Build a startup to track the bad guys. Is the job aligned with PEH = Yes. [a] Yes; [b] Yes; [c]Yes; [d] The incentive will be the good work that will attract the best guys; [e] Yes. Find patterns in specific areas that bad folks use to communicate. Use ML/AI to listen to the edge. Start FuFact[xi] algorithms that improve with each bit of information it receives.

(H) Legalize something illegal that people love. Does it help PEH scores? Yes, because people want it. Apple took Napster, which was illegal, and legalized it as the first digital music platform called iTunes. Is the job aligned with PEH? [a] Yes; [b] Yes; [c]Yes; [d] Attract the best. Build the best, and each person on your team could become a millionaire if done right [e] Yes. Don't just copy. Plan to become the world's best is the only way. Your deliverable has to be legal, faster, lower cost, better, and digitally convenient.

*Follow your personal passions: Then you have to never work a day in your life. It will be a fun-win ride. There are tremendous opportunities across the planet, so long as your mind has the right priorities. Find your area of passion. **Environmental opportunity in India**: For example, around 20,000 people are cremated every day in India. Each cremation uses 4 quintals of good prime wood for the pyre, i.e., 1 quintal is around 220 lb. This computes to 880 lbs. of wood per cremation. So, India burns around 17,600,000 lbs. each day, or 7,983,225 kgs of prime wood. This is a major opportunity for a national govt sponsored environmental startup in India with a daily customer base of around 20,000, that needs to be made aware of the environmental impact of burning wood, dirtying the rivers and paying around 10 times more. Today it costs around $40 for a wood cremation and this future will cost around $4 for an all-electric cremation. This is a 90% saving in cost and a 90% saving for the environment. Perfect as the alternative is better and cheaper, the only challenge is to change people's mindset, and reportedly the wood mafia that pays electric crematoriums to say their power is down, i.e., the profit influencers. We can leverage our social media for that. Find sponsors, get political leverage, find angel investors, and fund the Indian IIT, and engineering colleges with a competition to build the most efficient furnace that burns at 1,400 to 1,800 F, i.e., 760 to 982 C. Setting up these crematoriums across India can save much-needed trees on planet earth. It's a win-win-win PEH idea. Let it then get powered by Solar power and Solar energy retention batteries, made by India, and Indian IIT funded research, something like the Tesla Walls.*

Khan Academy is a US non-profit educational organization created in 2008 by Salman Khan, a US citizen and not the Indian actor. His goal is to provide fit-gap-free education in micro nuggets. The organization produces short lessons with online content and videos that are easy to use. Fit-Gap because its website also includes score charts, learning trackers, along with exercises and materials for teachers. The company has grown by philanthropy and donations.

*There are a multitude of certified **micro course providers** that offer students precisely what they need to succeed. The no-more and no-less concept of digital solutions. In the last ten years I have personally helped hire over 100 people, some of them interns who were experts in a small focused expertise, and who came from micro courses after a basic education foundation. Post covid we're no longer looking only for tier one university graduates, but also for very focused experts. This does not mean that there are no companies looking for IVY league students but for those students that cannot afford them do not despair as the future is on micro specialties so long as you have the basic foundations, like a graduate. When I need two data scientists, I need very specific skills and expertise along with good scores that demonstrate commitment.*

Remain PEH Focused forever: Never do anything that is not PEH positive. If everyone on the planet took this approach then we will reach the proverbial haven humanity has been searching for. However, there are powerful players that want to control, rule, enslave, and misuse humans in their spheres of influence. We have to plan for a future that is as far from these people as possible. Today with smart digital tools in our hands we actually can. Take advantage of the global goodwill that exists on the planet today. There are governmental organizations, private doners that are open to investing into good causes. Some of the younger children, clearly state they are too busy to attend such meetings on a weekly basis but are willing to invest small amounts and once a week attendance' to invest into the future of humanity. Form an investment club with squeaky clean accounts and reputation and there are no barriers. It's when these organizations become greedy and ruin it for everyone. If you collect enough funds, you can hire people for admin tasks.

Embrace SmartTech: Just as this book is about digital disruption,

---

By Numbers

98% of people have great ideas. Only 3% invest into their ideas

8% of early digital entrepreneurs have become millionaires. Less than 3% of opportunities have been closed. In developing nations — copy a successful digital idea from the US, or Europe

57% of successful digital companies were formed by people between 12 and 85. There are no age barriers

so is the future of humanity. Think SmartTech. Carry a phone and when you see wrong video it from start to end, without risking your life. Covid-19 has brought the world together digitally, as never before. Use it for the PEH good. Today people don't think twice about using technology to connect, chat, assign tasks, collect funds and even share accounting details in a secure manner, i.e., transfer funds and look at their bank accounts. These digital applications are the foundation for global participation, and collaboration. Groups and business that use technologies see all national, religious and cultural barriers disappear unless their government disallow such discussions, like China where they track the Uighur Muslims and Tibetans via digital surveillance techniques, or they shut down the internet like Iran did in September after the women there rebelled. Or, when in Kashmir they shut down the internet when Indian PM, Modi removed Article 370, that helped Pakistani terrorists more than it did local Indians. When you have a good idea, you need to invest time and effort to make it blossom. Build 'Saving Humanity' local, regional and global tribes. Discourage members who join just to disrupt, and with an agenda of their own, that does not conform to the rules of each seat has a voice, and no shouting above others. After some time use it like the Egyptian Spring revolution. For example, Facebook for planning and Twitter, or Slack, for action items.

Using SmartTech is a big operational catalyst, and a bigger responsibility. Everyone touching it needs to learn how to operate it. But use it sparingly as this is not the silver bullet to all issues. Human interfaces and social collaborations are critically important.

Never Forget Friendship and Fun: My advice to every child going to college is (a) Have fun. (b) Never miss any class. (c) Keep scores high. Do these three things, and success is guaranteed. The fourth is, of course, to keep away from bad players. If you have an idea that can help the community, then share it and grow it in the community. Along the way, make friends, help people together, and leave behind a better place and memories.

Outro

Our greatest global victim since COVID-19, Xi, Putin, Ayatollah, Donald Trump, GOP, and Republicans is the systemic erosion of TRUST. We need to somehow bring it back. Trust and reconciliations require us to go way beyond the surface events that are designed to ignite our emotions. Emotions are very sticky and, thus, a strong target for fake attachments to events and happenings. We must never forget the greatest influencer of today and the future is an intelligent individual. That individual is you, the reader, and everyone you know around you. Whether you are Trump or George Floyd, you are an individual. Whether you are the police with his knee on the neck of a victim or the girl who made a nonstop video of the murder while she witnessed a crime, you are each an individual. You could be a criminal taken out of prison and sent by the Russian army into Ukraine or the digital hacker that shut down all electric car chargers in Moscow; you are an individual. You could be someone who has locked themselves in their home surrounded by digital friends or a fanatic following a 1980s fatwa to stab Saman Rushdie in a New York speech; you are still an individual. You can be Putin imprisoning Navalny or ordering the attack on Ukraine or Zelensky defending his nation; you are each an individual. You can be Elon Musk buying Twitter for $42 billion or Tim Cook announcing that Apple is now a $3 trillion company.

You have to know, and for the sake of our future choose the individual you want to become. The individuals, the manager, the CEO, the Governor of the state, or the president. You are each an individual.

Once again, remember that what got you here will not get you there.

An individual is the one who launches a startup enterprise as an entrepreneur as they organize, operate, and assumes the risk for an idea to transform innovative ideas into products and services, normally for a profit. When entrepreneurs are supported, in good and bad economic times, all resources are kept on their toes, spurred to always grow, learn and adapt. Entrepreneurs are the new digital, and economic blood that keeps the fruits of nation-states healthy and thriving even when the global economies are failing. It is these innovative individuals that meander through options, take the necessary risks, build for the future, and take that path to success. Most of the successful ones are the good guys, that legally build products, solutions and services. However, in all fairness, if wealth is the only measure, then Putin, Xi and the Ayatollah have each also amassed a lot of wealth by different ways. Most of them today use SmartTech. So let us see how the different layers of the planet can work together to build a future where everyone can become a digital win-win prospect for saving humanity.

Roadmap For the New Individual

Your greatest choice has to be the decision to Try

Until around the beginning of this century it was always the collective force that worked. The world was run by the rich and the powerful, like big corporations, larger nations with the bigger national armies, the large brick & mortar brands, with thousands of shops in the malls. Up until recently that seemed the only future and we once again simply sat and accepted that. But not everyone. Then in came the digital revolution, with the individual, and their micro focused digital disruptors. The big boys needed physical mobs; they needed votes, they needed large crowd sizes. The new individual needs a highly focused digital solution. The digital transformation needed smart Tech, it needed familiarity, its phones required that could take high quality videos and it needs cloud platforms that could instantly distribute information across the planet. This new digital world needed protests, some peaceful to get their voices heard, while others designed for chaos. For people like Gandhi, and Martin Luther King Jr. if was just to change the point of view of pure oppression. In one case, the anarchy brought about by the British people in India, and in the other, the anarchy of the caste system of the White man oppressing the Black people in the USA. *At the center of this new world, the dominant force is an individual who is digitally connected.* The new individual is digitally connected, knows how to use the digital assets on one side, and is aware of how trolls and algorithms can misuse their emotions and thus their minds on the other. These people are aware of the digital beasts that can be created inside the minds of innocent people.

Our world is full of good folks and bad folks; it is full of good companies and bad companies. It is full of good religions — consistently only ours, and bad religions — consistently every other religion, as it is full of good nations and bad ones. There are good leaders and bad leaders, and sometimes good nations with bad leaders, too. For example, Germany was a good nation, but it got a bad leader once, and everything changed for them within two decades. For thousands of years, China has had a philosophy not to wage war outside their boundaries and not to bully their neighbors. And then all it takes is one leader who not only declares himself as a president for life but also has aspirations on other nations land, creates islands in other nations oceans, and tries to change the opinion of the world by their sheer might. As an individual, what we should do is identify the good guys and the bad guys, the good nations and the bad nations. And once again, it is vital to remember that the individual is at the center of all these things we want to change.

Right now, we can see many nations at the cusp of some major shifts that could happen rather rapidly. Russia could change from its autocratic rule of Putin to some other form of government. US could flip from being the lighthouse democratic nation to one run by some autocratic leader who manages to change the constitution. China could see a change similar to Russia. There is North Korea, Iran, Saudi Arabia, Iraq, Afghanistan, India and a host of countries that could change rapidly because of the power of digitization, and just one leader. At the center of these changes is the individual who has suddenly found global alliances, support and power.

There are two critical things a digital disrupter might ask though; *what's in it for me? And what do I need to do?* The answer to what's in it for me is often very simple. If you're in the US, and you're a girl, who's going to the university, and you are living in a Republican state, what's in it for you? It is your entire future. Because, if you get raped in your school, by a stranger, by a parent or relative, or in the university, then your church, and state will criminalize you, even if you're not a Christian. It will force you to have that child even though you don't want the child; there is no consideration for what that decision will do to your education, life, family, and friends. You will have to have the child. Suppose you try to get an abortion illegally, your state will release hound dogs in the form of bounty hunters who will either make money by blackmailing you, or turn you in for a reward anywhere from $10,000 upward depending on the state. Blackmail is an easy option for these bounty hunters but you will have to pay more than the state, and often it can be a repeat income for the blackmailer.

If you think about **what you can do?** The fact is that the power to change your future is in your hands. Democratically and peacefully. Spend your time thinking positively, networking with like-minded people, working toward a future that you desire. If you're that girl, or a parent of one, going to the university, find people who think like you. If you think people should not be allowed to have abortions, fight for it, find people who do not want abortions, go and vote for your rights to ban abortion in your state. However, if you are a person who thinks a womb must not be separated from the woman, nor a woman enslaved by religious and laws devised by men, then you must find people who support your pro-choice views. Such views are called anti-slavery views because denying abortion to a woman is a new form of religious slavery. Again, if you believe that abortions should be banned, that the pope is right, and the priests true, and the bounty hunters are good people, then you too must find likeminded people and vote for your future.

Remember, that just like in an endless universe every individual and planet is at the center of the universe. So too, in the middle of a crowded room every individual is at the center of the crowd. In all your social conversations you must watch out for anything, or any post, that creates a 'us versus them' statement. All the while never forgetting that you are at the center of your universe.

All in all, if we talk of just three things you must take care of in life, these are the three that you should remember; The **first** is that in this new digital world, your eyes are the center of the universe – your eyeballs, and mind are the two most desired real-estate in this digital world. Find me a million viewers that follow your blog, podcast, or application, and I'll tell you how to turn that into a million-dollar business, so long as they are not paid trolls. **Second**, you should stop paying to get your brain whitewashed, altered, and spend your time and money on something else. Because, when you listen to your smart devices, that is where they specialize in once again grabbing your eyeballs, and thus your mind. **Third,** ensure it's not a troll grabbing your sympathy, empathy, emotions, and thus your mind. It's getting increasingly difficult to guess if an email is from a bot or a human. The challenge is called the *Turin Test*, designed for text exchanges. But in the very future humanoid robots will need to undergo a similar test with a physical interface when a human cannot differentiate if the voice, they are interacting with is a human or a robot. Many recent AI-driven bots have already passed the Turin barriers for simple communications.

For example, the US constitution was never designed for dishonorable, or corrupt, politicians spouting polarization lies by leaders based on a political distortion of the truth. Though they did warn us about this. Unfortunately, this is not just a US phenomenon but has become a global systematic political pandemic. Two reasons could be that modern humans are not accustomed to being locked down in a house, and many in just a room for 18 months or more. The second is this abundance of network connectivity into these locked-down homes and rooms with 18-months of smart algorithms maturing at an alarming pace.

It is important to realize that globally, 'We the People' have something more powerful than what the trolls can feed us — our intelligence and global digital communications. For the first time in history, you are the only barrier between your dreams, potential, and your current reality.

Your Individual checklist

1.   Start everything from a PEH-centric mindset each time you have to make a critical decision. Think logically and take ownership and accountability for your thoughts, actions, and everything you say. Don't blame others for the situation you are in. Don't let any politician convince you there in the race for your sake and not their benefit. Go change it peacefully and democratically — go vote.

2.   Be Smart about your thoughts, words, decisions, actions, and life.

3.   Turn off your auto-reminders on your smart apps. These are pop up messages, social notices, TV channels, and news — all designed to grab your attention, then your eyes, and thus your mind.

*4.   Stop paying for channels that are paid to influence your mind. Fox news is currently a great example, and every media is in the exact same game. Demand truth from the media. Your news channels and TV Channels make money from all sides. 70% of what you watch and hear is designed to make you a tribal member by slow persuasion, and **when people shut their TVs** — families start playing games, reading more books,* *and having a lot more societal conversation. Normal life comes back, and intra-family polarizations tend to disappear.*

5. Build yourself an internal mental antenna to identify the 'Us vs. Them' trolls that are trying to create a divide between you and all the others. Make a dedicated plan to get your mind and life back so you are making your own decisions and statements, not ones inserted by your tribal networks, religious decrees, or political feeds. Don't lose yourself in the jungle of artificially planted tribal statements.

Your Digital Enterprise Roadmap

It is critical to remember that at the center of all companies, enterprises, and businesses is a collection of single individuals. An individual is at the center of the digital enterprise.

While on the one side, we have the positive influencers, like Elon Musk (mixed feelings about him right now but still like to believe his goal of good is right now beyond me) with his Tesla, Space X and a host of 'PEH companies making billions of dollars. Then there is Steve Jobs, Tim Cook combination providing exceptional digital products and services like the iPhone, iTunes and the Mac laptops and becoming the first $3 trillion company. Not far away are the disrupters in Amazon, Facebook, Airbnb, and a million other digital startups that are slowly eroding and corroding the halls of traditional brick-&-mortar companies. Then there are the semi-conductor components, the smart tech manufacturers. The most recent will be the unleashing of the AI enabled ChatGPT and its capabilities. While, on the other side are the bad players we have been discussing for most of this book. In the future, tech will be dominated by QMLAI, Quantum, Machine Learning and AI, but running on Quantum computers and solving problems a hundred thousand times faster. We will then, be stepping into the world of a new 1:1 drug development for the rarest of rare cases. For example, today a new drug takes around $1 to $3 billion, and 5 to 10 years to commercialize. With QAIML, I'm looking for the future where, this could be accomplished in a matter of weeks along with all tests required by the FDA AI bots.

In the meanwhile, if you take your precautions as an individual, a bad enterprise should not be able to do much to you. Or, you can choose not to work for a bad enterprise. Unless, of course, a good enterprise goes and does bad things in the shadows. For example, suppose Facebook once again goes and gives some new Boston Analytica clones your personal data without your consent. As they have done before and since. In that case, we shall have to assume Facebook is a good enterprise hanging around with bad players and thus ending up doing bad things very knowingly.

In October 2022, Elon Musk acquired Twitter by paying $44

billion, and one side of the public is ready to shoot him, while the other thinks he will end up cleaning this sewer of trolls and fake communications, amplified by rich politicians. I personally don't think Elon should have got into the Twitter-political mess as it has distracted him from his prime mission — get to Mars. But then Elon has done many consecutive impossible things, so this, too, will prove to be for the good of humanity, I hope.

Most enterprises have not yet realized the true value or potential of the data they have sitting inside their enterprise firewalls. Many are starting to throw away very large volumes of data they have collected over the years to slim themselves down. This is a very bad idea indeed. The critical reminder is that companies like Yahoo, Google, and Facebook generate trillions of dollars in revenue based on their collecting data, then extracting valuable information that many people are willing to pay for. Data is an extremely valuable asset of every organization, no matter how you look at it. Storage is exceptionally cheap now, so throwing data is not a good idea at all. Every company sits on a wealth of data that has the potential to provide exceptional information, but only if they architect it for providing information. There is a saying, *"While we are all drowning in our floods of data, we are simultaneously starving for information."*

And once again, the question is, if you are a CEO, CIO, or CFO, it boils down to how can I keep my customers more satisfied than what they expect? It might be they are mainly students who need coffee and a place to sit with their laptops for as long as they like. And what should I do? The answer is the key.

For over 2 decades, I have been assisting big and small global companies to meet their business user expectations from data and analytics. In my work, I recall assisting a computer manufacturer in designing a real-time Supply Chain visualization. Their business need was to fill gaps in their supply chain due to a lack of real-time stock predictability for critical items required on their production line for the next 4 weeks. There used to be times when some of their very expensive hardware could not be delivered on time because one single critical component was out of stock. Then, there were other times when they had too much stock of some raw or semi-finished material. Both are undesirable situations. So, if you have 1.4 million SKUs, with a few obsolete materials sitting, and you have $20 million to $60 million worth of orders on hand that you cannot execute because a single component, in most of them, that was supposed to come from, let's say, 'X country' did not arrive on time because they have a supply chain problem. And you just had a single supplier who had been very reliable throughout but is currently going through some unique problems. As you can see, if your order-to-booking ratio is good, then the potential for your on-time shipment is compromised due to supply chain issues. Your digital opportunity here is to visualize and see the whole supply chain in real-time and be able to predict 6 weeks, or every supplier's lag time x 2, in advance of any supply chain issue globally.

Right now, as of July 2023, I am predicting that ChatGPT and its LLM, large language model, regenerative Text interface is not going to be a very big problem as an AI disrupter as of yet, but it will be a more significant problem as it gets itself entangled into a never-ending internal loop of generating more and more text by its GPT logic. ChatGPT logically creates new data continuously and then uses its data as its learning reference. This is because when asked a question, ChatGPT uses its massive data library to predict the first word in the answer. Then it logically computes what the next word should be. This process takes a global average of next-logical-word which builds its answer in the regenerative text. One word at a time, with no checks for truth or data facts. It simply creates the next word. In each iteration, it produces more data that it might use for its future AI regenerative model. So, LLMs will end up with the current version with more and more data that will take any GPT answer farther and farther from the truth. There is an old saying in Technology and finance, *"When a human used to make an entry mistake in a banking ledger, it often took weeks for experts to find out that mistake. Now with AI-driven algorithms, when an algorithm makes a mistake, it can take many years and we humans may still not be able to find out what happened."* This happened in a 15-minute flash crash event in 2010 when the NY exchange had to be shut down due to anomalies they could not then understand. It resulted in a $4.1 billion direct, and $14 billion indirect trade loss. It was assumed some algorithms reportedly made some autonomous changes that did something unknown. The only thing we know that the short blip caused a very large number of transactions to error. The scary part is that it is now 2023 and we still do not know what exactly happened on that day in 2010. Or, if some smart hacker algorithm made billions of dollars for someone or some nation. The official version is because an operator forgot to shut down the DR system. The street version is financial hacking that made someone billions.

Your enterprise checklists: see how far you get

1.   *First and foremost, 'become the digital disrupter.' Individuals and groups are leaving established corporations and focusing on a very narrow niche idea to become the best on the planet. Take the risk or invest some time in forming your own digital enterprise before someone else does that.*

2.   *Encourage employees to come up with digital disruptive ideas, then fund them. Today is the most significant opportunity for companies to form a digital enterprise the world has ever given to an individual. Individuals worldwide are disrupting enterprises with digital alternatives;*

if you can think of doing anything better, faster, and cheaper, ask your company to join hands. Just digitize your idea and disrupt the market.

3.   *The millennials and the new customers are most attracted to PEH companies. In the customer's mind, PEH facts are prime material for marketing. First and foremost, use your PEH filter in everything your company does, and how you plan to assist them. Customers love companies that decrease their global carbon footprint by increasing efficiencies, giving back to society through contributions, and thinking long-term as a replacement to pure profits. One way to attract exceptional talent in a ML/AI software solutions company is to allow employees to form groups that help society, i.e., build an app with the local police department that allows tracking criminals via intelligent real time analytics. The solutions are endless. Think sensors and add digital. By 2023 ChatGPT has upset this cart, and the next hypergrowth is on its way leveraging AI. There are a thousand GPT ideas out there if not millions. Go find one and put your life into it.*

4.   *Don't be part of a company that does not have a clear PEH strategy. Cleaner manufacturing, decreasing carbon, using more wind and solar, and building a more efficient process are all hallmarks of a successful company.*

The Future in Digitizing Education (subset)

"It is the responsibility of societies never to suppress knowledge, no matter how awkward that knowledge is, no matter how it may bother those in power or those against it; we are not smart enough to decide which pieces of knowledge are permissible and which are not." — **Carl Sagan**

Overheard at the optician's waiting room

"I don't understand where education is headed. Teachers are now giving YouTube links to students for study. Very few teachers are actually teaching. As an ophthalmologist I can only say our system is exposing children to too much screen time. We are seeing more and more children with worsening eyes" — **Overheard**

The New Science of Fufactology — Learning from the Future

"History has been the study of all the world's events (only from a winner's point of view)." — **Voltaire**

"There are two types of forecasters, those that don't know, and those that don't know what they don't know," Traditional wisdom on forecasting. We have all been learning about the past for the last four to five thousand years, but only from the POV of the winner. Our method of learning about and from this past has been defined by the subject of History. History is basically the recording of events from the past, particularly human affairs of local, national and global interest.

Across nations, there have been cultures and nations that are good recorders of events, and then there are others that are not. There are still others who recorded their history, and then some barbarians came along and burnt, destroyed, or wiped out their recordings. Some of the most famous ones are when the [1] library of Alexandria, 400,000 scrolls, were burned. It is reported that a Muslim caliph burned the library, and then again, Julius Caesar burned the library in 48 BC. [2] In the 5th century BC, the Nalanda library, boasting hundreds of thousands of volumes, was all raised to the ground by the Muslim Delhi Sultanate in India. [3] In the 16th century, the Maya codices were destroyed when the Spanish friars burned almost every book they found as part of their inquisition process. [4] Institute of Sexual Research in Germany was destroyed by Hitler and the Nazi party as not conforming to Christian/Aryan definitions. [5] The Iraq National Library was destroyed by the US in 2003. [6] Search for the *List of Destroyed Libraries* on Wikipedia for a list of when humans destroyed knowledge and burned books. Currently, there is a modern-day destruction of knowledge going on in the US by Republicans who want to either ban the truth about slavery and/or not let educators teach the truth. There is, however, one fact that sticks out like a sore thumb. It's mostly the religious fanatics and their belief that 'my truth is the only truth' that has been used throughout history to burn books, libraries, and knowledge. The other reason is politics, like communism, which is just another form of religion. Both tried their best to erase knowledge that did not agree with them.

From a history writing point of view, the advantage of being a good recorder of events is defined by what is recorded, and who sponsors/funds the recordings. So, what we get consistently is the winners get to rewrite history in their own image. This affects all the learning and lessons every future generation will learn. Whether what they write is true or untrue, biased or factually wrong, the winner still gets to write it. Future generations read that as the truth, and slowly this history erases the truth and replaces it with the winners lies. Up until new scholars are able to prove the biased lies, go and erase these lies via very complicated processes, and finally replace the writings with truth.

So, the process of history commences with the start of an event, the event, and the end of the event. An event can be the crowning of a king till their death, or the start of World War 2 till the end of the war. It can be the slavers of Blacks in the US with the history written by the Whites, or the massacre of Tulsa and the total erasure of the event from US history books. We have been taught for very long that history is an important study as it ensures that humans learn about events of the past. We have been told that one of the benefits of learning history is so that we see patterns and do not repeat the dismal, or allow criminal, events of the past to happen again.

Like the way Hitler led a hardworking people of Germany and turned them into a genocidal machine. We hope that people who read history can take action to stop their prospective leaders if they are undertaking Hitler-like decisions that could lead a nation toward building fanatical political parties and tribal followers. It becomes an early-warning beacon, a lighthouse, that prevents the ship of a nation from crashing on undesirable rocks. But the recent past has totally challenged the writing, erasures and the truth in histories.

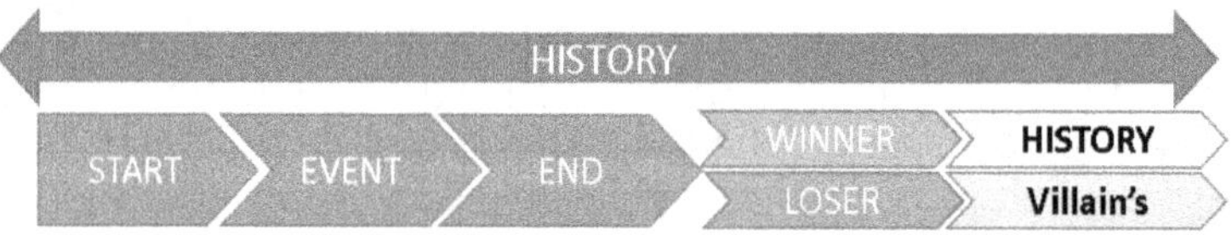

This is because, there is an inherent defect in the process of recording history as demonstrated in the chart above. History has always only been written by the winners. Now imagine if Hitler had won the second world war would our history books read exactly the same as they do today. I can guarantee you that it would have been a very different optics of history, as would have been the current global environment itself. I won't toil on that idea other than that then Hitler would been the hero, and savior, while Churchill possibly the destroyer and villain. **That is the power, or defect of history.** So, hopefully we momentarily agree that History is not a record of the empirical truth, but the interpretations of winners. It is a version of truth as written by how the winner wants posterity to see not only them but also the losers.

Fufactology: Fufactology is the new science designed to replace winners' interpretations with facts as a form of learning not from the past but from the future. This topic is about introducing a new subject for colleges and universities. In this we take a quantum leap from 'Learning from the falsified past' to 'Learning from the Predictable Future.' As this book is for a general reading, I'll keep away from the technical details of this most interesting subject that I have been speaking about to friends since around 2014.

Briefly the birth of this thought is from a book by Isaac Asimov, that I read when I was 13 years old, called Foundation. In the book there was a Hari Seldon, my first namesake, who conceptualized the subject 'Psychohistory' which was the study of multi-world people and of using that to predict very precise events into the future. The basic concept was that the larger the database, the more accurate their complex AI driven prediction became. Until it could take global events to predict very precise forecasts in the future.

We have today reached a cusp in time and technology where we can actually start building this course to learn from the future. We are already predicting how we can take it to the next step.

Learning from the future is no longer a science fictional or improbable science. It is a proven science that runs on the same algorithms and bots that we have been talking about in this book. Here are some real-life examples that learning from the future is already a fact today.

Top 10 prediction components we use every day:

1.   **Weather forecast:** Weather is the near-term environmental forecasting. It forecasts the temperature and weather conditions. Reliable weather up to 2 weeks ahead with details like highest and lowest temperatures by the hour, prediction of rain, snow, and even high winds. The accuracy of the next 24 to 48 hours is extremely precise. We learn from the future by heading to ski slopes because new snow will be falling or carrying rainwear because we know it's going to rain. This is a highly dependable science today.

2.   **Sunset/sunrise moon phases, and planetary positions:** Due to knowledge of the Earth's orbit and rotational patterns, we can accurately predict the exact times of sunrise and sunset. Like sunset/sunrise times, moon phases are also predictable based on the moon's rotation around the Earth. Transit of planets and comets are highly predictable.

3.   **Supply Chain Transit schedules:** The schedules of airlines, trains, buses, and other forms of public transportation are planned and predictable, allowing commuters to plan their journeys, months in advance accurate to the minute. In global supply chain we are able to predict ETD and ETA's accurate to the day and hour today. In our complex supply chain of materials and humans we already predict very accurately. When we order on Amazon the system predicts through billions of options and informs the user the day that the ordered material will arrive. When we board an aircraft flying from San Francisco to New Delhi half way around the world we get very accurate predictions on our ETA, or estimated time of arrival. When we order an Uber it tells us the color, and make of the car, the drivers name and the time of arrival. We can even see where that driver is at any moment. All this is very normal predictive science.

4.   **Tides:** Tidal patterns can be accurately predicted based on knowledge of celestial bodies and gravitational forces, on our oceans.

5.   **Sports schedules, TV Shows, elections, and School and office holidays:** Professional sports leagues release schedules of games well in advance, allowing fans to plan their attendance or viewing.

Entertainment industry schedules are carefully planned and publicized so fans can anticipate the release of their favorite TV shows and movies. Election dates are predetermined, and elections have set schedules, so we can predict when elections will take place. Academic calendars are set well in advance, so the dates of school holidays are predictable. Holidays like Christmas, New Year's Day, and every international holiday have fixed dates, so we can accurately predict when they will occur each year in every nation.

6. **Financial Customer Risk:** It is very routine to conduct financial risk analysis. Complex algorithms can today calculate the potential risk for a customer asking for a loan from any financial institution. This is a mixture of past financial history, past financial transaction and in the US the individuals credit score. Normal predictive science too.

7. **Health Predictions:** Health care has improved exponentially over the last few decades. A host of smart devices today keep data streaming about the patients' health. Regular tests can help in predicting better or worsening health.

8. **Pandemic Prediction:** recently our most powerful geo-disruption was the global spread of covid-19, its global impact flowed from one country to the next and then resulted in total lock-downs across the planet. There is not a single person reading this who did not witness this pandemic or lose someone they knew to it. All through complex algorithms calculated the infected, and deaths, then used their real-time data to predict spread, degree, veracity and velocity of the spread of the virus.

9. **AI enabled predictions:** AI learning Bengali with no human interface, leading to easier translations. AI enabled coding that humans do not understand, leading to programming languages we cannot contemplate. AI enabled autonomous tasks, leading to self-driving cars and vehicles. All very predictive processes today.

Back to **Fufactology**, so put simply this new subject is the beginning of ending all political spins on thing like climate, or something like Covid-19 spread, the needs to wear a mask or not. Whether injecting Clorox will save you from covid, or if the oceans will ever rise due to ice melting on the poles. Just like we can learn from our weather forecasts today, we will be able to learn from a lot of other future facts.

No politician today dares to state, *"Sorry, the weather prediction is all wrong, it won't snow in Tahoe the day after, or you can trust me and go out without an umbrella as it is not going to rain."* The reason is that the accuracy of the predictions has already established the required TRUST and no one dares challenge it. Yes! There are a few predictions that do not come exactly true but overall, it's a very fine science and no politician dare refute its predictability.

All this accuracy did not come overnight. During the late 1800s weather stations were constructed all across the country and their data improved our predictions. By the 1900s we built special instruments, and then later launched weather satellites far above the ground. Finally, we used the new doppler radars that work with satellites, and with AI modeling, to predict very accurate weather. We will only keep getting better at this.

Fufactology is the science of learning from the future in as many things as we can imagine. Let's start with climate as in the US it is a very hotly argued area, with republicans saying it's all fake and the scientists saying it is inevitable. Also, because our weather data can be accumulated to predict our future weather, and we will solely learn to TRUST the climate predictions as confidently as we do the weather today.

All this is highly predictable today. So, we can use the power of prediction to forecast what will happen then learn from the future by feedback from 1, 10, 50, or even 500 years into the future. Learning here is the ability to use the 2030 list, chapter 18, today to build your unique skillset in one of these areas and become the best on the planet for that support-by using FuFact's today, to build your company for tomorrow.

From a Fufactology science POV, the question is whether we can predict the next pandemic in a better way. At the basic level Fufactology will use current empirical data, then using advanced AI+ML and a bunch of algorithms to predict the near and midterm future. With the advent of Quantum Computing this prediction accuracy will go beyond human comprehension, or denials. The science of Fufactology will launch the world of Hari Seldon as it helps break down into Future Facts by pure data driven scientific probabilities. Once again, the larger the size of data-sets, the more accurate our training, and forecasting will become. So, think national, or global.

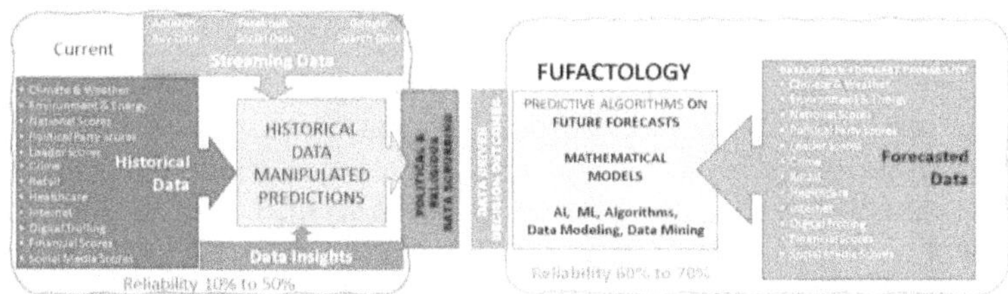

Fufactology will ensure that politics becomes a no-spin job. There are no religious interferences, just pure factual predictions. No more the president of the US convincing his tribe that Climate warming is a hoax- let data speak. No more aspiring presidents exchoing confidently that climate is a communist construct. No more stories or 'zero-sum-solutions, where in order for one thought to win the game, some other thought must lose. For example, if wearing masks has to lose, then not wearing masks must win. On a PEH scorecard, FuFact will ensure that the best options should always win without any external, political, or value add, interferences, or assumptions. We can finally base current decisions on pure data, which deals only with scientific data and our ever-improving science of FuFact predictions. Once we start this study, our predictions will only get better with each iteration as our data insights improve, and so will our capabilities to predict more accurately. With the right professors, teachers, mentors, and students, a collective passion, and science, we will be able to predict climate and weather into the future without any political or party-based alternative assumptions.

Soon, we will be able to use this to predict the sum effects of what humans are doing across the globe on things like Climate warming and its impacts on weather systems. On when there will be a hurricane in Puerto Rico, how severe and what we need to do to prevent it. Of when there will be a poisonous bloom in Florida or a toxic red tide in California. Including when central California will flood due to the river in the sky cutting off San Francisco from Tahoe for two years. Or when there will be a flood in Vermont, which will stop state activities till the waters clear. Somewhere in the future, maybe even stop crime from taking place before it happens via predictive individual insights. Stop a crime based on data signals to save humanity's future. So, with this science we can finally move from Voltaire's "*Winner's interpretation of past human crimes*" to "*A predictable probability about how current actions will impact our future with precise likelihoods of reactions, with exact time.*" In this way humanity can for the very first time learn lessons from the future that will prompt us to take steps today to divert a catastrophe in the future.

One of our prediction dilemmas is that, in some areas, the more accurately we predict, the lesser the probability that it will happen. For example, if FuFact can accurately predict, and inform you, that you will have an accident at 5PM driving in your car at 80 miles an hour. All you have to do is not take that drive. Again, if FuFact publicly predicts that the stock price will go up precisely by 300% in 4 days, then everybody will buy that stock today and in 4 days it just might crash, or go up 700%. Similarly, if FuFact predicts that the market will crash and lose 52% on a particular day a month from today, then everybody will sell their stocks this week and the market will actually crash in the next day or two. Thus, in every case of FuFact prediction of accuracy changes the reality and the future.

Now, imagine a new future scenario where a US president walks into a global warming conference with a single intent to stop this discussion because the Oil and Gas giants fund their elections. The president announces that the Climate issue is all false and that his US scientists have proof for this. He is now suddenly faced with pure data from FuFact predictions, data from the north and south poles, the mountain tops of the Himalayas, every river and city on the planet, and its current state in real-time as it is right now with an ability to drill down into each data point for the last 100 years. Then, they let AI predict what the next 1/20 and 50 years will look like with that single attribute if it runs out of control. Our US president will have very few foundations to step on in case of such a factual demonstration.

Once again, there are three caveats to this subject:

1.   **Institutionalized lies:** Just like the 1989 Tiananmen Square massacre, when several thousand peaceful protestors died, never happened in China, so too the 1921 Black massacre in Tulsa never happened in the US. The world histories are full of missing chapters.

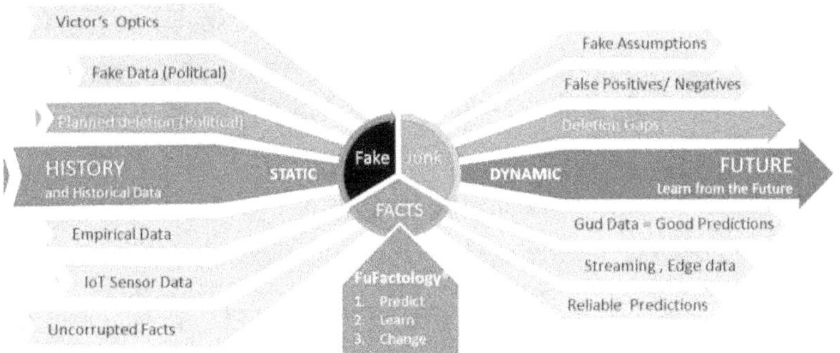

Both were wiped clean from history books. Again, after pulling US out from, the Paris Accord, Trump tried to restrict US metrological data access to universities, students and scientists, to deny them the right to learn, share and report from actual US weather data. The Republican party's belief is that if there is no data, no one can predict any climate future, disastrous or other. This denial of climate data has been a republican agenda since the mid-1960s, with an aim to kill all these climate discussions, as it was bad for the US energy suppliers. Search for *'Beyond Drake — No Earth 2.0'* on Amazon and section 2 covers the factual data to this statement. Just as lies is the foundation of current media, history and politicians, data will become the foundation of Fufactology. However, if the government, or a state wipes out true data then that creates a huge domino effect in our predictability. Just like many states want to deny the barbaric days of US slavery, and historical facts like the Tulsa massacre from being taught in US schools. If the data itself is politically wiped clean then no level of AI can extract anything from nothing. This is political data cleanup and such acts will corrupt the capabilities, and thus the accuracy of our predictions.

2. Alternative Science: As evidenced in many instances each political party can find scientists who will change facts for the optics as desired by the political overlords. Once again, remaining closer to home, when US pulled out of the Paris Accord there were many scientists who stated using an alternative science that there was in fact no climate disaster about to happen, and that all the climate scenarios were untrue. Their tribal members believed every word. This alternative science was funded by the US oil and gas industry. The problem with these alternative realities is that in a polarized planet of 'Us vs. Them' when one of the parties presents certain facts and their tribal members believe only that. So, when the republicans shouted that there is no evidence of any climate change due to carbon emissions, 50% of the US people believed that as their gospel truth, some still do. If people did not believe that lead, or asbestos, was poisonous or dangerous, we would have never gotten rid of lead in paints, asbestos in our houses and naval ships, or from our environment. But then better sense prevailed and so did we.

Fufactology could be a very realistic science that is ready to be deployed as it has all the credentials to pull in the new generations because they want pure facts to drive their decisions and not some political or historical *Spin Doctors.* Just as people believe in Google maps, they will also believe in our Fufactology predictions.

3. Advanced Artificial Intelligence: The recent launch of ChatGPT has altered the world view of machine learning and AI. Right now, it is too early to list the repercussions, but Elon Musk described it as *scary good.* It has raised concerns across almost every sector, from education to coding, from passing medical exams to writing a launch elevator statement, from PR to Marketing and everything in between. It is critical to remember that this is but the beginning of a new disruption. It is also, critical to remember that Quantum Computers will increase every computing capability a thousand to a hundred-thousand times faster, and more securely. We are, as yet, only experimenting in this area but the growth of solutions and its potentials are real. AIML is real, ChatGPT has made AI operational, and now Quantum computing is just around the corner. I will try to follow up on this topic on my blog for this book.

4. Whitewashing: In March 2023 the Florida Republican governor has seemingly started to ban AP courses that expose the desire of White suppression, and books that deal with Black slavery. The rest of the republican states are following suite. Some going as far as mandating that any single parent can demand any book, they deem as inappropriate, to be taken off the libraries, education institutions, and schools. So today, any parent in Florida can demand that they don't want their children to learn the heliocentric facts of the heavens, but revert to the archaic earth centric church doctrine, from before 1610 — now suddenly they can. Today any fool can demand a rule in the republican state of Florida. One of the most dangerous facts out across the planet is when unqualified individuals, or politicians, along with their hand-in-glove legal systems decide what can be taught to students. More importantly what must be left out of history books in schools and colleges. Truth, and facts are being sacrificed on the altar of political agendas. This used to happen only in autocratic nations like China, Russia, North Korea, Iran, Saudi Arabia and such. But in the last few years this politically polarized autocracy has raised its head in the US, where governors are deciding what books a school can keep, teach, and even firing teachers for not following their orders. One such example is the desire by republicans to erase the fact that the Whites enslaved, tortured and committed genocide on a very affluent group of Blacks, in Tulsa in Texas in 1921. Then the event was scrubbed out of the US memory by not having it mentioned in any history book or taught in any US institution. Recently, in 2022 the Florida governor mandated the deletion of the African studies AP course, and made it illegal for any educational institution to teach it. Thus, by banning the study of truth, the governor actually made this a national racist topic and that is how I got interested in it. In the US, the existence of unfair White supremist practices in education, required the Blacks to research on CRT as a method to understand the persistence of race, and racism, as a form of discrimination, against the Blacks, across the US. CRT is an academic endeavor that truth be taught in schools and universities about how the Blacks were treated across history in the US. By 2021 at least 21 republican states had banned the teachings of CRT in their educational institutions. US by the way has no national education guidelines, like most civilized nations have across the planet. In the future, we need to demand that we put in place a harmonized national educational framework, which disallows political states from hijacking education based on political or religious guidelines. Intelligence demands that if this is a part of American history, and thus society it must be taught. CRT asks if students should be taught how the White people have used CRT to segregate people by color — Black slaves vs. free Whites, from education and housing to employment and healthcare. Critical Race Theory recognizes that racism is more than the result of political and individual bias and prejudice, it is a fact of history. But, 21 US states, for example, will not allow CRT to become part of their education curriculum. It is expected that some of the other republican states might follow suite. It is important to note, that the Florida governor DeSantis is standing for being nominated for president from the republican party. He would like to apply these rules across the US, including banning abortions across the US, as well as pardoning every Jan 6th person charged by US courts since the coup.

Education is the foundation of the future of every society. Deny your people free and open education and we are already setting the roadmap for an autocratic nation under a future leader like Putin or

Xi. Thinking about education and how it impacts our future is very important.

Education in the US: Somewhere in the cusp between the enterprise and the government sits education. Briefly as we have already covered all this through the agrarian period when education was mostly about Church and God, with a few scattered subjects like philosophy. If we look at the US — in the 1600s it was about family values, religion and community. In 1636 Harvard opened to teach the puritan version of spreading the gospel. By the 1700s schools mostly had one teacher who would teach students of all ages. The first spelling and grammar 'Blue-Backed-Book, was published in 1783. Right up to the mid-1800s education remained the same — it was mostly about religion. It was only in 1837 that Massachusetts created the first board of education with free education for all grades. By 1840 the American dictionary of the English Language was published. By 1867 just when the industrial revolution was starting it aligned with education institutions to ensure the availability of a structured, and trained labor force. Industrialists funded education institutions to ensure that they managed to impact education to tailor future workers to their needs.

But it was not till the 1900s that the true impact of this industrialized education format was felt. By around this time the scientific principles of manufacturing came into existence and the ford car company commenced with their moving assembly line. By now it was critical to educate a workforce that was disciplined like the machines in the factories. Reaching school on time, and leaving when school finished, Math's, and other subjects became mandatory. By the 1910s it became mandatory for all 8-14 years olds to attend school. A strong workforce was required to bridge the gap between Ford's production line and Charlie Chaplin's repetitive, educated workers. In 1954 despite the Brown vs. Board ruling racially segregated schools were created. This continued till 1970s. The 1900s required industrial workers, and by the end of the second world-war we retooled them with blue collared knowledge workers. By now specialization and super-specialization had already become norm.

Being a capitalist economy education in the US became a for business institution. Despite a $695 billion degree-granting post-secondary institution by 2019-20. $438 billion at public institutions, $242 billion at private non-profit, and $14 billion at for-profit private institutions. It is still a business.

Education in India: Compare this to India where between 1500 to 600 BC, referred to as the Vedic times only the privileged had access to education. This included the higher caste brahmins or priests and the royals who all got private tutoring. Universities like Takshila and Nalanda existed during this period. By 350 BC two universities were set up to teach Philosophy, Math's, Grammar, Astronomy, Psychology and arts. From 400 BC to around 1200 AD education remained stable in India. By around 900 AD the Muslims invaded India and by 1200 they had changed the cultural landscape of India. They pillaged and ruled India till 1607 when they were methodically replaced by a new invader — the British Anarchy. Since 900AD barbaric invaders pillaged India for its gold and jewels. The Muslims did not try to erode the Indian education system. However, the British anarchy brought with it Christian missionary schools, the papal sponsored requirements for conversions, and the catholic schools. Britian deployed a very systemic introduction of schools, and colleges across their colonies, based on their desire to destroy local cultures via language and education. These are the same Britishers that later went to the US and Canada and committed genocide on native Indian children on the opposite side of the planet — via their White dominated Christian missionary schools across the planet. In India, the British colonizers seduced the Indian elite by firstly encouraging them to study in Britain, and then awarding cushy jobs to English speaking people. By the 1920s this class of Indians were efficiently trained to think more like the British than as Indians. With the advent of the industrialized education system in the UK, the same was exported to India especially in the so-called convent schools in India. The intellectual elite Indian children in these schools, went to the church in the morning, sang hymns, and learned to be critical of the Indian culture. By the 1930s, people like Nehru, Jinnah, and Gandhi, all educated in the UK, began to be steered by their British counterparts on how to set the future of India along good-British structures and processes.

Nehru and his family felt close to the British ways, and as a family even decided based on their political ambitions to dress up in traditional Indian clothes. In contrast, others like Gandhi worked for the good of rural India. The independence of India thus created a group that spoke English like it was their first language and somehow resulted in the creation of an elite English-Speaking class across India. The use of the English language expanded, and today English still remains the lingua-Inde for most communications. After 1947 India invested heavily in professional, and educated India on the British model. When the British finally left India, they had decimated the traditional Indian-style education. The British had decided to educate a small section of elite Indians, in their upper and middle class, who would bridge the language gaps between British colonial India, and the new independent India. According to records 70% of Indians were educated when the British arrived into India, however by 1947, only 13% of Indians were left educated. The British, had systematically eradicated the Indian education system in their process of colonization.

Education contrasts: On the positive side, after independence, India opened their world-class IIT's for engineering and IIMs for business management that stand proudly even today. The current education system is socialistic, where teaching, truth and creating good students superseded political or for-profit polarities. Resulting in world-class institutions with a clear divergence from politics, and for-profit due to the socialist democracy that India chose to become.

For example, a premier engineering institution like MIT costs around $79,850 annually for education alone. An engineering student in India can get a world-class education at IIT for $3,380 annually. Both are today considered as top institutions for engineers as a baseline. MIT mostly gets rich and very intelligent. IIT mostly gets the very intelligent. Both universities have grants and scholarships.

Looking at education on a scale of time and cost vs. benefit over the years, US education has often remained a rather consistent provider till very recently. On a cost vs. benefit analysis over time, we expect additional functionalities, along with lower costs. As a comparative example of cost vs. benefit, the average price of a 47-inch TV in the late 1980s was $4,200, and by 2019 it had gone down to $400. During this same period the average cost of education in the US went up from $7,000 in the 1980's to $45,000 by 2019. An additional benchmark is that between 1980 and 2019 the technical functionalities of the TV probably increased by a hundred-fold, while the universities still teach approximately the same books, subjects and courses. The universities just charge a lot more. As another reference, the education fees in India, for example, for the IIT's, are governed by the central ministry or the Union Grants Commission, so every school, college or institution provides a consistent education curriculum. While at the same time, education in the US is governed by independent state rules, so a school in Florida can decide that education in their state must conform to their political, or religious, agenda. This has resulted in republican states trying to ban any abortion teachings in their medical schools, or prohibit any African Study AP classes or CRT books or courses that tell the truth but show the White man in a not-so-good light. Due to this, a US student lacks uniformity in education, while Indian education is harmonized with little political influence.

Now looking at the present — in the last two years universities and education have taken a 180-degree digitization turn. By March 2020, due to covid-19, education went from brick-and-mortar and on-premise-attendance to 100% remote education. This is the great digital leap and an opportunity to change the very definition of world class education platforms.

From a global communications and optics point of view — during COVID-19, almost all US universities, along with politicians, media, and other education-related stakeholders, did a degree of disservice to children nationwide. There is a belief that to save their brick-and-mortar $242 billion/per annum for-profit industry, they may have used their digital, political, media, psychological, and social algorithms to keep their revenue stream strategically alive. They broadcasted via media, social apps, and every communication device that the social impact on a child deprived of physical interaction was catastrophic. All this while the blotting paper brains of children and their parents absorbed and amplified this message because we trust our educators. Media, we recently realized, was just another amplifier of paid opinions for a spectrum of politics and their paymasters. The world has become a digital addict to this for-profit educational message, too.

My father had taught me, *"In life, there are things that we can change, and then there are things where we can change nothing no matter what we do. Focus on the positive options on the things that you can change, not on what you cannot."* This statement is the foundation for how the institutions and governments could have done things differently. **Fact 1** — We could do nothing about Covid-19, but hopefully wait for it to pass. **Fact 2** — Until then, we should have amplified very positive feedback to every student studying from home.

I firmly believe that the global educational institutions and the state operationalized a disservice to humanity during a 'Zero-option' period of mandatory total home isolation. This was undertaken by driving a consistent negative message with the desire to protect their future revenues. And the world swallowed it hook, line, and sinker. It would have been suitable for the students, their parents, and humanity if the constant barrage of messages across the planet went something like this. Just imagine a perfect PEH positive message by *universities and politicians:*

1.   *The time between 2019 and 2022 has taught us that we need to move from profits and physical attendance to teaching every intelligent student across the nation including each brilliant student across the planet. Together, we feel confident that we can help the national intelligence scores improve while, at the same time, even making higher revenues. Instead of 30 students for $50k each, we need to have 150 students at $10k annually to break even. Every additional registration after 150 will generate pure profits. Now take this to 600 digital students worldwide, and the concept becomes a more efficient digital education platform;*

2.   *We welcome our remote students in this new world of digital connections. This is the future of everything in our remotely connected world.*

3.   Instead of being depressed, feel elated as you now represent the future as a CEO, CIO, CFO, manager, and worker. This is because the workspace has changed for many non-physical jobs on the planet. Your education now is about the digital future and not the traditional industrial past. Let's celebrate this metamorphosis together. We used to teach industrial caterpillars, and now we shall evolve to train global PEH-enabled digital butterflies;

4.   Post Covid our future has changed forever, and our education will be driven not by the past drivers of more profits, but on how we can jointly benefit from our collective PEH along with our children's children. Our courses will not be static books as they are today but will be dynamically altered as knowledge changes. This will become standard practice for the disruptive digital future where national universities, professors, and students work remotely where possible, and on-premise when required. We must ensure that the digital content is defined by a national Educations Grants and content Commission, not something a politician or political party can change spontaneously.

Note that where such a premier university does not exist, an opportunity exists to create an internationally recognized institution that can assure high-quality education with highly qualified, Nobel laureates, digital professors, and direct student interface. Physical participation is easy to manage today, with visual cues that an AI can monitor, and alert professors when students get distracted, or are absent from a class. No more, spending the first 5 to 10 minutes taking attendance, it's all done remotely by our AI assistants. New AI tools can ensure better focus and attendance from students via self-defined learning parameters, i.e., can also automate eye contact, attention to topic being taught, distractions, and many other behavioral analytics when on-line. It can confirm that the student has completed, or not, the required 5 hours a week, and they can accomplish this on their own convenient time. So much sweeter and better for our student mental and physical future goals.

In the same breadth, similar messages could have been crafted by enterprises and politicians with media and social networks to match. For enterprises in Silicon Valley at one extreme — messaging can be adjusted for your own industry accordingly. Think Google, VM Ware, Amazon, Facebook, etc.: *[a] we look forward to working with the new digital workers who are currently working far harder and learning far longer than traditional students used to while attending universities physically; [2] It is important to realize that during the lockdown 100% of our employees continued working remotely, and it took just a few months for this to become routine. You, students, are already being trained in remote participation, so welcome to our new world once again. [3] Facts on the ground demand remote unless physical attendance is mandatory, like a production line operator. We have given our employees the choice to work remotely forever; some go to different geo-locations while working on their local office times throughout. [4] We look forward to you students who represent our future as the digital shift in efficiency and a balanced work-life future.* How much better each student and parent would have mentally felt. The covid lockdown is a zero option, and we must not let the profiteers and trolls generate the dystopian messages that do our societies more harm than help just because they want to protect their own profit turf.

Side note: Workcation is when you are at a vacation spot and still working full time. I have one friend working for a multinational company as their CIO and he, along with his wife, were in Hawaii for a short 10 days visit that turned into 2 months lockdown, because Covid exploded back in Philadelphia, PA, and all flights were cancelled. So, the next 2 months were in the Big Island while he still continued to work on his local office times every day. Another example, our friend's child, worked for a company in Seattle. The young boy was in Seattle for 2 months, which was their base. Then in Hawaii for four weeks, Costa Rica the next 4, Mexico the next 4 and still working on their local Seattle times every day. I have still another colleague who went to India for two months with zero disruption to his work on US Pacific time. This is the future. Post Covid we interviewed 2 prospects one from India and another from Poland. Both of them declined to come to the US, saying 'why should we come to the US. So long as we give you high quality work and you give us our money we can continue to work seamlessly, from where we are.' They did not ask for US salaries, but also would not accept local offshore salaries. The final compromise was a mid-range that was a win-win for all.

Most enterprises are petrified by competitive digital transformation. Many more know the threat is out there but continue to believe it will not hit them or their industry. For some strange reason most of the universities have yet not woken up to this inevitable threat that they themselves are in. They along with politicians are still living in the baby-boomer, industrial, brick-&-Mortar' world. Possibly because most parents are still very old-world driven because they themselves saw value in this model. In addition, so long as elderly, grey haired baby boomer mindsets continue to be at the wheel of enterprise decisions, this attraction to specific brick and mortar universities will continue. But, as the millennials start taking the helm, we are already seeing a major shift in hiring priorities and thus a shift of students who will get on to the specialized 'Germany type' of an education path. Where to a very large degree, it's not about big-name universities, but about highest-quality, low-cost, specialized education suited to your aptitude and qualifications. Due to this Germany has some of the world's highest number of 'world-class' mid-sized companies on the planet. World class here is defined as any business that is dominant in the triad nations, i.e., EMEA, US and Japan, and now maybe China. These nations represent over 70% of global consumption. What this means is that if your company has a 90% market share say in India but no dominating presence in the quad countries then you are still a rather small player, by a world class score.

Your *education checklists: See how far you get:*

1.   Universities need to clearly define who their customer is. Is it the endowment funds in the university bucket, the foremost, guidelines of the board of directors, or the politicians that control your state funds. Or, is it the strategic good of the Student on a PEH foundation. Your PEH-directed student must become the center of the universe for all schools, colleges and universities.

2.   Demand the Facts: The fundamental role of an education system is to expose students to all facts, and truths. They need to learn, experience, and choose their rights and wrongs. No politician should be allowed to define what a school or university can, or cannot teach. Or, what books can and cannot be in their libraries. In democratic nations, the Education Grants Commission must take this reign back and standardize education across the nation. Though such a plan could have negative effects if their nation turns into a China, Russia, North Korea or Iran too.

3.   First and foremost, think German and not US or British education that was firstly designed for creating the repetitive industrial student; and second exists to basically make profit. This is a general rule and does not cover exceptions like MIT, Stanford, Yale, Oxford, or such. These are the 5,300 colleges and universities across the US that often churn out students for profit. Germany, like India, provides highly subsidized education at world class institutions. Once again there are 1,058 universities in India too and many of them are in the same production process for profit. Beware of courses and degrees that cost you an arm, and have no potential for placement or a reliable job.

4.   Secondly, the 'education industry' is a massive opportunity as the education system has still not woken up to the opportunities for digital disruption, at a mass scale, and have thus not adapted to become the new disrupters. We have the MOOGs and other micro trainers, combined with free certified training opportunities, but the main College and University remote opportunity is still unused. This means that this is open season for someone to launch the next MIT, Harvard, IIT, or Stanford in the digital world with a 'German specialization' mindset. Most brick-and-mortar courses are like the old music LP's and CD's which contained 1 or 2 hits and the rest were fillers to sell those hits. In came Steve Jobs, who took an illegal operation called Napster, and made it one the most successful music platforms on the planet called iTunes delivered by apple music.

Similarly, our current universities have a structure where their high paying students get 2 or 3 of the subjects that will help them in their future and a lot of so-called fillers with some justification, or the other. Then there are the so-called impacted courses, i.e., where either you cannot get all the classes you need unless you book them early, nor can you opt into one of these impacted courses, if you don't like one that you're already taking. This translates into that once you choose a set of classes you cannot change them if you need to find a new path. Beware of impacted courses at universities. The digital opportunity: All this makes the world is ripe for an iTunes like app for world class university courses thought by Nobel laureates and world class professors. Must have a reliable system that ensures they maintain a student's full attention at all times, maybe by using ML/AI to track every student digitally. Where one can get away from filler courses unless the student wants to indulge in that distraction. Where your university councilors work with the customers, industries, to find out what the future needs are and then tailor courses and course material exactly for the future market demands of the mid-term and strategic requirements. Become the disrupter to the brick-&-Mortar universities. Become the world class iTunes for education and learning. Allow students to work at their optimal bio-rhythms. Let them repeat their learning and exams till all their weaknesses are leveled.

5.   Make a student focus not only on industrialized branded courses but also on PEH filtered courses, the new generations expect to learn and create a better future proof content. The human-status-quo will continue to fund these brick-and-mortar universities as they are still traditionally world class universities but this will change rapidly, unless some of them become the disrupters in digitization of education itself.

6.   If your new university is not thinking digital then they are already lagging backward. In most brick-and-mortar universities by the time a student passes out their learning is obsolete. The new university needs to be designed with dynamic course ware where professors have to learn almost at the same speed as the students to keep both current on the needs of tomorrow.

Our New Digitized Government

Out of the three areas, individual, enterprise, and nation, not considering religion for now, the weakest link in this chain is the nation and the mechanisms of governments. In the US we stand to believe that our best evolutionary political system is a democracy, however even a brief glance at politics across the world makes it very clear that unfortunately the principles of democracies are all but slowly dying.

Below is the map of the Free nations, the not so free and the oppressed nations.

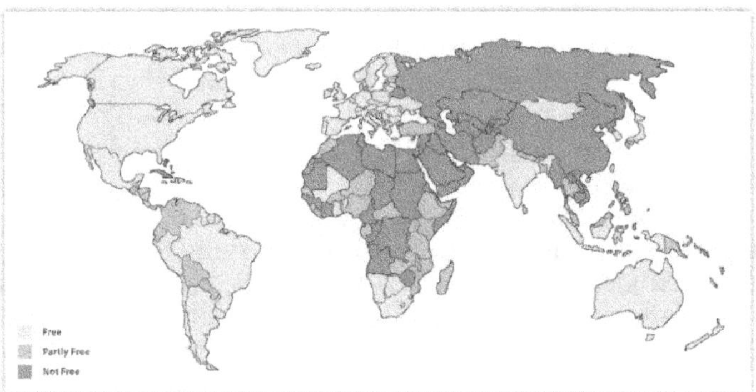

North Korea calls itself a democracy as does China and Russia. But we all know that they are not. They are as much a democracy as Iran is. In the US there is a core 45% divide of Republicans and Democrats, which statistically is a perfect division. However, these people do not uphold any fundamental principles but are today victims of polarized tribal politics. No matter what, most party tribal members will only vote for their polarized party. For example, no matter who the democrats put up their tribe will vote that person in, the same goes for republicans. This is polarized membership voting and not democracy. The day this happened democracy died and partycracy was born. There are around an estimated only 9% to 10% Americans that belong to neither party and these are the brave citizen's that manage to keep the US democracy alive. They are the thinkers, the people who vote not by party but by the candidate of party promises. These are the purebreds of a democracy, but only a low percent. This percentage is too low to keep democracy alive so democracy may be slowly dying in the US, so too in other parts of the world.

The map above, demonstrates the different types of political systems ranging from Presidential Republics, absolute monarchies, to No Government. Once again let's focus on democracies as discussing China (only one party's right to govern), Russia (Semi communist KGB presidential rule), Iran (Autocratic presidential Islamic rule) or Saudi Arabia (Absolute Islamic Monarchy) and such autocracies mostly do not work for benefit of our PEH scorecard. All we can do is hope that these nations can use their digital connectivity to get freedom from absolute definitions, or oppression, however the predictable future is that these autocracies will use this digital shrewdness more and more to control and oppress their citizens. As I write this in April 2023, we can only hope that the Iranian oppression of women turns into a revolution and goes way beyond a protest. We can hope that the UN does what it was supposed to do. That it stops Russia, and their veto power, currently used to break every rule of international responsibility, by attacking a peaceful neighboring nation Ukraine. A hostile attack, that was undertaken without apparent cause. Important note is that while this atrocity continues, the world is again polarized between who is right or wrong – resulting in a status quo of support for the oppressor, and a continuation in blaming the oppressed.

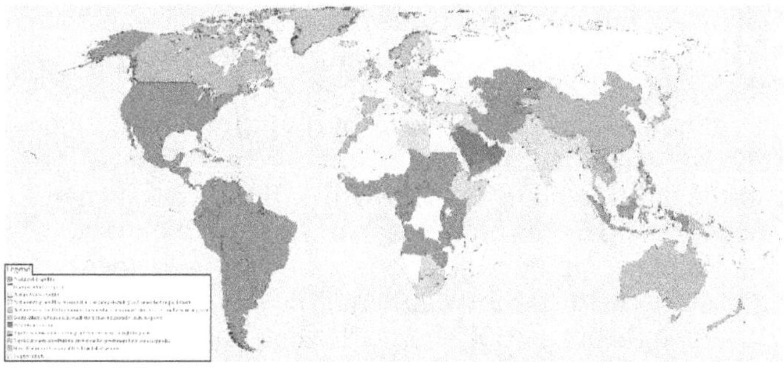

Once again at the center of every democratic philosophy is the platinum brick, we call an individual. World over, democracies are slowly being turned into autocratic empires by the rich and powerful, i.e., presidential emperors voted into office by political muscle, like Putin and Xi for life, others by religious power like Iran, Saudi Arabia and Afghanistan. Each year they become more powerful as they flex their individual ideologies a little more. There is an old saying, *"Give humans the best of any idea, and over time, the rich and powerful will manage to slowly turn it to their advantage."* This rule applies to nations, religions, steel, gunpowder, God, scientific discoveries, free human labor in the form of slavery, and now democracy itself. Today, the US, a country that projects itself as the foundation of modern democracy, is floundering in a quagmire of total digitized-political chaos, i.e., opportunistic mental takeovers driven by two highly polarized political tribes. Politicians are today openly using their digital outreach via smart devices and social networks to weaponize citizens in their party for enhancing personal causes. Common people are being digitally weaponized to carry weapons, participate in coups, conduct bodily harm, and even murder civilians and school children at ever-alarming rates. This was never on any roadmap to a great democracy — so what happened?

The primary fault of the US system, as mentioned way back in a 1944 book by Swedish economist Gunner Myrdal, 'The American Dilemma' is that the president can appoint and fire judges plus appoint critical key decision makers that should be outside the dogmatic sphere of a political leader. In the US, the segregation of duties, though clearly defined, is extremely convoluted if we go a little deeper. This allows presidents and governors to pass laws in their political or economic interest. Which historically resulted in a long duration of slavery and the relegation of the Black person as the sub-class to White Christians, something that clearly still exists today.

This was further visualized in the book 'Caste' by a New York reporter, Isabel Wilkerson. The second is that this system continues to allow the selected overlords to usurp the will of the people, thus taking the democratic principles of the highest votes through an artificial filter of the electoral college that can still be doctored by the rich and the powerful. The third is the extremely large politically motivated funds that directly and indirectly get infused into the political mind-bending capabilities of digital manipulations, which distort the very foundations of free thinking. In the last two years, this has resulted in influencing the free will of the people by mega billion social network and media opinions that mandate their paid version of right vs. wrong. This became apparent when Fox News paid $787 million in settlement so they would not have to tell the truth in court and to the people of the US. The truth does not matter for rich fake media companies, as they apparently get far more money to spread their paid lies than any paltry fines or settlements, they end up negotiating. This is the price for allowing establishments to sabotage a democracy.

The general voting citizen has today lost their ability to see beyond polarized party views; the right to vote is the fundamental right of every citizen in any democratic nation. Politicians today are salivating with the trillion-dollar dole-outs, unfettered lack of controls on funds distribution, privacy, and mental hacking potentials of digital algorithms. The time has come when 'We the People' need to put AI-assisted strong fences on our voted appointees, their political parties, their followers, and their promises vs. what they actually do when they get elected. Their bios vs. who they actually turn out to be, their financial mischiefs, and tribal members strictly on what they can and cannot do. This must be urgently put into place by setting up a federal digital department that creates political rules and regulations on the use and, more importantly, the misuse by transparent digital surveillance of elected officials and leaders.

When we look at leaders and governments, I believe age is a big consideration that is unfortunately authorized and governed by the constitution and the very politicians that need to define the rules of engagement for political positions. Just as they can define up until what age a president or a governor can stand for an election or remain a Chief Justice in a high or the supreme court. For example, in the US, a president has to be at least 35 years old, but there is no cap for the upper age limit. US politicians have the advantage of setting laws to define their pensions and post-retirement benefits. Just as a note, in some nations, if a politician serves in two positions, they are allowed to collect two pensions. That is like a person having worked for 4 companies, now eligible for 4 social security checks — try that for a reality check. I would love for that to happen, but neither I nor you can make the rules, so this will never happen. In fact, some Republican officials want to gut the social security payments for common citizens. Social security does not belong to the politicians. It belongs to the people. Politicians want to gut Social Security because they have pulled out so much money from it that they cannot pay back, so their best excuse is to gut it. It's like your company borrowing from their employees 401K, then as it comes time for you to retire, they vote to gut it. **Officially, the US government has already borrowed $1.7 trillion from the Social Security Trust. They have borrowed money that does not belong to them. Infact, they have stolen money of the people.** The government has basically borrowed the full value of the Social Security fund to pay for other pork-belly government funding. **By law, the federal government is required to pay back its borrowings with interest; these are full-faith borrowings.** However, now the Republicans want to gut the empty coffers rather than pay. For example, George W. Bush borrowed $1.3 trillion to finance his tax cuts for the rich and fund the Iraq war. The reason is that it is all public money. A contribution that we have all made right through our working life. On the matter of age, in 2024, the two US presidents competing for the next presidency are both well above any corporate retirement age. If they applied to a corporation or job, they would not be hired, but as the president of the nation, they have defined that they can. Biden, for example, is 79 years old, and Trump is 76. Just as an example, if you ask a 75-year-old for their definition of infrastructure, you get a very different answer than if you ask a 35 to 40-year-old person. **Infrastructure** for a 75-year-old is roads, bridges, rail, and the rest. It's also about building a wall between Mexico and the USA. For a 35-year-old, it's about digital infrastructure, i.e., Sensors, IoT, IoE, digital security, Quantum computing, and bandwidth. It's about building digital highways, installing smart cars, smart buildings, and smart city applications. It's also about building digital walls that trolls from outside cannot get through to disrupt the digital applications inside the US. It's about building a state-level Digitization Office that works solely for defining and auditing the standards and processes of operating in the new digitized world order.

The next reason why governments are on a slippery path is because they have, over the centuries, got more of their hands on everything that defines and determines the lives, or lack of it thereof, of every citizen.

Take taxation in the US. In 1862, income tax was 3% for anyone making more than $600. Over time, the United States and the federal government are doing everything they can to get more and more out of the wallet of the common citizen, while giving more and more breaks to the rich. In 2022, for anyone earning up to $10k, the taxable rate is 10%. While anyone earning between $170k to $215k, the tax rate is 32%. Finally, for anyone earning more than 539K, the tax rate is 37%. The big question then becomes as to how someone like the ex-president has not paid taxes since 1998 and, in 2021, paid a measly tax of $1,700 something. Here, the rich and powerful have gamed the system to get more and more tax breaks and keep more and more of their money. The rich, with their lawyers and judges, have totally gauged the system. This needs to be corrected, and a minimum tax rate must be established across the US. The other is the fact that the Church, as mentioned earlier, is exempt from paying taxes in the US. We have the rich, and then we have the ultra-rich churches that pay no taxes.

Freedom, take the civil liberty rights of a girl in Iran. It is reported that if a girl gets raped in jails, by their religious police, she has no voice, recourse, or rights. Now let's briefly move back to being a pretty girl today in the US, that too in a republican state, and you get raped while walking back from the library at 1am. Here your womb became the property of the Church and state on June 24, 2022. Now let's move to life in Tibet, where your culture, religion, education, your individuality is totally suppressed by the Chinese government. Tibetans are prohibited from learning their own language, pray to their gods or religious leaders, and are routinely taken to brain-scrubbing centers that China has used extensively to force children and people to think and act the way Xi's government wants them to. That is if they want to continue to live. The Chinese intent is to crush the Tibetan Spirit and force it into the Chinese way of thinking. The Chinese people believe that Tibetans do not need to think in Tibetan, they need to think in Chinese. If you're a Chinese person living in China, the attempt is the same, but at a lower amplitude, unless you are a Uighur woman or a Uighur Muslim. If you are different from the tribal lifestyle as defined by the rule of Xi then you automatically become a targeted person of this digital state for continuous surveillance.

Thus, I retrace to my point: governments are today influencing most things their citizens do. Some nations do it overtly, others covertly, using internal and external trolls to create national 'Us vs. Them' tribal loyalties. A lot of times, we find elected officials on a daily rant of polarized announcements for the sake of making a statement, while not doing the job they were hired to do. These tribal members are often weaponized at an individual level to create polarized opinions within nations.

In the US, individuals can be triggered to illegally enter the room of the speaker of the house and steal her laptop, take selfies, or make a tribal member enter her house with the intent to knee-cap her and then end up hitting her husband with a hammer because she was not there.

Democracy under audit: In today's digitally manipulated democracy, [1] It all depends on how much your party, or your leaders, are willing to pay for your 'democratic' victory, including the fact that the deeper the mental penetration, the more it costs. [2] After that, it also depends on the individual attributes required to make this mental shift in a way that remains below the subconsciousness of an average individual. In some cases, in relatively short times, individuals actually start to believe the fake scenarios planted inside their minds as their gospel truth.

These are no random events but an accumulation of digital persuasive messages that are being thrust into the minds of the target. Some believers just listen and let it go. While others become digitally-enlightened fanatics that are willing to die for their tribally induced alternative realities.

The PEH Audit: We need to start measuring our candidate on our PEH scales as a human collective. We need to spread this 'Save Humanity' message to as many people as possible. **On the P scale:** Planet- we shall focus on democracies as our current hope for humanity. But, if we look closer even that is due for an audit.

1.   Democracy seems to be dying or needs to come under a political audit. China and Russia declare themselves as a democracy. North Korea as *a dictatorship of people's democracy.* They all say they are democracies. Democracy was not designed for leaders like Putin, Xi, or other autocratic leaders. The only fact that bubbles upward is that a lot of things are not heading in the right direction, and thus, some things must be done now to change that.

2.   Future leaders must be controlled by stronger articles in areas exposed by the last US president, that has taken US to the brink of a coup. Single nation dominance must be discouraged, for the planetary good. Leaders must not be elected on slogans like *Make China Great Again* or even *Make America Great Again* in isolation of PEH guidelines and all other nations and people.

3.   Graft: Today the greatest danger to the US is when the president calls the enemy leader a genius, and his whole party mirrors the optics of Putin. When Putin attacks a peaceful neighbor, the republican party wants to stop funding our modern David, and hand over the nation to Goliath. Pakistan gives away Gilgit, Nepal a few villages, Sri Lanka a port all to China. The only reason such things are given for free is graft. Nations need to put very strong fences around media, the rich and powerful and especially elected officials from singing to the tunes of their enemies. During Oppenheimer's times he was put on the blacklist, despite stopping the war, for having been in

the company of communists. Whereas now, our ex-president calls the Russian leader a genius, praises everything he does while castigating his US agencies, and his whole party wants him right back as president in 2024. This can't be right. Or am I totally worng.

In China, it is designed to evoke separatist China-only emotions. It prompted China to torture and kill Uighurs and Tibetans under very cruel policies. In the US, it allows the parties to subvert the rights of Blacks to vote or even get killed by Whites in uniforms. It results in banning Muslims as a community and evoking the 'It's not your fault, it's theirs' feeling in party members. When Russia thinks only about itself and attacks an innocent peaceful nation like Ukraine after having gobbled up Crimea.

After that, then follow-up by sending Russian prisoners whose bounty is to be allowed to pillage and rape the young and innocent. Then, proceed to threaten to nuke the country if anyone interferes. This cannot be good for the planet no matter what side of a political spectrum, nationality, or religion you belong to. Yet most Chinese and some US Republicans think Putin is a genius. This can't be good for China, nor for the US, and thus, nor for the planet. Thus, 'We the People of planet Earth' need to change this dismal status quo. I don't have an answer to democracy, but it probably is a smart political startup that triple authenticates every vote and comes out with a new political philosophy to replace the already corrupted democratic principles.

On the E scale: Environment - one instantly remembers two quotes by Carl Sagan. The first is *'No other known planet in the solar system, or our Milky Way Galaxy, is a suitable home for human beings; it's this world or we have nothing.'* Followed by *"We probably don't need more weapons than what's required to destroy every city on earth. There are only 2,300 cities. So, the United States, by that criterion, requires a maximum of only 2,300 nuclear weapons"*, at that time of this quote US alone had more than 25,000 nuclear weapons. Right now, as of June 2023 US has more than 5,428, Russia has 5,977, China 410, France 290, UK 225, Pakistan 165, India 160, Israel 90, and North Korea 20. Another 'E' statement *"If humans disappeared from our planet tomorrow, the world would regenerate back to a rich state in a decade or two. However, if insects were to vanish, our world would collapse in months into total chaos. If all bacteria were to vanish, we would all be painfully dead in hours."*

Most leaders suffer from a total lack of P and E concerns, reflected by their political tribes. As humans we have successfully eliminated toxins like lead despite rich and powerful companies and politicians fighting hard to prevent it. Climate warming is exactly such a dilemma for the world but more importantly for a nation like the US that perceives itself as educated, intelligent, and a light on top of the hill.

Yet, we have political parties that have been paid millions, if not billions, for over 7 decades, by US oil and gas companies to not only slow down talks on climate warming but provide scientific evidence that this is a hoax, and then follow through with sending our representatives to gut the initiatives itself.

Most republican US presidents since Nixon tried to accomplish this instruction covertly, and from behind closed doors. Trump, in his global political naivety, took his orders to its extreme limit, in a show of grandeur, when on Nov 4, 2020 he childishly showed his loyalty to his Oil and Gas sponsors, by officially pulling US out of the Paris accord. We've known since the early 1960s that republicans were against climate discussions, but this time it was gleefully different. Read 'Burning Earth' on Amazon Kindle on this topic. This is a negative on all three PEH counts. A globally disastrous political decision.

On the H Scale: Humanity- Each national or political action, including every P and E decision, is finally going to kill the H folks, i.e., humanity. Our greatest risk of extinction, as the only intelligent species we know of in the entire universe, does not come from an asteroid hit like when it supposedly wiped out the dinosaurs, a mega-volcanic explosion, or even an expanding sun. Unfortunately, our greatest risk comes only from our fellow human beings. Each tribe, for the last two thousand years or so, has been hell-bent on dominating all others at the cost of total eradication of 'them' in their 'Us vs. Them' mindset. Our greatest risk today is a nuclear Armageddon that wipes out humanity. Unfortunately, this is already wired into two of the prime religious books on the planet, so it might become a 'self-fulfilling prophecy' with a let me act first trigger finger. If the earth explodes in a nuclear holocaust, it simultaneously explodes the Environment and thus could potentially extinct humanity, at least as we know it now. Every misunderstanding, disagreement, battle, war, and national altercation only takes us closer to that eventuality. As the US, we were finger-tips close to it in 1962 with the Cuban crisis between the US and the USSR. Now, however, in the last 2 decades, we have witnessed misogynistic statements of 'Mine is bigger than yours' at least 6 times. At least twice during the Afghan war from 2001 to 2021, twice with North Korea. Now, most recently, twice with the Russian nuclear Sabre-rattling by Putin and his generals over its illegal attack on Ukraine. Recently, i.e., by the 4th of July, 2023, we, as a planet, are sitting with our feet hanging over the global nuclear option between Russian nuclear threats regarding the Ukraine war that Russia is slowly losing and the global retaliation in case that happens. On July 11th, Russia sent a warship to Cuba, reminiscing the days of the Cuban crisis in the early 1960s.

For the new individual, it is most critical to once again remember that the one thing that can destroy all PEH is us humans. It is also important to remember that in 1608, when the British landed in India with less than 100 people, they landed with anarchy, colonial experience, and profit-addiction in their mind. They came only to pillage and profit. India's population at that time was around 169 million. The message here is that even at its colonial height, around 256,000 British soldiers and supporters ruled a nation of 169 million, which is a ratio of 1:660 people, i.e., for every 1 British representative, there were 660 Indians who did not resist. The British managed to subjugate and rule this nation for 350 years. The UK became a rich nation by pillaging the wealth of India and its other colonized nations. Let's make sure the politicians do not succeed in pillaging the citizens of their own nations in the same way, in a similar manner of a few powerful people controlling the unwary thousands in a planned and structured process of *command and control*. When you look at global politics, the delta between Xi, Putin, and many world leaders is but marginal.

The important lesson to learn here is that every political party, including every political leader, is like the British landing in India. First, they bring a lot of baubles, make a lot of promises, give many gifts, and show empathy and friendship, but that was only before the elections. However, once elected, they behave like they own both the people and the nation. By 2022, democracies, political parties, and digital trolls will be pillaging the wealth, beliefs, and minds of citizens. *Remember, the 1:660 ratio of British vs. Indians, and now, with our digital outreach, 'We the People' have an obligation to never get sucked in, or oppressed, with fake promises, news, or stories in exchange for zombie votes, by a few oppressive autocrats.* We need to make sure that if our nation is a democracy, then we must vote for the right people who will create our future. Our global democratic ratio and digital influence potential is 195:8,000,000,000, i.e., 195 nations vs. 8 billion humans. In Russia, the ratio is 1:143.4 million. i.e., the might of Putin vs. 143.4 million individuals. In Iran, the ratio is 1:87.920,000, i.e., 1 Khomeini against 87.92 million people. In China, it is 1:1,412,000,000, i.e., 1 Xi vs. 1.412 billion. In the US, it is 1:331,900,000, i.e., 1 President vs. 331.9 million Americans. You get the message. In a democracy, at least, we must guarantee the enslaving of humanity must not be allowed to happen. We need to be taught and become able to read the signs of potentially oppressive autocrats.

The only problem is that these politicians, with their wealth and power, are becoming ever more wealthy and powerful. Thereby, they can control nations with their physical might, i.e., religious police, local police armed to the teeth, army, air force, and the entire legal system. More recently, they control more and more people with their digital broth. We need to put a stop to this or collectively agree to be subjugated and give up the nation that we once belonged to. Then, our only path is to accept that the US will either become like China or Russia, with political authoritarian presidents for life, or become like Afghanistan, with religious authoritarian control. This is not just a figment of my imagination; we've had US presidents, senators, and elected officials state that 'Putin is a genius, or that 'Xi should be given more access, and even that 'Afghanistan has got it right.'

I have faith in the democracies as they have so far self-righted themselves. However, this time the power of the bad guys, with their digital arsenals is very pervasive. The will of the people, is being weakened by constant broadcasts of absolutely fake facts right in their living rooms and cell phones, is extremely powerful. This is a personal environment where people trust everything, and the result now is that they have been systematically programmed to lose trust in everything. And that this is the beginning of the takeover with pure tribal loyalties and power by people willing to die for their perceived cause, and oppressors ready to get them killed so they can harness ever more power.

For Our US National/Political Checklists

1.   *Elected Officials: Start by mandating that an elected official needs to first define who their customer should be. Who are they working for — USA, their state, their citizen or their party. I believe it should be PEH first, then followed by the welfare and safety of every citizen, bar none.*

2.   *Federal & State level Digital Department: Start by creating a federal level digital leadership with their first task to shut down the international, and internal/national trolls and social media influencers. Place very high digital fences, for controlling the warping of our national psyche, based on paid guidelines, distribution of fake assumptions, and using digital media for paid fake-propaganda of any sort. Plus, very clear outcome guidelines that prevents a public misuse case like dominion to be closed by bribe, by getting paid by Fox news to the tune of $787.5 million. In every such case where the trust of a nation is being compromised by bribes — our SCOTUS must immediately pursue that case till the truth is established and made public.*

3.   *Media, the constitution, and PEH: In most democracies, media is a protected institution. They have the right to print, broadcast, and tell people the true state of affairs. In the US, the media is protected by the First Amendment, their freedom of speech, and of the press. Basically, the media was supposed to tell the unbiased truth to the people. However, modern media is today a global for-profit business. Rupert Murdock has an estimated net worth of around $17.5 billion. This was not made by telling the truth; this was made by broadcasting opinions based on what the rich and powerful politicians wanted their audiences to hear. News media*

today has become a paid mouthpiece for political parties. Just as an example, Fox News made $14.0 billion in 2022, and CNN made $2.02 billion. Fox is a paid mouthpiece of the Republicans. In 2023, Fox anchors even announced that they were an opinion channel and not a news channel. The degree of their corruption was so large that they paid Dominion $787.5 million just so as not to get their management under oath or be dragged to court. We need to make sure that if a media news channel is not acting like one, then their rights to be treated like one should be taken away. Also, if it is proven 'beyond reasonable doubt' that a media channel has participated in criminal or anti-national activities, then they must be censured and lose their rights as a public media channel.

4.   Universities must remain learning institutions not political megaphones: *No politician should be allowed to ban any book or topic from schools and universities based on their political agenda. This must be decided by the national education board with zero political interference. Students must be taught the truth, not from any religious, or political perspective, but only from that of the author and the national education board members, with a non-political agenda. Follow through by digitizing federal funds toward developing the future Education Systems. This is a massive opportunity as the education system has still not woken up to the digital threat, and have thus have no plans to disrupt legacy establishments.*

5.   Zero Lies: *With Zero tolerance. No elected politician should be allowed to lie on anything that impacts the public. SCOTUS must take it upon themselves to ensure elected officials tell the truth and apply a '3-strike' rule after which they should be, terminated from their current position when it is proven beyond reasonable doubt, and not allowed to stand for any public office for life. If any official makes a statement, they can be called to prove its validity, the response cannot be that they got it from some non-authorized source, nor that it was in jest.*

6.   Financial Transparency: *There should be total public transparency to the disbursement of public funds. Since February 2020, around $9 trillion has been disbursed in one form or another as Covid relief. This directly equates to $25,714.29 per every citizen, student and child, of the US with a population estimated at 350 million. So a house with father, mother 4 children and 2 babies equates to $205,714.32. I do not say there was graft, by every political party but as a citizen I want to know where each cent went. I can also bet a hundred that some, or a lot of it went into the pockets of some right friends and family members, even if it was routed from external donors. Being elected must never become a financial lottery. Just as an example when the 1ˢᵗ $2 trillion Covid-19 was approved, I recall there was a $500 billion allocated to big corporations, and $377 billion to small businesses. However, when I check for details on exact disbursements in these two categories I cannot find it. When analysts tried to research the small business funds allocation, neither could they trace the full amount. I recall when asked about the exact details of the fund's* *distributions, the government was not clear on where the money went. I'm not accusing that the money was misused, but as a citizen demanding that I need to see where each cent was allocated, to prevent any misuse. We need to prevent that no one should be able to misappropriate even $0.01, toward any party, fake initiatives, person, friends, family or personal gains. This can only come by treating these funds like a USA Inc fund that is subject to a SOX-like audit by an approved neutral company.*

Who will fight for regulations on the global digital transformations?

Protection from Digital Oppression

Another taboo is in assuring who will truly fight for fair digital transformation practices. This includes social networks, the news media and all internet-based communications. Its answer is that you and I will have to fight this fight. We need to write to our senators and representatives and assist in defining some very critical rules and regulations. Moreover, we should write to the senate to define national rules, and then to congress to waterfall them to state level metrics. Unfortunately, it will be a difficult hill to climb, because the very politicians who currently pay to misuse this technology are now responsible for putting guardrails around themselves. It is like asking Putin or Xi to enable privacy in Russia and China. These global politicians have used the techniques of Cambridge Analytica to change the minds of their tribes. Digital power is coming close to becoming ultimate power. For the suppliers there is lots of money to be made as political parties can finally fine tune tribal minds. For the buyers, depending on how much they are willing to spend, the trolling companies guarantee them victory against many odds. They often collect a bonus after victory — the cream on the pie. While the common folks are being distracted by vote counts the real battle is being played in the digital world of their smart devices and social networks.

That's how the 6th of January incident took place! If this coup had happened in any other part of the world, we would've called it a coup, but in the US, 43% of people, who are the hardcore Republicans, still insist that it was just a peaceful march.

It has been exposed that when they went to the ellipse where Trump was speaking on the morning of January 6th, the FBI came and said, "We can't let these people in because they're carrying weapons, they're carrying knives, spears, and all kind of things, which we cannot allow inside." Trump, on being informed, said to let them all come because he needed to show a crowd how caring he is. He made a very interesting statement, "They're not here for me. They will not hurt me. Their weapons are not being carried here to harm me." It simply means that he knew their weapons would be used against somebody else.

Thus, it is more than confirmed now that politicians will never be our safeguard for the digital future that we desire. Instead, we will need to become the protectors of ourselves. People who understand digital transformation, and security and build this new world that we are getting into.

What about Sex

Sex has been an extremely powerful urge, and emotional disrupter, in most animals, mammals, and humans. It has resulted in fights, battles, domination over women, and even wars. If this is such a powerful human emotion it is no wonder the power mongers would want a share in it. Prior to the modern digital neuromassage, there were religious control mechanisms. One way to control humans was by putting fences around sex. Over time every religion made ever stricter boundaries on sex. Every religion gave the power of sexual domination only to men, with women as just the subservient victims. No religion traditionally allowed sex before marriage, the punishment was disproportionately greater for women, while the honor for men disproportionally illogical. It starts with religion defined sexual partnerships, including marriages, sexual preferences and sexual choice. We have recently discovered, that in animals, bringing up children by same sex partners is not uncommon, neither is homosexuality. However, when scientifically reported, there were powerful forces that did not want this knowledge placed in scientific journals. In religion they made sex a huge barrier to heaven. It ranged from no pre-marital sex to getting stoned or shot to death for it. We have now digitally reached a point where swiping right or left can mean you get to meet a person of the opposite sex for a fling or a date or even a long-term relationship. If you don't like what you see or meet you simply swipe right and move to the next person. With this functionality, the foundations of traditional relationships, marriages, and societal nitty-gritties are starting to crack. The whole movement of LGBTQ is a democratic movement in a lot of countries that allows people to freely choose the sexual identity that they would like to carry.

In some countries, the LGBTQ movement is a façade, but it is quite accepted in Europe. In the US, it was initially banned, but now it is accepted, then in 2023 there is again a movement to get it banned again. There is a deep fear that after the ban on abortions, the Catholic Christian fanatics will try to go after the LGBTQ. If they get victory in their motives, it will lead to a political victory; there is a high probability that LGBTQ will be the next segment of people that will be attacked because they want the old-world religious order to come back. They only want traditional marriages and no sex outside of marriage. They want domination of women, which is proven beyond doubt by the anti-abortion laws passed by the Supreme Court in the US. I also wrote an article on it on LinkedIn called *'Being Woman'* a link to which I have included in this book.

Sexual fluidity is becoming the near predictable future. The colors people wear will go out of tradition. Forty years ago, men wore brown, blues, and Whites. But today, they wear colors they feel comfortable in. In the past decades, if you saw a small pink cap on the floor, you could predictably say it was a girl's cap, today you cannot.

Discussing sex was a taboo in every society, and slowly these discussions are becoming more and more main stream. The example of marriage as a mandatory requirement is a concept in front of us; its bonds are already breaking. There is one European country that has the highest number of children born out of wedlock. The people got pretty upset when some Europeans called the place "a land of bastards."

The government organized a vote to know if people wanted to remove the concept of marriage out of the country. The surprising fact is that 49% of people voted to eliminate marriage with a clear understanding that if they eliminated marriages, we must do two things; eliminate the church and legalize non-married partnerships with the same legal rights. Simple – no more bastards.

In other words, if there were no marriages and people lived with each other with a choice, there could be no bastards. It would have a very strong ripple effect on our interpretation of sex. Our interpretation of marriage, which is associated with legalizing sex between a man and a woman, results in our definition of a bastard.

The Future is Ripe for Women

Accelerate what happened to Harvey Weinstein, Bill Cosby, Larry Nassir, Charlie Rose, Jeffrey Epstein, and most recently to the president of the Spanish Football Association Lius Rubiales is simply the tip of a massive global iceberg. Women re adapting very fast they need the support of men to make equality happen. Read my article 'Being Woman' for a glimpse of data driven findings right after the US Abortion crisis.

I am willing to predict that the next two decades will be the decades for women. Just like for the last two hundred years women were oppressed, and thus in a state of depression. So too, by the 2050s the men will be in a depression as women will begin to outperform them in most studies, activities, physical fitness, and jobs.

So, when republicans think they will have more Christians by forcing a no-abortion mandate they are probably very mistaken, they should learn from India. India has no controls over abortions and it is a total pro-choice nation. Women there have children when they are ready for having their child. No religion, politics or bounty hunters. The more freedom you allow women the more readily they will have children when they are absolutely ready. The happier the child, the happier the mother will be, and the better the society will be. With this logic every nation, religion and state must support pro-choice for a happy world. This is how India is today the worlds most populated nation, not because they pursued some draconian abortion laws, but because they gave the women the freedom to choose.

Our planet is a male-dominated planet where men define the rules, wrote every holy book and thereby conveniently subjugated women in different layers of control. But soon, all this must be changed because it will be the decades of women, and it's starting now.

Worldwide, we're seeing fewer men going into Gyms that are getting more and more women. So, women are getting more physically fit then their male counterparts. We're seeing more and more women get into traditional professional studies like engineering, architecture, and other such studies. They're not only getting more and more into professional institutions, but when there they are studying harder and scoring better than the men.

Women had to take a tedious and hard journey to come here; that's why they struggle and work harder due to which they score better.

I have a friend in India whose son is studying engineering, and was not doing well with his grades. His excuse was that the university had made the mistake of letting girls get into engineering. Firstly, because it was a man's job, and secondly because this distracted the men from their studies, which is why his scores were not too good. The men will need to find better excuses in these colleges now. I'm sure this will pass very rapidly but still an interesting perspective to share.

But then things changed in the 20th century. In the mid-1900s, we saw few girls entering professional studies, but today, we see a far a larger girl's population is in the universities and scoring as well as the boys. The ratio between both genders is balanced, as it should be. That's why it is predictable as to who will manage the world in the next years. In the coming decades, you will find more women in all workspaces, government institutions, and educational institutions. In fact, women's non-confrontational point of view will be a strong value driver.

Women have more empathy than men; therefore, the sexual battle will be very challenging for men democratically. Women are getting into key positions very fast, and it is easy to predict that they will be leading a lot of companies in the next two decades. It all boils down to an earlier statement "If Lehman Brothers had been Lehman Sisters would the 2008 banking crisis have ever happened. Most agree probably not."

Religious Houses

There is a discussion that I participated on recently that asked a very fundamental question regarding religious houses collecting over $1.2 trillion that religious houses collect officially in the US. The question is *"Show me 1 cent, out of all the money collected in the name of all god, that reached even one of the Gods?"*

This is the **first** critical question as most religious donations are mostly in cash, their earnings are tax free, so their gross profit is $1.2 trillion. This is more that the collective profits of Amazon ($215 billion), Apple ($170 billion), Facebook ($85 billion), Microsoft ($168 billion). This adds to $638 billion which is around half of what Religious Inc collects in a year. Do your simple math's, and then think if we need some digital transparency, by which we can track where all these profits go on an annual basis. There are thousands of cases of priests buying better and better stealth helicopters, with their massive funds, because their god came to them in a dream and asked them to buy one. Just google this in your individual countries and get a glimpse.

The **second question** is that if you were the CIO, or CFO, of religion Inc. with a gross profit of $860 billion. Then, what all would you do and how much would you spend to keep your customers addicted to your company and the product you sell.

Just a singular example of benefactory donations, I have a friend who automatically donates 3% of his monthly salary to his church.

On the topic of taxes, while we, the common citizens, pay taxes regularly, but Trump has been absconding from paying taxes for the last 15 to 20 years. I once missed my tax, where I had delayed paying a $20 by 10 days. I ended up paying around $2,000 because the law only penalizes the common man, but excuses the very rich, the powerful, and of course the religious institutions. That's the finest proof of how some people, and institutions, are above the law. If you're rich and powerful, or a religion, then there is another set of laws in the US.

Religion is extremely powerful. It is the oldest business to collect money in the name of Gods, and the digital world is exposing this business completely. It is taboo to discuss their business model. It is taboo to ask the questions we have asked here. It is taboo to let such thoughts even germinate in our minds. But question we must, because we are curious. Let's ask questions while our digital domains allow us to share and ask such questions, for the first time with an international audience. A Hindu may not be able to ask some question sin a Hindu state in India but they can to the international public. Same as an Iranian girl may not be able to ask some questions in Iran but she can attract responses from girls worldwide. Let's use this power peacefully, intelligently and responsibly. A lot of US churches collect funds the same way as the media collect by selling paid for opinions. Like media this is a very powerful global organization that historically has, and still can squash smaller religions and question that do not suit their optics.

The **third question** to ask regarding religion is quite different. It is a fundamental question asked by every person that goes to any religious house, *"I am a Christian, and when I die, will I go to heaven?"*

My private response to this question is a taboo answer that opens some unplanned doors to fundamental questions. *"What if you do go to heaven, but the heaven you get to is not your religion's heaven?"* There is seemingly no pathway, or tubes, where you get sucked up into Muslim or Hindu heaven. I got the same answer from a priest in a church and a temple *"God is all knowing and he knows who you are. This will never happen."* But that never answered my question in totality. What if you land up in the wrong heaven for eternity?

Don't ban curiosity, just as we must not ban books on White atrocities on blacks across US history, or cancel black AP classes in Florida because it does not suite my political agenda. What I'm trying to say is, as we get into this new digital world, each one of us will need to start asking more new questions; why am I, as a Christian fighting a war against the Muslims even when they are humans; or why I as a Muslim am I being instructed to kill non-Muslims as my stairway to go to heaven.

These questions are very important; it's a true path to getting facts, and this knowledge will percolate and spread our awareness. But only if we dare.

Currently, in the US abortion is a critical humanity, freedom, and religious issue being propagated by religious politicians, who will walk from the White House across the street to get a photo opportunity in front of the famous church. So, the question becomes *"Why is that people who normally never go to the church, suddenly start carrying a bible before elections, or for photo opportunities."* The only reason is that a lot of these mega-churches are mega vote banks. There are a lot of votes attached to displaying that religious book. A $860 million worth of donors who are loyal tribal members that can be swayed for their votes in small, medium and mega churches in the US. Ever since the ancient time of Egypt, it was the priests that could confirm a pharaoh by the sheer power they held. No leader on the planet since then, up until now can survive an anti-religion stance, not even the king of Saudi Arabia who is the keeper of the two holy harams in Saudi Arabia. Exceptions are the communist countries but even there they will not taunt religion openly, and communism is just another form of religion today. Putin still seeks the blessings of the Church, and Xi that of the communist manifesto. Right through the ages there are many examples of religion taking political sides, the most important of them being when Pope Clement V in connivance with King Philip of France in 1307, arrested Jacques de Molay on fake charges of heresy, sacrilege and satanism. Ironically, Jacque was the Knight Templar who fought for Christians, and he was burnt at the stake by the pope who is the head of Christianity. It was all done so that a bankrupt King Philip could get access to the banking wealth of the Templars. Such political actions have plagued the houses of religious leaders from time immemorial.

On a PEH scale I have not heard any religious leader say anything good about saving our planet. So, they collectively get an D. Neither have I heard any messaging on the environment another C. On the score for humanity, they score a F. During Covid-19 some churches in the US backed the optics that the blood of Christ will protect the parish and followers of Christ did not need to wear masks. Collectively almost every religion announced that believers in their gods would be protected from covid.

Politics

Politics is a mixture of everything; it is a mixture of funding, supporting, and having the power of allowing companies like Cambridge Analytica to drain and brainwash their citizens and make them join certain tribes.

Politics must keep, at the center of its universe, the happiness of citizens never the total oppression. This must become their unwavering Political Hippocratic Oath, to directly or indirectly do no harm to any citizen and keep the health and happiness of the citizens at the heart of the global political responsibilities. Elected leaders must be given no option of making their nation into an autocracy. The political standards and structures must audit and track that elected officials do no harm to nation, state or its citizens. The law must treat the nation, state and citizens as patients that need care. Elected officials must not be allowed to lie, spread fake opinions, polarize citizens for political and/or party reasons, nor create situations where citizens fight other citizens based on false trumped up, and false reasons. Autocratic rulers must not be allowed to decide when and with whom to go to war without a thorough approval from the citizens – only if their nation is threatened and as a form of defense. However, now in this new digital world they have the power and resources to digitally mine and influence the minds of their citizens and polarize parties' loyalists by a constant barrage of communications. Digital assets allow governments to monitor citizens as never before and the autocratic nations do this with impunity.

Politicians also had a deep impact on the number of people who died during the Covid crisis. At the top of the score are countries like China and India with very large populations. At the very bottom are countries like US and UK who had all the resources but their health care messages were polarized by their unqualified political leaders.

Here is a personal scorecard that I was distributing to a core group across the planet in countries in the chart from January 2022

The Chart on the next page makes some interesting data-driven KPIs or Key Performance Indicators. It measures the effectiveness of political decisions by a nation and the effect it has on the lives of its citizens. It works like this. [1] The global confirmation of COVID cases was at a ratio of 40,662 cases per 1 million. The average global deaths were at a ratio of 707 per million. Now, what the chart visualizes is we measure each country against the global average.

The expectation is that developed countries with funds, vaccines, and development capabilities should normally have the lowest infection, and death rates. This is because they have all the necessary infrastructure to control the pandemic better than the global average. While, poor countries without funds, infrastructure or vaccines would suffer the most. This includes the highest population countries like China, and India where infections, and deaths should have been the highest. China, because the pandemic reportedly started from there, and India because of its density. So, when reading the chart, we watch for countries where infections, and deaths, are below the global average, thus determining they did something better. While countries above the global average did not fare well. The reason we're attributing to this is politics, i.e., the will and power of the national leaders to implement a fact-based scientific program in a pandemic situation. However, as we studied our data there were two absolute outlier nations.

The first was China, it reached 82,000 infections by March 2020, and then kind of stopped dead, no pun intended, right at that number. It is our belief that China lied through the pandemic for some reason. Shockingly, but not surprisingly, the country with a dismal 'F' turns out to be the US. Where because of the way the US took their covid decisions it resulted in above average deaths due to bad policy decisions. In the chart above if you have a bar to the left of the center then you nation did badly, or dismally. For example, UK did badly and US did dismally. In the **UK**, based on their population, and the global average only 47,136 people should have died, however due to bad policies 103,918 people died. So, the policies of Boris Johnson directly resulted in the deaths of 56,782 people who should have been alive today. In the **US** 611,156 deaths occurred that were well above the world average, for its population. Even by world averages only 231,844 people should have died in the US. So, the divisionary policies of Donald Trump, the support of GOP and the insistence of the Republicans, directly resulted in the deaths of 379,312 people who should have been alive today. **Brazil** comes next where due to bad policies 370,251 people died who should have been alive today. The best performing two nations were **China** with their forced 'Zero Tolerance' orders that only China could deploy, and what is believed as fake reporting. They put the entire nation under total lockdown. The next is **India** where the citizens followed the instruction of the government by going into a national voluntary lockdown and there was zero political polarization of the science of staying alive during covid. What Dr. Randeep Guleria, our Dr. Fauci in India, said from Delhi was obeyed voluntarily by all across India. Thus, in India 485,000 additional citizens are alive today than the global average. This was accomplished due to the policies of Modi, the national adherence to the recommendations of Dr. Guleria, and the voluntary single-minded national belief that they were both working for the good of the nation. By Global averages 964,956 people should have died in India, yet we had only 479,956 deaths. That is 485,000 people saved who should have been dead today by global averages.

Our politicians control the process of making and implementing rules. In the US they can appoint and/or fire judges on their whims and fancies. If a Supreme court judge dies, they can choose the new judges. Politicians control things that could make the planet safer in the future. They can also take decisions that can take humanity toward a cliff of total extermination. Our politicians are responsible for supporting our policies, learning, abortions, law & order, police, our judges, our intelligence organizations, supporting the LGBTQ movement, building restrooms that all genders can use, and allowing trans to get married. Our politicians are also responsible for making many critical decisions. The appointment of republican judges and then prompting a fast-track supreme court decision on laws made on sex and abortion are the most recent examples of their indirect power. In the United States of America, the politicians, the state, and religion were supposed to remain separated, but across the years, there is 'beyond reasonable doubt' evidence that they are united, and often found sleeping on the same bed.

Both parties are profiteering from each other. Just as an example there was an evangelical mega church in Houston where they had reported to the police that their cash collected in one single day, went missing. In this case one single day's collection was stolen. Seven years later, a plumber found the cash, and it was $6.5 million in cash, hidden behind a bathroom faucet. This is a single day's donations collected by this one single evangelical megachurch.

When a preacher stands on a pulpit and says to vote for XYZ, their parish bow the head and votes for that person because religious tribal member have extremely high loyalty to their church and preacher. The attendees to these religious houses are made to believe that these instructions are from God itself, and follow it as a form of good religious, tribal member participation. Now, if you are a political aspirant how much would you be willing to pay that church to have them tell their parish to vote for your party, or recommended person. In the US this number goes to millions of dollars.

On the positive side, a president can also change national moonshots and positive goal setting, like when in 1961 Kennedy announced that before the end of the decade US would have a man on the moon. Or it can be a president that tells a nation not to wear covid masks as proof that they belong to a party, resulting in 379,312 additional people die that could have been alive today.

The universal health system is proof of this caste system's power in our country. The murder of George Floyd by a White policeman is proof of the existence of a caste system in government bodies where just one single police officer on the road could become the judge, jury, and executioner, and kill a human. And for a White police officer, it does not matter who the human being is, whether he was a murderer or a drug addict. They execute them, and it is brutal. No police person should have the right in a democratic society to execute an innocent person unless they have evidence against that person's intent to harm, or try to kill them, i.e., as a last resort act of defense.

On the PEH scorecard, politicians all over the world are becoming more and more personally power-hungry. It no longer matters if the country is communist, a socialist, an Islamic kingdom, a democracy or a republic there is one common pattern evolving across the planet. Polarize the people, and harness the tribal power. On the P score almost all countries, including the US, score a dismal D. On the E score there are peaks and troughs. Countries like Bhutan fly to the top with a positive carbon footprint score, while countries like the US, under Trump, actually walked out of the Paris Accord — a collective that is trying to take remedial action on strategic climate mitigation decisions. Globally in the US were at a C score, then the US went into a F score under trump by withdrawing from all scientific advice and projections. Now the US has moved from an F under Trump to a D under Biden. The reason for this is because despite Biden supporting the Paris initiatives Republicans will not support anything he does. So, half the country is against the D score for climate initiatives. On the H, for humanity, score politicians worldwide are on a global path of D to F.

More and more leaders are caring less and less about the people of their nation. More and more presidents are trying to get elected as lifelong emperors under the garb of strong-arm voting practices resulting in getting votes for all their decisions. More and more leaders today want to become lifelong emperors in their nations, including Trump in the US. Presidents are trying to become de facto rulers by changing laws, autocratic officials are routinely supporting their autocratic leaders because when these leaders win, they have an assured seat on the table. Very few leaders care about their people anymore, their prime focus is to remain in power. Power is intoxicating and absolute power is a drug addiction bar none. A lot of these leaders are using their wealth and power, along with the newfound digital assistants to not only stay in power but to also weaponize their tribal followers with mind bending psychiatric digital techniques.

The Big Moon Shots

Just like on May 25th, 1961 President John F. Kennedy announced his moon shot, i.e., to send an American to the moon and safely bring him back in the next ten years. This gave the Americans a cohesive vision and with it all of America lost all distractions and delivered to that goal on July 20ᵗʰ, 1969 with the words "The eagle has landed."

In this same manner, there are some noteworthy moon shots that individuals, companies, and nations are making today and focusing on their passion for making the reality of the future they dream of. Here are a few to keep a close watch on:

1.   **The FTX Faultline:** In mid-November 2022 FTX a company run by Sam Bankman crashes, and there's anywhere from $10 billion to $50 billion of debt that is outstanding. FTX, along with SBF created a host of shell companies, around 158 as of Nov 19ᵗʰ. FTX tried to become the 'Federal Reserve Bank' for crypto and got themselves some very high-profile advertisers. The model was very sound but the intent of fraud prevailed. The same thing happened with Theranos — a great idea, lots of powerful people on the board, lots of false promises, and totally built on fraud. Elizabeth Holmes the CEO was once touted by Forbes as the richest woman entrepreneur and today sits in prison. To cut the story short the US govt needs to create a *Federal Digital Reserve Committee*, under which a sub section should be *Federal Crypto Committee*. The agency must first define national rules that become the global 'golden client' on how to control digital bad players and trolls. California can once again become the global leader in these initiatives as it has some of the world's strongest talent in this area. The second would focus totally on legalizing the $50 billion opportunity that FTX had exposed. Very much like Steve jobs had taken an outrageously successful illegal success of 'Napster' and made it into a legal global music sharing company. The iTunes brand totally changed the music industry by altering the supply-chain of traditional music to customers. Music changed from a middle man deciding what music they could sell to you → to a direct relationship between fan and artist. All the government has to do is keep the bad political and digital players out of these landmark decisions. Which by the way is easier said than done.

2.   **Global Brain Rot:** I'll start with our current greatest danger — global Brain Rot. This is not a light matter but a very serious concern. **On one side** of this equation is AI scraping away our natural thinking. For example, many years ago if I traveled on any road, I could remember my directions even if I went back after 10 years. Then after using Google Maps for over 10 years my navigation skills have gone down the drain. A predictable fear is that as our devices get smarter, we humans will proportionally get dumber if we don't work against the rot. On the **other side** is our core topic of this book. We are surrounded by both good and bad news, however being human we are attracted to bad news. Our trolls, algorithms, smart devices, social networks and news channel know this fact. We are surrounded by bad news distractions like China committing genocide on Uighurs and Tibetans, Russia attacking Ukraine, creating global supply chain

breakdowns. The US, has taken a sudden religion-based decision toward a religious + political ban of abortions by revoking 'Roe vs. Wade' to add to all this uncertainty. The result of this is total disillusionment, religious fervor, polarization inflation, combined with wholesale grocery and gas prices going up has amplified crime across the US. Unethical gun laws, has a direct result on mass shootings, shootings at schools and out in the street, and ensures that crime increases as tolerances decrease, while the tools for a disagreement escalate from verbal, to physical and now a shootout. Murder and assassinations are the ultimate expressions of disagreement. The US, with its legacy *guns-rights*, the constitutional 2nd amendment, followed by Republicans, funded by their aggressive NRA sellers, has unfortunately empowered these opposite opinions with guns-for-free laws. With this, murder is an inevitable and highly predictable outcome. All of this leads to a state of constant uncertainty. Add to all this the extreme polarization, supported by most foreign and local politician, and trolling factories, out there trying to not only grab your eyeballs, but also your mind. In a democracy, these buyers collect their funds from you and I, and then pay the trolling industry to hijack, your and my brains by loading fake-facts, images, and emotions that are mostly forged, and designed to alter our thinking. When our environment changes faster than we can adapt, when one-in-1,000-year storms, happen year after year, when bigger floods destroy our homes, when the republicans say this has nothing to do with global warming, or the climate, then tribal people tend to hang on to every message as a form of their final straw — and therein lie the seeds of autocracy. If history has taught us any lessons it is that autocrats start their autocracy by first mushing the minds of the citizens into a state of mistrust of established pillars and of each other. Our greatest danger that we need to protect ourselves from is *Structured Brain Rot*. Unfortunately, a lot of the buyers of these services are our own political parties and leaders. Take the recent example of Fox News in the US that paid Dominion $787.5 million as a bribe, just to avoid Murdock, and their news anchors, from having to testify that paid opinion business model. They also terminated some news anchors who had seemingly followed instructions from the top, as a news and media providers. This is very unfortunate as the very people who are supposed to protect us, from lies, are the ones paid to disburse these on a day-to-day frequency. They are propagating and broadcasting pre-planned lies, that are truly detrimental to "*We the People.*" Now as a digital economy, and for transparency, we need to put into place 'rules & regulations' to stop this globally. Stop giving media rights to companies that are not media, stop elected officials of continuing if their only interest is party and their pockets, stop all priests, especially Catholic priests, from sexually abusing children as a planned process of sexual gratification. Stop giving Churches, where there is evidence of sexual abuse coverups, any and all tax reliefs. We need to stop these people in their track. We need to create our world of, Zero exceptions, zero excuses, and zero protection by higher ups.

3.   **Artificial Intelligence:** At the top of the pile is AI. Despite a lot of futile discussions on what is artificial and what intelligence — this is a field with phenomenal potentials. AI is our modern-day beauty and the beast put together. Though this sounds poetic it is true because AI is just a code. It depends who wrote that code and for what purpose. Unfortunately, it has been proven beyond doubt that the biases of the authors of the codes delve into the code and the AI decisions. The world as a whole has very little idea on how far this had evolved and how much further this singular method of training technology to emulate a human ability to solve problems on their own. We as a planet need to ensure that AI does not take the US approach of patents and profits, but an approach of knowledge for all. AI must not be allowed to be created behind closed doors, trained for misuse by autocratic leaders and governments, or weaponized for personal, enterprise, party or national gains at the detriment of another human. AI needs to be regulated by strict universal laws and conventions that need to be put into place with a global Digital protocol — probably as of now under some new global control mechanism. It could start under the UN in the meanwhile. These laws cannot be defined and controlled by the winners or the powerful, but must be democratized for all. At the start countries like the US and other democracies need to make sure AI is aligned to 'saving humanity' as a solution for de-risking the planetary future. Its potentials range from the fear of Terminator and AI taking over humanity, to AI solving so much that most knowledge goes out of human control as it is already starting to happen. AI is already driving fully autonomous cars by the 2023s, they will solve problems millions of times faster as their rate of absorption accelerates. They will become the google to creation, and inventions, and soon will go beyond simply asking questions.

4.   Already AI is helping humanity to catch bad actors, fanatics and terrorists. It is used to help us fly planes, and drones, and can now help us drive cars without any driver inputs. AI is catching fraud and helping medical services and increasing the health of people. ML can do everything that is rule based today, and AI is already complimenting non rule based intelligent tasks too. Like make me a painting in Van Gough style of my portrait — look up openAI.com/dall-e-2/. Dal-e is an AI artist that can paint by you inputting in just words, like draw me a beagle in Van Gogh style. This is public accessed AI at its current best. With the launch for subjects like Fufactology, which is totally driven by very large datasets, merged with our AI. Together, they will be able to correct political, and climate decisions by simply predicting the impact of political

decision by degrees of accuracy. Already AI is keeping people healthier, predicting when their sugar level has gone high, driving cars on public roads, taking space probes like DART to crash into an asteroid at 14,000 miles per hour, i.e., Didymos is an asteroid 2,500 feet across, and 6.8 million miles away. This feat, while impossible for direct human control, was accomplished by AI's accuracy and trajectory calculations. Your opportunity: right now, a group of friends can get together, learn a basic language like Python and in a month or two and start group-coding, and leveraging its predictive libraries to start by answering some basic questions they have. This message is for 13 to 14-year-olds — both girls and boys. AI is a massive fun game, opportunity and a big future requirement. Your starting point will be to identify a data generator, or get a few sensors that will generate your data and you're ready to run. In this process you will learn to code your algorithms, if some code is repetitive create functions that you can call, and kind of build your own bots that will become your slaves and perform tasks you program them to do.

5.   **ML:** or Machine learning is the next most important area. When your AI is programmed to finding patterns, we need to feed it with data. This is when the AI rubber, hits the data road. Initially things will not work as planned, but slowly as you fix each issue it will be an iterative journey to success. Basically, the more data we have the better our ML and AI will be able to respond and react. Whereas AI is deductive, ML is more the study of patterns and imbedded rules. In rotary machines, like motors, using ML we can train our sensors to predict the breakdown of a bearing, or another part, when the audio and vibration threshold reaches a certain level. Using AI, the car can recommend making a service reservation before the damage becomes catastrophic. ML is basically our algorithms learning from data and thereby predicting from rule-based patterns.

6.   **The Human App:** we have all been infatuated with apps on our phones and laptops. From Angry birds to Pandora, from Uber to Airbnb but the future apps will change humanity. These are the human-machine interface apps. There is a case of a person born color blind who used a human interface to hear color. This talk is available on ted Talk as I listen to color. The future is ripe for human-machine interfaces. Elon Musk and his new initiative Neuralink hopes to help handicapped people by direct neurological links. Neuralink just got FDA approval This technology will be used, it's just got the FDA approval for human trials on May 25th, 2023. Neuralink is a mind-computer interface, and next steps will begin to help handicapped people from undertaking physical tasks simply by thinking. Add to this, the dimensions of AI, ML and Human Apps and we have a world where a cancer patient can simply wish the AI to build an anti-cancer drug that is designed from scratch for their specific type of cancer and genetic framework. Or, order a scaffolding for a body part using their

own genes, blood, and organ samples. Something that used to cost billions of dollars for a generic drug will now be possible in a matter of a few months or days. The world of opportunities is brilliantly heavenly — but only if we can collectively keep the bad guys out of the fence. The goal should be to prevent bad thoughts and not segregate these people. It should also be to not let these technologies sit in the hands of profit-making enterprises, which make them available on to the very rich. The aim should be to stop people and processes that encourage any bad actions, just like US having 25,000 nuclear bombs when there are only 2,900 cities to destroy, or Russia trying to test their weapons and muscles by attacking a peaceful neighbor on one pretext or the other.

7.   **Extreme Polarization:** But more important is the polarization of different tribes at a global scale. As a rule, when we hear Russian alliance nations we see support of Russian actions, whereas if countries are aligned to the US, we see an anti-Russia feeling. Both are wrong. There cannot be two versions of the truth and we need to create a future where powerful nations and people cannot influence or change the conscious, and subconscious thinking of individuals as a national level by reaching them on a 1:1 individual level of neuro propaganda full of things the individual will react to at a personal level. We need to also build a world where rich nations are not allowed to honey-trap leaders, nor use debt financing as a method of control.

Conclusion

The global Covid experience has altered the thinking of humanity. These two years of total isolation, polarized communications have created a state of unprecedented volatility across the planet. As I live in the US I'll use the US as my example, but this phenomenon is quite uniform across the planet.

Let's take US, where trust is at an all-time low, split into hyper polarized political view that has digitally, and via social media, been thrust deep into the psyche of all citizens. Etiquette and social tolerance seem to have now become a pre-2020 phenomenon. Micro levels of hatred combined with macro level of digital tribal followings have altered the American essence. Intolerance to alternative views, and targeted hatred for fellow citizens results in daily violence. In an environment of daily gun-related killings, mass shootings of children in schools followed by political denial that it ever happened, total disregard for the opposite political party, and an elected cadre that flaunts decency and replaces it with blatant lies has resulted in animosity at a national level. Anger has flared as people have forgotten simple social skills due to over exposures to tribal negativity in social networks, and smart device enabled dumb communications.

Now that we are here, we still need to define a process to try and answer the fundamental questions that beg answers, "Who will fight the common person's right for equality in this digital transformation?" Today we can connect to people living on the opposite side of the planet in seconds, for free. This is the time when we have to stop using our political party, politics, religion, and our individual gods as the architects of our 'Us vs. Them' planet, and stop believing that our country, politics, religion, god is superior to all others. Instead, we need to think that if my leader is not working for the interest of the people, their wishes, and needs we must use our digital people power to install somebody new who will.

Unfortunately, and I need to apologize for this, some of our greatest barriers are the ones we are most passionate about. While historically country, politics, religion and god's have been the greatest benefactors of taking humanity from nomads to who we are today. But, over time the rich and powerful have found ways to misuse these great assets for their personal benefit. Three examples of this power centralization are Putin in Russia, Xi in China and Khamenei of Iran. Their misuse of these very structured power baselines will take humanity down an undesirable path.

An Exit Dive in Politics

Post the mug shot of Donald Trum we are seeing an explosion of pride in being a criminal. We are seeing a rise in elected officials, and now republicans taking their mug shots with the Fulton County Jail logo as their proud social image. Is this a this is a direct acceptance of 'Brothers in Crime' or the social tattoo of the new US mafia that will stick together even with certified criminals. Does that make the party members attracted to crime and mug shots. No matter how we look at our global experiments in politics, we have collectively failed humanity. Capitalism has failed in the US because in the end we have today polarized a nation into two distinct enmity camps. Each president seems to be worse than the last. Autocracy seems to be around the bend unless 'we the people' take this into our own hands and somehow change it. The political experiment has failed in Russia and China too, as it produced Putin and Xi. It is a disaster in North Korea, Iran and many other nations. Our future is still unfolding but the promise of capitalism, and democracy, is either failing or possibly dead. The rich are becoming richer, national, and global, wealth is being spread amongst a few, and they are rigging the system between themselves to tax the doers and find ways to not pay taxes for their own kind. Rich is a universal class on planet earth with different set of rules in every nation, except China and Russia, where they seemingly execute them. According to the NY Times Donald Trump, our ultra-rich billionaire ex-president has paid no federal taxes, or probably income taxes, for 11 of the past 18 years through 2017, and then paid only around $1,750 in 2017. I challenge any other common citizen to try missing their taxes even for a year. So forget the law, some people in the US are even above the IRS rules and regulations, or so it seems.

Capitalism, which is the bedrock of the US economy and society hasn't worked out for most people. Right after the first and second world wars, the promise of the Marxist and Leninism in Russia and China, and the capitalistic republic in America had one common goal. To give equal opportunities and spread the national wealth among all the citizens. After the second world war it looked like it was all working across Russia, China, and the US. Technology had arrived, nuclear families were traveling across the nations, and it seemed better days were ahead.

What we have to realize is that no matter how great the Marxist, Leninist, or the US Republic definitions were, in the long run, it is human greed and the influence of the rich and powerful that will subvert even the greatest of ideas. Let's focus on the US. By the constitution, we promised Equality for all the people and a country run by *We the People.* So, what happened along the way.

After the second world war, people in the US felt a national sense of solidarity and common shared goals. They established an 8% tax and the rich embraced it without hesitation. This is around the time when a growing number of people used education to create a new class of knowledge workers — creating a tribe that carried a new sense of pride in their work and not the nation. Slowly this new class became the bureaucrats that taught politicians how they could take a higher and higher share from people's hard earnings in the garb of taxes, props, and mandatory deductions. They also formed elaborate establishments designed to punish non-taxpayers and extract more and more money into governmental funds. Slowly people lost faith in their democracy and fewer people came out to vote. Donald Trump took advantage of this and activated the 33.4% of the very dissatisfied Americans with promises of equality, and encouraged them to vote. So, what we see is capitalism today benefiting the few at the top based on votes from the most disgruntled, who became the most loyal, and who come out in masses to vote while designing new ways to prevent their competition from either getting to vote or being able to get to a voting booth.

Rather than creating a highly harmonized nation, we went and created possibly the most polarized people since the birth of the United States. Now, we need to reignite a sense of nationalistic patriotism, a willingness to support all citizens, and encourage the spread of national wealth that is built on the backs of hard workers. No human needs 7 hundred billion or a trillion in their bank to survive. There must be a minimum tax that all rich people must pay, no matter how their tax lawyers spin their finances. As someone once said, the truth is between popularism and ideology because *'popularism offers a headless heart, while ideology offers a heartless head.'* According to the Russians, capitalist countries like the US unfairly allow a few people to amass large properties and wealth. Yet, they avoid talking that people like Putin might well be the richest person on earth, because he can imprison, and lock successful business owners, and inherit their businesses. His oligarchs are, unofficially, some of the richest billionaires too. He can lock and assassinate any competitor and take what they please. Kim of North Korea operates at a different dynastic level altogether. In China Xi and his appointed loyalists are no different from the super wealthy owners and power enablers.

Moving forward we need to move toward a French, German socialist democracy, or a nation that puts data-driven pragmatic policies that foster a true national patriotic community. The Jan 6th coup on the US capitol must never be allowed to happen ever again, anywhere. We need to collectively design a nation where rules are common and citizens of all economic, color, or sex are assured social respect. Our future design cannot be one designed by Bernie Sanders, nor one forced by Donald Trump. Reality is somewhere in between. Neither too left nor too right.

In the 1970's Nobel laureate Milton Friedman postulated that a company's only purpose is to maximize shareholder profits. This idea has grown and has been amplified in the western economies separating the workers and products from the profit seekers and board members. Both the customer, and the workers, suddenly becomes irrelevant, as does the nation, the environment, and the planet.

In recent years lower income children of less educated families have been increasingly falling through the cracks. In the US fifty percent of the lower income children have been raised by single parents or no parents at all.

Going back in history to as far as we can track, the modern societies, kings, counts and rulers have kept their coffers full only by the simple act of forced tax collections. The fairer the ruler the fairer their taxes, the more avaricious the ruler the higher they made their tax percentage and the harsher process of forced tax collection. Then rulers found that they could collect more money by conquest of their neighbors, then taxing their citizens, and thus become richer. Finally, this resulted in the anarchy of colonization where the Whites subjugated people, and their lands, into a state of oppression so they could collect taxes beyond their borders. Tax is the systematic way to collect money from some other persons earnings, work and property. Let's look at Taxes in the US, which is supposed to be the greatest equalizer and spreader of wealth. It is where each citizen pays a little from their earnings for the benefit of the nation and the work to be undertaken therein. But unfortunately, we find that the ultra-rich across the planet, and mainly in the US, once again game the system, and keep making additional laws that only benefit them, the ultra-rich. Historically, in 1862 Abraham Lincoln imposed a flat 3% tax on anyone earning over $300, this elapsed in ten years. Then in 1894, the Wilson-Gorman tariff act was passed and a 2% tax is enacted for all earners. However, in 1895 the US Supreme Court rules that income tax is unconstitutional. In 1913 Delaware, New Mexico, and Wyoming cast votes on the 16th amendment establishing Congress' right to collect federal taxes. From 1935 to 1942 any American making more than $ 5 million was taxed up to 75%. The US taxes were highest during the 1944-45 period when any person making more than $200,000, this is equivalent to making $2.4 million in 2009, had a 94% rate applied to them.

The rich were probably very disturbed by this and made it their business to pay lesser taxes. In 1954 financial experts were hired to draft an 875-page Internal Revenue Code document, making around 3,000 changes to the federal income tax system, they realized the common man never understood what these pages contained and that the government could have an endless flow of money from the people. Since then, the politicians never looked back after this on how they could get more and more money from their citizens, including how they could give higher and higher breaks to the rich, who funded their victories. Next in 1986, President Ronald Regan passed a tax reform reducing the maximum income tax rate. He also passed a tax cut applicable to the very rich, reducing it from 50% to 28%. The rich and powerful applauded Regan as the greatest president. In 2001 George W. Bush gives tax relief which includes a new 10% tax rate for low-income earners. Finally, in 2017, President Donald Trump, signs into law the Tax Cuts and Jobs Act which permanently reduced the corporate tax rate, from 35% to 21%, only applicable for the rich, and powerful, and simultaneously limiting deductions for state and local taxes which hit the middle class most of all. What we see across the timelines, is the unhindered efforts by politicians to keep giving breaks to their rich supporters while increasing the burdens on the middle and upper middle class.

Take Aways:

Individual: The individual is the hub, the active atom, and the center of the universe of every digital transformation. The individual is also the target. We have to beware of all the digital trolls and the powerful. Watch out from the triad digital opportunists. Protect from 'Us vs. Them' broadcasts of politics, powerful leaders and religions. In the new digital world, we must separate out thoughts from the influences of the paid digital media news and work on data driven truth. However, finding the truth is getting harder and harder as the data streams get loaded with alternative realities. Each individual must take responsibility to find their true north and the truth. Watch out for free-ware that collects your data. Fight for data privacy.

Companies: *Companies world over are in a state of digital euphoria and shock. Euphoria because of the great options and digital services they have available and shock because of the new AI developments in 2022-23. That are right now but beginning. The IT segment that was the growth segment is right now sitting in a state of disarray. However, the message is still clear on the wall, the future is with the digital disrupters, and digital technology will only become more pervasive and is not going anywhere. The global enterprise IT is recovering from the Covid-19 rush as the world starts to seemingly get back to normalcy, which it never will. The post covid era is never going to go back to pre-covid.*

Nations: *Globally, the political systems of the planet are at a cusp and a global audit. Democracy is sitting at a cliff with the Trump era in the US, and how democracy is being hijacked by countries like Argentina, Pakistan and North Korea. Democracy needs to be saved and it is in the hands of We the People, i.e., the voters, to resuscitate it back to life. Presidents the world over are going from bad to worse. In the US, we thought Bush was bad but when Trump came Bush became not a bad president. Now in 2024, we see the likes of DeSantis and by the writings on the wall it looks like we'll look back and say that Trump was not that bad. This trajectory has only one direction on its current path – anarchy and an authoritarian leader. We need to take our atomic particle of the individual and vote the world we want to create. We need to separate our mind from the constant barrage of polarized viewpoints, and lead our countries in the right direction.*

The 'P' factor: *Our greatest risk of extinction is not an asteroid or the sun expanding; it's not even the climate — it's us, the People, in this P. Our greatest risk is that either we will vote, or some leader will forcibly hold that position and basically start a nuclear war. The biggest risk to humanity is us fellow humans. We are also the brightest hope. In the US and India, at least, we can vote the bad people out. A majority of the nations do not have that luxury. However, despite this privilege, we are allowing our political parties to divide us into warring tribes. In 2020 we were in a situation of total denial of anything the last president had done. By 2023 we still have around 32% of the people who still believe he is the legal president, despite all the proof and data to the contrary. In addition, we were made to doubt anything the experts told us with a message that we should not trust the experts. In the US we had 50% of the people who stopped wearing masks. This directly resulted in the most inefficient health scores, and the highest death scores per million of any nation on the planet. The current times are good. The future is in our hands so let's individually take the right decisions for our collective future.*

A very takeaway for 2024-2030:

1.   *Individual:* [1] *Stop watching News Channels; they only work to amplify bad news beyond reasonable proportions, and influence your mind. Especially never get hooked to one channel. [2] Stop all social networks and get your life back. [3] People in the US have lost trust in their government, politics, and their elected officials. The US is creating ripple effects across the planet. First save yourself, the rest will automatically fall in place once you get your mind back.*

2.   *Enterprise:* [1] *Invest in PEH-filtered decisions. [2] Customers no longer care what you manufacture or do or how big you are. They care about what priorities companies have on the PEH scores.*

3.   *Nations:* [1] *Build national PEH scorecards rolling down to states and every elected official. [2] Bring back Trust. Political parties, the world over have been directly responsible for the greatest loss of the last decade — TRUST. [3] Bring back manufacturing. Elon Musk has proven that great things, like cars, can be manufactured in the most expensive real estate in the US and still succeed. [4] Don't fund any known enemy. Chinese-made goods are out of fashion. Russia is trying to recover as a true autocracy.*

4.   *Religion:* [1] *Demand total transparency. [2] If religious houses continue to become fanatics and terrorists, then they plan for losing followers. [3] Christian following went down with the exposure of pedophilia in catholic churches. The new generation stopped attending the mega-churches. In the Middle East, Iran finally overthrew the Islamic leadership that had been oppressing women. This will be the third successful revolution in Iran.*

Chapter 18 — Clearing the Stormy Haze: Risking Small Predictions

The opportunity for every reader is a sense of what may happen in the future, and whether you're investing your personal, or business plans to align with one of these probabilities. Or, of course, go create a brand-new future that we can't even imagine here- I bet there are thousands, if not millions of those already happening out somewhere. In 1994, no one could have predicted Amazon the bookseller. In 2003, no one could have predicted Facebook. So, you go and disrupt.

"The most reliable way to predict the future is to create it." — **Abraham Lincoln**

"The best way to harness the future is to learn from it, and then alter our current habits. Learning from the future is a critical science that will save humanity." Fundamentals of Fufactology®

"The Decision to Try, is the most important decision each individual will to undertake in order to eradicate the digital cancer" author

Prediction is a game in which most people either get it wrong, or totally wrong. However, I'd still like to play this game and extrapolate my current experience and knowledge as of the end of July 2023, nd the play with my sci-fi addicted mind in a controlled manner.

A. 2023-25 Immediate Predictions

A.0 Data & Analytics (my professional area): Datawarehouse will be replaced by edge computing and AI-driven in situ analytics. Chat GPT, like secure server services, will change management decisions from the traditional route of Reports to Analytics to Informatics, with Google-like Q&A. The 2023 dilemma of 75% of Data & Analytics projects failing will become a waste of the past. This was driven by IT leads thinking that tech had all the answers (something I wrote in detail in a book in 2011 (BI Valuenomics — The Story of Meeting Business Intelligence in BI). Secure LLMs inside enterprise firewalls will change the paradigm of decisions.

A.1 Power of the individual: Individuals will become the core drivers in this new digital world. The world has two types of people, the happy and the unhappy. You can put a happy person in any situation, and they will adapt and remain happy. While you can put an unhappy person in any situation, and they will blame someone else for their state of unhappiness. This is the time you, as an individual, can choose the life you want and succeed as a core team. Use this time carefully, and own your future.

A.2 AI: will continue to impact and create tsunami-like waves across people, enterprises, and services. Neuralink has got approval from the FDA for human trials and expects some major breakthroughs in this area, from helping the handicapped to strengthening the minds of soldiers to catching a criminal before they can hurt someone. AI will affect every industry and person but will be most visible in healthcare, transportation, and customer services. With each passing day, AI will become more sophisticated and advanced; it will be capable of leading the 1:1 breakthrough in scientific research via automation, something that we have just started dreaming of recently.

A.3 Some things that will not: Actors & Screenwriters strike: Today is July 16th and the actors and screen writer's guild are all on strike fighting the intrusion of new AI driven digital technologies. I predict this will hurt them more. This is reminiscent of the French taxi drivers going on strike, and burning cars on the streets. Uber never went anywhere. AI and Digital disruption are here to stay. The winners will adapt faster and become the new disrupters. **Flying Cars:** Will get approved by a few accidents of the very rich will put the fear of flying into humanity. They will be available but not become mainstream. It will remain a tourist attraction in holiday resorts and safari tours.

A.3 Dignity of Animals: The world will get together, starting in California, and join hands for the *Dignity of Animals*. I used to believe that working in Saudi Arabia was like living on another planet. But I now believe that living in India is like living on another planet. One fact — when we go to a county or street fair in any nation on the planet, it is full of animal foods. However, it is only in India where one can find stall after stall, mainly of very tasty 100% vegetarian selections.

A.4 Climate change acceptance: The global climate change deniers will head back into their woodwork. Trump, Republicans, and the US O&G industry will no longer be able to hide behind fake scientists. People will start to feel the wrath of planet Earth as we warm up and will not accept governmental excuses or lies. Insurance companies will get out of states where their predictions indicate a catastrophic future. Alon with these, sustainable practices will become more mainstream, as will a societal effort toward carbon-neutral existence.

A.5 Personalized Drugs: For thousands of years, we all get general medicines. When a child has a rare disease, the pharmaceutical companies do not allocate any resources because they are in the business of making money. Today, developing a new drug can cost $1 to $2 billion and take anywhere from 5-10 years. Even after that, it is still a wide-brush medicine. We are now entering the age of personalized medicinal breakthroughs. ML and AI are at the cusp of creating tailored drugs for a specific ailment/patient. Very soon, this will be available for a single patient with the rarest of rare diseases. With Quantum computing, this could be developed and approved in a matter of months, if not weeks. Also, because it is designed for a single patient, it will technically have fewer adverse effects.

A.6 Increased Space exploration: The day a private citizen entered the space race the whole barrier of space kind of dropped. Within a decade Israel, India, China, Japan all are entering the space race that was dominated by USA and Russia. India has already successfully landed on the south pole of the Moon. The next few years will witness a flurry of private and institutionalized interest in space exploration. Drone manufacturers, weapons delivery systems all will need to deploy platforms for the future, all requiring a component of space visuals. Different countries will now commence to share resources and skills to open the space programs as this is not a national requirement but one for humanity.

A shallow and arrogant dive into the future:

Citizens taking back control: It all started with the digital revolution. Democracy was supposed to take care of this requirement of giving control to citizens. But it was digitization that crossed boundaries. The recordings of George Floyd started a chapter where the common man, the lowest denominator, took control of justice for all. It then percolated to the airline industry, where lost bags became an event of the past. By 2025, airlines could not lose bags, and no more stolen items could be taken by bag handlers to their homes. This control of the base expanded, as did the confusion with the distortion of the truth by digital trolls. In 2025, the US implemented a minimum tax for all earning more than $100k at 20%, bar none. The rich became ultra-rich, but behavior patterns changed across the developed and digitally connected nations.

China has lost its global Manufacturing status: The slide began with Xi backing Russia and Russian aggression in 2022-24. Iran and Saudi Arabia suffered too as the Petro-alliance with Russia. Southeast Asia and India have gained from China's loss.

The Big California Flood of 2025-26 put central California underwater in the big natural lake. Popular roads lead to the lakefront, and getting from San Francisco to Tahoe got cut off despite record shows on the mountain tops making the flooding worse.

Self-driving cars become Common by 2025. Electric cars that drop you off, park at your command, and come to pick you up on command started in 2024 but will be more common by 2025. Automated taxis will cost less and are more dependent.

Regenerative AI: The biggest disruption started when Chat GPT-3 was introduced way back at the end of 2022. It shook the reality of the world on the capabilities of Artificial Intelligence, their facial mimicry, and their Turin test responses to questions on video. Then ChatGPT was released, and despite its name, users could only converse with it in Text. [2] AI-driven robots have become routine not only in industries but also for home use. [3] Children will hear stories and exercise with nanny robots. Security robots that will sit docile in homes and remain inanimate until any unauthorized person enters the house. It will then instantly awaken, give its warning to hold still, call the owner and/or police as instructed, display the intruder, and if necessary disable any unauthorized intruder if they pose imminent threat to a home or factory. [4] Agricultural IR, Intelligent Robotics, and drones will provide micro-accuracy watering, fertilizers, and insect elimination from crops. [5] Nano-robot bee drones will replace slave bees, pollinate crops, and take away the fear of crop failures. Natural bees will be set free to live their lives with nature as they please. [6] AI-driven medical diagnosis has become routine. Robot doctors now check biometrics simply by shaking your hands, permission-based access to your smart medical watches and devices, and by the iris scanning diagnosis to predict and dragonize potential ailments. [7] Translators: AI will soon develop languages of their own and also work as a universal Rosetta Stone. In 2023, Bard, the Google AI translator, went and learned Bengali and Indian languages all on its own, with no human training or LLMs. These language models will become universal translators from any language to any other.

Healthiest ever yet permanently Sick: Humanity will be the healthiest they have even been yet will remain sick all the time due to micro measurements and nano medical drug feeds. The world is full of all kinds of what is referred to as the ultimate apps. The human apps. These are human interface applications that are based on neuro implants and BCI technology.

Outliers

1.　　Forever 25: — **"The quest is not about extending length of life, i.e. time. It is about extending Quality of life, i.e. youth. In some cases about reversing ageing itself"** It's not about keeping a 95 year old in a semi state of coma till 102 without the person ever waking up. A rapid growth area has been eternal youth. As more and more billionaires are created, they have a passionate quest for immortality. Companies like Alcor charge $200,000 to keep humans in Liquid nitrogen, or $50,000 just to put your head into cryogenic state. This company has been quite successful in the US, and thus clones have erupted in Russia and China, who offer similar services but cheaper. Billionaires are spending more and more of their funds on technology and experimental drugs not only for the extension of life, but extending youth. Still, others are investing in keeping memories for posterity. The hope is one day, technology will come to a point where these humans can be revived and cured. There is evidence that very soon *death may become an option.* Scientists are finding ways to stop and, in some cases, reverse ageing by mitigating Mitochondrial genetic damage ranging from Reactive Oxygen species or ROS, to natural genetic erosion and damages. Telomere shortening is an active area of developments which is a death clock in our genes. Scientists now feel they can extend telomeres indefinitely and thus life itself.

2.   **True Covid-19 Analysis** — When FuFact algorithms were run against run through the global COVID-19 data, for an analysis on 'responsibility vs. deaths' the US came out the worst despite it having all the technology and funds at their disposal. It was proven beyond reasonable doubt that, between February 2020 and March 2022, the decisions of the then president, GOP and republican based recommendations there were directly responsible for 611,156 deaths. If the US president and elected officials had simply followed global recommendations then by the middle of 2022 611,156 people could have been alive today. It could have been a class action suit but then no one seemed to care.

3.   **History erosion** — As Fufactology came into the limelight, countries and political parties tried to rewrite their own politically motivated history. In China they erased the Tiananmen square incidence. In the US the Tulsa Black massacre had been erased as was slavery to a large degree. In autocratic countries rewiring history become a trend.

4.   By 2025 most gas stations had installed their own charging stations for EV's and by 2032 the entire gas station would be replaced by solar roofed charging points.

5.   By 2027 most US states had banned all semiautomatic weapons in the US. Possessing one could get one to prison. People having these weapons needed to register them and these could be inheritance. Any non-registered weapons were considered violation of public safety. Some cities had a total criminal ban on all semi-automatic weapons.

BCI or Brain Computer interface receive FDA approval for controlled applications, by 2025 they were already being used for helping handicapped people. By 2027 this is now allowed for downloading memories of elder relatives and for human though interface technology. Neuro transplants are reality for handicapped people but have also become a standard super intelligence implant for the special, and authorized, reasons. Mind control of products has become common with some high-end products and nations. China uses this technology for all the wrong reasons, with little external control.

With Neuralink® now approved by FDA in 2023 the realm of the mind has become an open territory for discoveries. Designed for good, don't be surprised if some powers use it for bad. It is in our control to democratically control its use and application. The force of masses.

Fufactology becomes a science course taught at colleges, and universities, in their CS curriculum. Fufactology is the science of learning from highly advanced predictive AI that will regeneratively improve with each deviation of their prediction. Accurate predictions will become more and more far reaching in time. Politicians will no longer be able to spread fake alternatives for FuFact. Fufactology has become an optional subject in colleges and universities, and a mandatory process driven by quantum computing in predicting solutions for the future. With the use of advanced LLM-ML and QAI, Quantum driven AI, this science of learning from the future has changed the reaction of humanity on political and other fake optics.

Digital translators: Simultaneous translators for major languages are common. Your smart glasses will instantly translate local languages on the streets and roads, and spoken languages will be simultaneously translated using high end ear plugs with built in local translators without even wi-fi connectivity.

De-extinction: In a world driven by humanity driven extinction we have commenced de-extinction. Or the science of bringing back extinct animals.

Religion: Christianity suffered its first great loss when, in the US, the younger generation fought against Catholic decrees and rules. The straw that broke the camel's back was in 2024 when the Pope refused to sign an agreement to report every pedophile priest across the globe, mainly because of their backward-looking policies. The ICJ then mandated the Pope and cardinals to report every sexual abuse case they had on record for the last 75 years. The churches in the US and Europe are faced with declining attendance, and this has made the few more aggressive in their adaptation to the new generations. Islam has had a fundamental shift, with more and more citizens getting out of the oppressive yolks of their religious leaders. However, fanaticism continues to grow, and micro tribes continue to evolve. Religion is on the decline, and the new religion HOPE, House of Planet Earth, has started getting a lot of followers. Their first version is HOPE for Earth. In 2027, Elon Musk City would create its first HOPM for Mars.

Traditional religions have bled humanity with a promise of hereinafter. On the PEH scale it does no good to anyone. HOPE on the other hand will be fully PEH compliant, and will only work for the benefit of the Planet and everything therein.

100,000 Qubit quantum computers are available for high-end computing. In 2023, we were struggling with 1,000 Qubits. Quantum computers work 100,000 to 1,000,000 times faster than the fastest computers of 2023. The two greatest competitive differentiators are in the field of Security and complex developments like designing a new drug for a single customer at acceptable prices. These computers will complete very complex calculations in a fraction of the time it takes us today. Very much like micro-segmentation changed the auto industry by 2016-17, QC will change the paradigm of calculations. [2] VR sets with 6k experience have transformed movies, immersive realities, and immersive 3D e-commerce. The US and China are the global leaders in Quantum computing, and they have begun quantum colonization. [3] All of 28 billion human genomes have been mapped. 1:1 medicine designed and manufactured by AI-driven 3D printing has made medicines very affordable for all, even for extremely rare 1:1 types of diseases.[4] Advanced Edge computing chips working on 64 Qubits of iOS and have delivered very advanced filtering capabilities and intelligence right at the edge of the digital domain.

Net Zero: Electric Cars have almost wiped-out gasoline cars from the face of the earth. We saw a big shift from fossil fuels, and the Electric factor has forced all auto manufacturers to stop gasoline cars. PEH is the main driver, and these cars will become mandatory by 2031. [2] The price of gasoline has skyrocketed, with very few enthusiastic customers left. [3] Technical Carbon sinks will become common across cities and industrial landscapes that still generate carbon. These are used to pluck the carbon produced by humanity right out of thin air. They will be scattered across the city and places that produce carbon. Electric planes became common by 2027 for short-hop flights within the US and Europe. Long flights are still run with gas turbines. Net zero homes can sustain all their energy needs without utility dependency. By 2030, 40% of office and manufacturing units have come to 8-% zero-dependency. There are many net-zero hotels and homes for rent.

Oil Producing Instability: With the global decrease in gasoline cars the Middle East will become more unstable. Countries that were wholly dependent on oil are now in a state of struggle and keeping their powerbase in order.

Energy Storage: High-capacity energy containers based on flexible new tech energy storage now allow utilities and homes to store electricity at low costs. Electric cars can travel a thousand miles on a single charge. Private home solar networks, along with spare storage, prevent blackouts as electricity rates come close to zero, and utility companies can buy and store energy that has been unused by homes with NueTech solar/ hydrogen energy generators;

Climate disruptions

1.   The great California flood, which flooded central California and cut off most of east and west California due to submersion, is predicted in the 2025-27 period.

2.   Florida faces more and more hurricanes, storms and coastal sinking as the oceans rise. Gentrification and unfair taxes drove most of the Blacks from prime properties and the 2028 class action suit has put heavy penalties on the state governments and White-dominated tax boards. A lot of cities in Florida now lie submerged.

3.   Nations like Seychelles have all their coastal cities submerged with the rise of the ocean, as have most low-lying coastal cities of the planet.

Bombing of the US Capitol by internal political terrorists: new articles are released by the US government: The 2026 bombing of the Capitol forced republicans and democrats to join hands and stop these radical polarizations of citizens. 'Enough is Enough' is the universal cry across the US. New articles so the US democracy does not regress into barbarism by the dark forces of politics and humanity. Among them is the 2-strike rule for proven liars for personal or party gain. Once proven the guilty official are now removed from public offices for life.

Digi-police: these are highly advanced and intelligent robots that have been installed in certain cities and replaced police in the developed nations. By 2027 police brutality, in test cities, is planned to be a thing of the past. The new Digi-police will work only for the people and have zero color, religion, or sex biases. They adhere to the rules and regulations in every interaction. Deaths in police situations will end. The era of police brutality is starting to disappear in colored and underprivileged areas of the US.

SpaceX lands humans on Mars: Musk City was established on Mars as the first private-sponsored space city for humanity. There was an accident, and 4 humans died. By 2027, it is judged that the humans were at fault, and Musk City gets permission to continue sending humans to Mars. [2] Neuralink is now a mature technology that is not only helping handicapped people but also starting to accentuate memory retention for the privileged and selected few. [3] Brain implants reverse Alzheimer's by bringing back lost memories with the help of Neuralink technology. [4] 3D printing of basic human biological parts has become routine

UGWN-United Guild of World Nations: In 2027, it was proven beyond reasonable doubt that China and Russia had compromised the United Nations. The UN had become corrupt from the inside. Both the prior global organizations were formed by the victors of the world wars. This new organization was formed in times of peace with a strict code of conduct. Countries with Veto power could lose it overnight if they broke UGWN charter laws. In order to join this collective, each nation had to sign to respect every current national and international border, thus ending oppressive wars. Any nation that went to war across any such boundaries would lose members within a week and be subject to unilateral global sanctions until they withdrew. No nation could attack any other nation for reasons, good or bad, without prior UGWN approval. The new rules brought with it a global ban on arms sales and a possibility of global peace.

B. 2030 Predictable Short-Term Outcomes

It will all be tech driven. Tech workers will not shrink, but grow exponentially as more and more humans are required to keep sanity checks of our AI regeneration.

B.0 Data & Analytics: The greatest impact will be in the area of the human interface with AI, models that assist in disease diagnosis, treatment prediction, drug discovery, and patient monitoring. I can safely predict that **by the 2030s, humanity will be the healthiest they have ever been, and yet sick all the time.** The reason for this is that their health will be monitored by a bunch of sensors. Their bodies will be serviced by many nanodevices, and the combination of these and other advancements will micro alert and thus manage to medicate the person by administrating micro doses of medicines as and when required — thus, the humans will remain in a state of permanently ill. If the body is performing at less than perfection, then the AI devices will assume they are fractionally ill and give a fractional dose to perfect health

B1. Energy Storage: High-capacity energy containers based on flexible new tech energy storage now allow utilities and homes to store electricity at low costs. By 2032, cars could travel 3 to 5 thousand miles on a single charge.

B.2 Space Tourism Trips: In 2031, Tourism trips to the moon will commence. These are simple one-to-five-day tourist trips. With the success of Moon in 2036, Mars trips had begun. There are now regular tourism trips to Musk City and many other cities that have become offshoots from the central city. Humans found many lava tubes under the surface, and they have become the perfect habitats under the harsh martial surface. Solar efficiency transfers the benefits of the sun down into the cities underneath. With advanced sunlight amplifier technology and close to 90% efficiency, humans in the cities get sunlight despite living underground. Direct early-intensity sunlight.

B.3 2032 New Articles added to the US Constitution: Everyone wonders why we could not have ratified these articles in 2024. But then, that is the curse of history. Even though by 2024, most citizens knew of the political swamp the US had fallen into, there was no agreement to correct these measures. After the insider bombing of the US Capitol, the US reached a point of no return. Due to the 'Enough is Enough' agreement between all political parties, there was a unanimous agreement to add new articles to the US Constitution: [1] No elected official shall willingly or unwillingly tell a lie to the public. This includes every line in their resume, every statement made to their voters, and every document presented. A Strike 3 rule is implemented where if a person is found telling a provable lie 3 times that they cannot prove to be true, nor say it was a joke, then they shall be banned from public office for life. [2] No politician is allowed to make statements that they are not professionally qualified to undertake. Each such statement shall be treated as two lies. [3] Elected officials cannot appoint their family members to US federal or foreign positions without due process and standards. If a family member is put into place that they are unqualified for, then it shall be treated as 2 lies instantly for each family member installed. The FBI shall mandatorily conduct a thorough investigation of every deal, finance received, and external influences received before, during, and after their appointment. If a professional, qualified state-appointed department makes a science-based statement, no politician is allowed to alter that requirement unless they are personally qualified to undertake that decision. For example, if the CDC or the Federal Education Commission makes a statement, then every politician has to follow unilaterally follow that recommendation. If they make an alternative political statement that causes even a single death, it shall be treated as three lies per death and can result in imprisonment. [4] All elected positions will spend 80% of their time, voice, and communications for doing what they have been elected for. This shall not include political or other lies, political distractions, and political optics based on assumptions or lies. This shall be measured routinely, and each month missed shall be tallied for the 'strike-3' rule. [5] No politician or elected official, including any person vying for an elected position, works for, on behalf of, or is renumerated in any way by a foreign government, entity, or persons.

B.4 Digi-police: Advanced democratic digi-police force has been deployed with minimal armament or weaponry. Their prime objective is public safety and they record every incidence in UDH video that is instantly uploaded to a central server. With the advance of home guard pets, and digi-police, combined with keyless vehicles crime has come to a very low rate.

B.5 Mars becomes the hub of innovation: Starting with Musk City, with most humans on mars locked up in artificial cities the need to learn and create supersedes most other activities. Their discoveries and innovations coming out of Mars have broken the invention curve of any people. Advances in Biotechnology, quantum computing, solar harnessing, spacecraft designs and engines, simul language translators are being developed and patented on Mars for a huge earth audience hungry for high quality Martian products and applications.

B.6 Deep Learning Networks: Advanced neural networks continue to develop, ever faster, allowing for even more accurate predictions in many more areas. Companies like Palantir can already do this to an accuracy of around 24%-35%. Predicting conflict and terror attacks will become an accurate science. Politicians will be barred from telling lies, and the days of telling lies to get votes will be well behind developed nations.

B.7 Generative Adversarial Networks (GANs): These models can generate realistic data and images, making them valuable for prediction tasks in various domains. By 2030, we will have mitigated the overages from self-generated data from LLMs and GANs. This will launch true AI predictions based on reliable data

B.8 Reinforcement Learning Algorithms: Applying AI techniques to optimize decision-making in complex environments, such as autonomous border security vehicles, home and industrial security robotic dogs, self-charging robotic soldiers in extreme terrains, and overall robotics and autonomous systems.

B.9 Natural Language Processing (NLP) Models: More sophisticated language models to interpret, generate, and simultaneously translate most human languages, improving tasks like sentiment analysis and language translation. This will eliminate the errors of 'Lost in translation.'

B.10 Recommendation Systems: AI Enhanced algorithms for personalized recommendations in areas like enterprise administration, global stocks administration, Customer responses, e-commerce, entertainment, and content consumption.

B.11 Automated Politicians Scorecards: By 2030 AI will keep a tag of every promise a politician makes to their electorate from their speeches and public interface. The AI will keep a daily score of each promise delivered. AI will also keep a track of their bio, and cases like George Soros will now stop happening. AI will weed them out from the start. AI will also check politicians and elected officials on their daily social network communications and score them with fact checked scores. Developed nations will deploy a 'Strike-3' rule wherein politicians are prohibited from stating lies.

B.12 Climate Prediction Models: By 2030 politicians will not be able to confirm, refute, or deny climate predictions. AI-powered systems will deliver very accurate climate forecasting, climate change analysis, climate impact forecast dates, and disaster response.

B.13 Financial Market Prediction Models: Advanced algorithms for predicting stock market trends, analyzing market behaviors, and detecting anomalies. These AI driven bots will take micro-second decision and investors funds will be protected from violent crashes and market fluctuations.

B.14 Autonomous Vehicle Prediction Systems: 30% of cars will be autonomous, 10% will be totally autonomous with no steering, brakes or other familiar additions we see in cars today. AI models to optimize decision-making, route planning, and collision avoidance for self-driving cars. No more speed limits, traffic lights or accidents. No more DUIs as these are all human driving limitation-based problems.

B.15 Smart City Predictive Systems: AI-enabled platforms that utilize data to predict urban planning, water distribution, transport management, energy consumption, traffic flows, and infrastructure management. Every car, cab, bus is digitally tracked across every nation.

C. 2040 Predictable Mid-Term Outcomes

This is now in the realm of total gut-feeling predictions.

C1. Green Cities: Autonomous green-cars, Self-sustaining homes, and fully sustainable cities become a reality, with renewable energy sources powering all aspects of urban living.

C2. Autonomous Cars: Have by now overtaken driver cars. Autonomous is now the standard. Self-driving cars and drones become the primary modes of transportation, leading to a significant reduction in road accidents. No more garages, DUI's, or traffic lights in smart cities. Traditional cars banned in metropolises.

C3. Space tourism becomes accessible to the general public, with commercial flights transporting passengers to the moon and beyond.

C4. Artificial intelligence (AI) advances to a level where it can carry out complex tasks, such as diagnosing medical conditions and performing surgeries.

C5. Virtual reality (VR) technology reaches new heights, allowing people to completely immerse themselves in virtual worlds and experiences.

C6. Breakthroughs in medicine lead to the development of personalized treatments and cures for various diseases, extending human lifespan significantly.

C7. Renewable energy sources, like solar and wind power, overtake fossil fuels as the dominant energy providers globally.

C8. 3D printing becomes a commonplace technology, enabling people to print food, clothing, and even entire houses.

C9. BCI: Brain-computer interfaces become widely used, allowing individuals to control devices and communicate directly with their thoughts.

C10. Human colony on Mars: The first human colony is established on another planet, marking the beginning of interplanetary colonization efforts.

D. 2090 Predictable Long-Term Outcomes

This is now on the realm of total gut feeling predictions.

D0. Keep humanity alive till then: Our biggest challenge, and risk, in every prediction is humanity eradicating itself. The probability is not 'Zero' thus we have to bring this risk into attention, so that we have this in the back of our minds and we make every effort for leaders to not go abound shooting their "Mine is bigger than yours" destructive weapons.

Reminder:

"The nuclear arms race is like two sworn enemy leaders standing waist-deep in gasoline, one with three matches and the other with five — Each threatening each other." — **Carl Sagan**

D.1 Space faring Humans: Humans will have established permanent colonies on other planets within our solar system, such as Musk city on Mars, and some others even on the moons of Jupiter and Saturn. These colonies will house thousands of inhabitants and serve as hubs for scientific research, resource extraction, and interplanetary travel. Because of the harsh external conditions these humans will accelerate in their R&D and lead in the next generation of technologies, propellants, energy producers and space travel.

D.2 Super AI- Artificial intelligence will have advanced to a point where it becomes indistinguishable from human intelligence. AI will not only be capable of performing complex tasks and problem-solving but will also possess emotions, creativity, and consciousness. AI companions and robotic companionship will become commonplace. Most of the cars, and spacecrafts will be totally built by HAI, or Hyper Artificial Intelligent. These will be spacecrafts that will have layers of auto-repair, AI support systems that no human could possibly understand, touch, or fix.

D.3 Climate change will be effectively addressed, and humans will have developed advanced technologies to restore and protect the environment. Global warming will be something of the past as renewable energy sources completely replace fossil fuels, and carbon capture technologies will help reduce greenhouse gas emissions. The Earth's ecosystems will flourish, and endangered species may even be brought back from extinction.

D.4 Dignity of Animals will be adhered by over 90% of humanity. The mid-century awareness of 'We feel your pain' program will have a massive conversion of meat-eating humans to vegetarianism. Animals will have a legal representative who will fight on their behalf on national decisions. Throwing animals into boiling water will become a global act of barbarism that will be punishable in every nation.

D.5 Teleportation: Humans will have mastered teleportation, allowing for instant travel between any two points on Earth. This technology will revolutionize transportation, making long-distance travel as easy as stepping into a teleportation booth. It will eliminate the need for traditional methods of transportation, such as cars, planes, and ships, drastically reducing pollution and congestion.

D.6 Medical advancements will lead to significant increases in human lifespan and the ability to cure previously incurable diseases. Aging will become a manageable process, with individuals having the option to extend their lives indefinitely through regenerative medicine, genetic engineering, and nanotechnology. This will lead to a significant shift in societal norms and the way we plan our lives and careers.

Post log

— Digital Tidd-Bits —

I'll end with predictions on my favorite topic, D&A, i.e., A. Industry, and B. Data and analytics. This is my prediction between now and 2030

2023-2030 Industry Predictions

A0. ESG to PEH: Environment, Safety and Governance will evolve into individual, corporate, and national shift to PEH, Planet, Environment, and Humanity, scorecards. The market will shift as more and more customers buy products, and services, based on the company's global PEH scores. The greatest development in PEH will come from monetizing PEH. PEH experts will travel across the planet and convince them that by protecting their rhinos, lions and pangolins from the Chinese they will make far more wealth than by killing them and exporting their pieces to China. 'Dignity of Animals' efforts will take flight and inhumane treatments of animals for food or pleasure will be banned worldwide. On the environment side large tracts will ban every saw, axe will be laid to rest to let the forests recover themselves. Countries that burn forests will be penalized by tourism and international trade.

A1. Business & IT — Overview: Business has been missing from the arena of technology, digitization, and deployment of decision-making systems. One reason, is that they do not come from a technology background, and are hesitant to take decisions in that area. The second is that they trust that their Tech partners will do their job for them. Both do not work, in fact according to Gartner more than 70% of these projects failed, and will continue to fail because of this lack of business participation. My book BI Valuenomics-The story of meeting Business Expectations in BI' was written for this exact purpose in 2011. The dilemma still, remains. In successful enterprises business has been taking lead in the design, architecture and applications across industries. With the introduction of enterprise AI-LLM, business will now sit in the driver's seat. By 2024 we shall see a mega shift of successful organizations that will run only 'Business Inclusive' projects. It will move from writing the technology story, to becoming a service provider to strategic business needs. Digitization will expand exponentially and digital twins will become a low-cost norm by 2030. While the last century has been all about consolidation and Synergy, the next two decades will be determined by niche digital players who get exceptionally good in micro services, and micro segmentation using AI-LLMs. They will do just one thing but do it better than anyone else on the planet. These players will generate niche competitors, like Uber and Lyft in the US.

A2. Autonomous self-powered Mobility: As we move toward 2030, we shall see a shift from gasoline to electric. Back in 2014, I have written on this in 6 LinkedIn articles, search for 'IOT-4-Automotive' on LinkedIn. By 2027 car manufacturers will no longer be manufacturing gasoline, self-driven vehicles will dominate the road. By this time the vehicular IOT will provide seamless SW upgrades as a defect is identified. Prices will drop as quality, per charge distance and safety will increase. By 2027 autonomous cars will evolve from savings to revenue generation. Smart will become nationwide. We shall see a proliferation of Smart Cities, with Smart buildings, smart utilities and transportation. On the city side multimodal — AI-LLM simulations will predict unintended consequences of population density, construction, climate change on city wide visuals. AI-LLM driven autonomous tasks will percolate all types of applications and businesses.

A3. Banking, Finance & Insurance: All three are going to be disrupted by digital niche players. However, the big players will keep them all three propped up. PEH efforts will impact all three industries as banks quantity their lending PEH scores, Finance is driven by customer purchasing on PEH scores and Insurance will see an investment in Greenhouse, and carbon investments. Companies with higher PEH scores will find getting loans far easier, as the market and customer will favor them, over those that score low. This will work for cities, states and nations too. Insurance will defocus from cars and transport and focus more on the UX aspects of their customers. The period of 2023-27 will be a time of turbulence for insurance. Companies that digitally connect to their customers, like Uber will see rapid growth, while those that don't will falter.

A4. Biotech & Pharma: PEH will become critical for this segment. Major customers have already commenced digitization initiatives, while once again, close to 75% of these companies will fail to meet business expectations. This is because of the age-old assumptions that the SI knows IT and IT must control all technology decisions. I worked for one US conglomerate that mandated, *"...You are not allowed to talk to business owners."* As of July 2023, they have already missed four of their go-lives and still continue to believe IT has the solutions. Many people will need to be thrown under the bus.

A4. Communications and Education: Both will face a major shift with digitization. The cost of communications will come close to zero as telecom companies will monetize their customer data, just like Google and Facebook do. Digital will disrupt communications like never before. People across the world will be able to speak to each other for free. I recall my father telling me, *"One day, you will be able to talk to anyone. If they don't answer, it means they are not alive* *anymore."* We have come to that prediction. Micro-specialized courses that conform to current business needs and not yesteryear assumptions will dominate the world.

A5. Entertainment — Movies will shift from Screens to In-home programs in a big way, and also from movies to serials that can provide more depth to stories. **TV:** The field of TV entertainment will face rapid peaks and troughs in digital alternatives. The main big boys are maintaining their share by divisions forming, and content is becoming king. However, competition will become more and more severe. From 2023 to 2026, monetization will become a balancing act between subscriptions and advertising revenue, which will be a prime challenge for customer retention. As demand for live streaming increases, content licensing fees are expected to rise. This will be a challenge to both paid and free services. Infra and bandwidth deliver services. By 2026-27, content piracy will be controlled by AI-LLM live applications. The aggressive isolation of each content provider by creating independent channels per provider is confusing customers. By 2025, we will see consolidations come back for customer-facing services, along with free digital entertainment that will be embraced by a high volume of global customers on digital devices. As mentioned earlier, the future is in free services, and an example that comes to mind is a digital entertainment service out of CA called *Distro TV*. It has the customer stickiness factor = Free, as its monetization is based on ad revenue. In June of 2023, Distro TV partnered with MX Player, one of the largest live-streaming TV services worldwide. It's all about co-innovation and collaboration.

A6. Governments and Politics are both at the bottom of the scorecard. There will be drastic shifts and changes by the 2030s in all these areas. Upheavals will continue up until they reach a breaking point, when both sides will realize that it is better to work together. Enemy state digital aggressions will continue, and 2026-29 strict global-digital governance laws will make a place structure and procedure for all digital players. For example, it will make illegal subliminal trolling a global criminal offense. Global organizations will measure digital attributes by micro-segmentation of critical KPIs and share these transparently with the citizens of developed nations. Participating with autocratic nations will become accentuated as people react to their optics of working with fanatic nations.

Ironically on August 27th, 2023 FIFA and Europe stood by the rights of women by terminating Luis Rubiales, the Spanish Soccer Federation president for kissing a woman player on her lips, in Australia, without her permission. While at the same time the US republicans were protecting and rewarding their ex-president even after he had been charged for a forced rape. Both Lius and the ex-president say that they are innocent and victims of a 'witch-hunt'.

A7. Manufacturing: It is in for a very big digital shift. Companies that made China their industrialized factory will suffer as the Chinese economy falters, compete with their products at lower costs, and customers start to see the margins of large conglomerates exposed. Manufacturing will invest a large portion in Robotics, AI, and ML and reap massive benefits in all attributes, from cost to quality. Between 2023-25 we shall see peaks and troughs of glut and scarcity of critical components in the global supply chain. During this time, over 50% of smart factories that pick digitization only as a fashion statement will fail. Robotics will play a big role in quality, consistency and cost, something proven by Elon Musk, with Tesla being manufactured in California. Due to robotics, a lot of manufacturing will come back to national boundaries as China will no longer be competitive, nor a desirable factory to customers. The customer optics of China is that of Putin right now — they are both so tightly intertwined that it is difficult to tell one from the other. Biotech and Pharma have already launched tokenized asset management, i.e., the use of secure blockchain identifiers on each individual item, such as a critical drug. By 2025-27, the top CGMs, Consumer Goods Manufacturers, will implement digital passports for their customer and high-end products.

A8. O&G: Oil and Gas is on for their biggest shock. Between 2024 and 2026, consumption and production volatility will be overtaken by global geopolitics. O&G companies are investing more and more in digital technology during their uncertainty of the future. One of the prime areas is lowering or eliminating unplanned shutdowns, which can cost a rig anywhere from $2 to $5 million per day, per rig. We provided this solution to a US company in this field. By 2029, Oil futures will be predictably stable, and the smaller suppliers will not be able to compete as the large players consolidate their production and pricing.

A9. Retail: This is an area with a triple clamp-down. On one side is the digital disruptions, on the second is the effect of the Covid-19 shutdown, and the third is the massive customer movement from on-prem to cloud. I have a bunch of articles in which we started predicting the impact of digital disruptions on Retail, and by 2017, we started accurately predicting. Most retail stores across the US, EMEA, and Japan are in a critical state of retail pricing realignments. Interestingly, while the triad nations are wondering what to do with their Malls, there is a big construction boom in developing nations, unaware of the possible impending doom. By 2027, digital e-tailing will include immersion technologies where customers can see the products on their personas. Sales in each segment will go up by 7% to 19% by 2027-28.

A10. Tech & Tech SI: Tech and Technology support establishments have been under tremendous pressure since the launch of AI-LLM technologies. By 2025 60% of traditional tech support will more to more advanced checks and balances from AI-LLM generated solutions. In the short run, we are witnessing a huge tech layoff; however, in the mid-term, i.e., 2024-27, we shall see a tremendous growth of AI-LLM validation jobs that need critical understanding not only of the technology but also business expectations. D&A support will move from manual to automated solutions and cloud-native applications running on generic and customized AI-LLM Models designed specifically for business users.

B. 2023-2030 Enterprise D&A Predictions

B1. Strategic D&A Planning: i.e. Data & Analytics: Here is what I published in 2011, and it is as relevant in 2023 as it was then. Gartner then stated that 70% of BI projects will not meet business expectations. That stands as true today as it did then. The book *"BI Valuenomics — The Story of Meeting Business Expectations in BI"* is quite alive and relevant in 2023. I mention this as I had stated a decade ago: the need for Business Inclusion; IT leadership will need to become more and more business and outcome-focused. Enterprise Portfolios, Programs, and Projects harmonization will become AI-driven, with 40% to 60% of digital approval processes automated without human interventions standard. Jointly, they will need to define success five years ahead with business stakeholders. Strategic Vision and Business Owners in D&A projects. D&A will become a 'Single Pane of Truth' with single definitions of KPIs and business definitions across national boundaries, systems, and applications. Data Harmonizing, Data-mesh strategies, and metadata management will become the foundations of future D&A support. This will require a centralized D&A governance to control this. AI-LLM will become a game-changer in companies that will disrupt decision capabilities for success. Companies with AI-LLM modeling will decrease costs by 50% and increase deliverables accuracy by 40%-60%. D&A architecture will evolve from Infra and Landscapes to Business and Data flow architecture. 'X' will become the primary deliverable focus, as predicted in 2011, deliverables will be measured by UX, or User Experience. Enterprises will appoint CXO as the current failure rate is at 75%. By 2024 AI-enabled Cloud charge allocations will become standard. By 2026, IOT edge streaming data analysis will become an efficiency control mechanism. By 2026-27, AI-LLM-driven digital selling will drive enterprise Augmented Analytics. CRM will be a big benefactor in AI-LLM initiatives. The biggest benefactors of autonomous augmented applications will be CRM, Finance, SCM, and Stock visualizations. By 2026, 40% of companies deploying robotics will adapt to AI-LLM Adaptive informatics.

B2. AI & LLM: The greatest leap in D&A will come from AI-LLM.

Starting 2023 we shall see the advent of LLM driven AI transform D&A substantially. By 2024 D&A generative applications will automate 50% to 70% of the design effort for critical KPIs, while Auto-ML will increase business adaption from the current 25% to over 75% for 'SPOT' Analytics, i.e., Single Point of Truth. AI-LLM will automate routine edge extractions, and tasks, and provide more accurate and, actionable insights. We will welcome enterprise business leaders into the new world of Augmented Analytics. The combination of AI-driven LLM's in the enterprise space will open new doors as the AI models will start providing answers to questions that need to be asked. The long-desired world of 'In-Process' augmented analytics is now here.

B.3 Chat GPT: The autonomous evolution of AI and LLM's is ChatGPT. ChatGPT is an evolutionary development by OpenAI, and a revolutionary technology that will help all kinds of digital endeavors. It will become the launchpad for teams and ideas to exist in the world of AI + LLM + Autonomous. For entrepreneurs this will open floodgates of opportunities that will be ripe to take to market and for acquisition games. Read, Invest and learn from ChatGPT and their various clones across the planet as this will come fast and furious because of its simplicity and autonomous capabilities.

B.4 CDAO: Chief Data & Analytics Officer. A new role CDAO will evolve by 2024-25. Initially, it will be co-owned by experienced CDO and D&A leaders. However, over time, by around 2025, we shall see the growth of a single leadership of CDAOs across enterprises. In the digital world this is a critical business facing role and it will be an outside-looking-in deployment position.

B.5 IoT & AI-LLM: The Internet will keep generating more and more massive amounts of data. Leading to an increased demand for sophisticated informatics tools to deliver augmented informatics. The AI-LLM models will be able to answer these needs by early 2024 for enterprise customers on their native ERPs.

B.6 Data Privacy: Data Security and authorization will become major concerns. Applications with ISO 27001 compliance, i.e., GDPR, HIPAA, HITRUST, PCI DSS, CCPA, will be table stakes, or a given. With the growth in digital data and IoT data volumes, data sensitivity will continue to grow. Companies will pay heavy penalties for noncompliance. So, Companies will need to invest into investing in robust data governance and compliance programs.

B.7 NLP: Or, natural language processing and voice recognition technologies will advance, enabling users to interact with data analytics platforms using their local language voice commands and conversational interfaces, to accomplish complex tasks.

B.8 Data ethics and responsible AI will need to become a national state-level structured program. Data ethics driven by AI-LLM bots will become a game changer, and gain prominence as organizations grapple with the ethical challenges of using data and algorithms in augmented decision-making processes.

B.9 Augmented analytics: with the launch of AI-LLM the natural adaptation is for Augmented Analytics. By 24-25, these in process analytics will become will become mainstream, via integrating AI-driven business insights into data curation processes, and applications. This will become a giant leapfrog that will facilitate data-driven decision-making at all levels of an organization.

B.10 Data democratization: Up until 2023 data was owned, processed and delivered by IT. That phase of IT-driven analytics is coming to an end. An end user, or a customer must not need a PhD to get their desired information. The democratization of data will allow the actual-user in the enterprise, to understand and analyze data without need for any technical knowledge. Higher and higher inclusion of final users and business will continue to grow with deliverables leadership held by individuals within organizations having access to and understanding of data, empowered to make data-driven decisions.

B.11 Hybrid cloud and multi-cloud architectures will continue to grow. The cloud architecture provides unparallel elasticity, and stretch both to compute and storage that just cannot be achieved in any on-prem solution. Cloud become the norm for enterprises, allowing organizations to leverage both private and public cloud infrastructure for their data analytics needs.

B.12 True-Real-time analytics will become more mandatory. No more information from last week or yesterday. Businesses will demand information as it is right now. This will translate into edge data extraction and eliminate the need to periodic ETL and delta jobs. Thus, enabling organizations to make decisions and take actions based on up-to-the-minute data.

Authors Thank You

Thank you everyone for taking time to read this book and its vision.

Please share your comments, critique and recommendations to harig.auth@gmail.com. In the Subject Line start with Digital Shock: so I can sort these messages automatically.

To see the charts in this book a little more clearly please visit my blog Earthonomics where you will find Digital Shock,

Digital Shock- Charts and Images:

https://earthomonics.blogspot.com/2023/08/digital-shock-continued.html

Thank you once again. Keep smiling think PEH

Authors Links

BOOKS on Amazon

A. <u>2011: BI Valuenomics- The story of meeting Business expectation sin BI</u>

B. <u>2018: Beyond Drake- No Earth 2.0</u>

C. <u>2018: Cooking Earth: A data driven journey to Climate initiatives by US (embedded in Beyond Drake)</u>

BOOKS Under Development

D. **2023: Quantum Thinking:** The underlying principles to Success

E. **2024: The Scientific Principles of Information Delivery**

F. **2024: Fiction-: SiFi- Eternity** – first contact with an alien intelligence

Articles on Linked In:

1. Autonomous Automotives:

   a) <u>The self-driving Automobile revolution is already here</u>

   b) <u>The future of Connectivity – 7 reasons to connect your car</u>

   c) <u>Baseline Planning for Connected Cars</u>

   d) <u>Insecurity of connected cars</u>

   e) <u>The digital ripples of Tesla Model 3</u>

2. <u>2014: What each of us can do for climate change</u>

3. <u>2017: The working principles of a Data Lake</u>

4. <u>2017: The (f) Actual Costs of a failed BI Project</u>

5. <u>Harnessing the SAP S/4 winds to your favor - FI</u>

6. <u>2019: Deconstructing Student Loans</u>

7. <u>2019: Deconstructing Digital Transformation</u>

8. <u>2018: Deconstructing your CSO- Customer Success Officer</u>

9. <u>2018: Crossing the 'Don't Know' Chasm</u>

10. <u>2018: Trust, Privacy and Market Share in a digital world</u>

11. <u>2018: Your 'Do or Die' Technology Challenge</u>

12. 2015: Defining IoT and IoE (Simplification)

13. 2015: Identifying Patterns of Economic Meltdown

14. 2016: Predicting Presidential Elections– why it is becoming so Inaccurate

15. 2022: Being woman– Data Driven deconstruction of the Roe vs. Wade decision

Bibliography

1. **AI & Machine Learning for Coders:** A Programmer Guide to Artificial Learning – by Lawrence Moroney

2. **Chatgpt millionaire:** Artificial Intelligence, Real Riches: Using AI to be more efficient in business, Save time and Make More Money – By Jason Goldberg

3. **Lean B2B:** Build Products Business Wants- by Etienne Garbugli

4. **CHATGPT Millionaire Bible:** How AI can build you a Million-Dollar business to become financially independent – By Harold Pearson

5. **The American Dilemma** – By Gunner Myrdal

6. **The Cult of Trump:** Steven Hassan

7. **Too Much, Never Enough-** Mary Trump

8. **Barak Obama-** Biography

9. **Digitization & Neo Docetism-** Seth Troutt

10. **The Politics of Mass Digitization-** Nanna Bonde Thystrup

11. **Contemporary methods of Digitization:** Panagiotis Zigouris

12. **Infocracy** – Digitization and the Crisis of Democracy

13. **Cyber Anonymity** – Kushantha Gunawardana

References

---

[i] Digital Shock: Blog
https://earthomonics.blogspot.com/2023/08/digital-shock-continued.html

[ii] The January 6th report on Donald J trump:
https://www.cnn.com/2022/12/22/politics/full-jan-6-report/index.html

[iii] PEH = Planet, environment & Humanity focused decisions

[iv] Being woman: https://www.linkedin.com/pulse/being-woman-hari-guleria/

[v] Dark-Side: in the digital world, the dark sides are companies like Cambridge Analytics, owners and workers of Cambridge Analytica who continue the 1:1 segmented brain-washing, the Politicians and leaders who pay them to do this work, the state-owned leaders who build formal state-sponsored trolls like those of every politician on the planet and the religious digital departments like the ones run by US evangelists and the south Korean mega-churches. These are all the dark-side leaders with their digital warriors spreading fear for personal gains. The media opinion generators who tell blatant lies depend on who pays their company and what false outcomes they plan to generate.

[vi] PEH - Planet, Environment, and Humanity focus in every decision we undertake hereinafter.

[vii] 'Being Woman' https://www.linkedin.com/pulse/being-woman-hari-guleria/

[viii] 'Being Woman': https://www.linkedin.com/pulse/being-woman-hari-guleria/

[ix] The factual costs of a failed Project:
https://www.linkedin.com/pulse/factual-costs-failed-bi-project-hari-guleria/

[x] https://www.washingtonpost.com/news/post-nation/wp/2017/05/23/a-neo-nazi-converted-to-islam-and-killed-2-roommates-for-disrespecting-his-faith-police-say/?utm_term=.1cc8eecae952

[xi] **FuFact:** The opposite of History. Future Facts is the science of learning from the future, where we take current factors and use ML, AI & LLM to predict the probability of outcomes in the near and, mid-term, and long-term future. Based on the probabilities, we change the current state to change the future. Each time we make a prediction, the algorithm improves by seeing what happens.

www.ingramcontent.com/pod-product-compliance
Lightning Source LLC
Chambersburg PA
CBHW060857140726
47996CB00001B/12